Springer Complexity

Springer Complexity is an interdisciplinary program publishing the best research and academic-level teaching on both fundamental and applied aspects of complex systems—cutting across all traditional disciplines of the natural and life sciences, engineering, economics, medicine, neuroscience, social and computer science.

Complex Systems are systems that comprise many interacting parts with the ability to generate a new quality of macroscopic collective behavior the manifestations of which are the spontaneous formation of distinctive temporal, spatial or functional structures. Models of such systems can be successfully mapped onto quite diverse "real-life" situations like the climate, the coherent emission of light from lasers, chemical reaction-diffusion systems, biological cellular networks, the dynamics of stock markets and of the Internet, earthquake statistics and prediction, freeway traffic, the human brain, or the formation of opinions in social systems, to name just some of the popular applications.

Although their scope and methodologies overlap somewhat, one can distinguish the following main concepts and tools: self-organization, nonlinear dynamics, synergetics, turbulence, dynamical systems, catastrophes, instabilities, stochastic processes, chaos, graphs and networks, cellular automata, adaptive systems, genetic algorithms and computational intelligence.

The three major book publication platforms of the Springer Complexity program are the monograph series "Understanding Complex Systems" focusing on the various applications of complexity, the "Springer Series in Synergetics", which is devoted to the quantitative theoretical and methodological foundations, and the "Springer Briefs in Complexity" which are concise and topical working reports, case studies, surveys, essays and lecture notes of relevance to the field. In addition to the books in these two core series, the program also incorporates individual titles ranging from textbooks to major reference works.

Series Editors

Understanding Complex Systems

Founding Editor: S. Kelso

Future scientific and technological developments in many fields will necessarily depend upon coming to grips with complex systems. Such systems are complex in both their composition–typically many different kinds of components interacting simultaneously and nonlinearly with each other and their environments on multiple levels–and in the rich diversity of behavior of which they are capable.

The Springer Series in Understanding Complex Systems series (UCS) promotes new strategies and paradigms for understanding and realizing applications of complex systems research in a wide variety of fields and endeavors. UCS is explicitly transdisciplinary. It has three main goals: First, to elaborate the concepts, methods and tools of complex systems at all levels of description and in all scientific fields, especially newly emerging areas within the life, social, behavioral, economic, neuro- and cognitive sciences (and derivatives thereof); second, to encourage novel applications of these ideas in various fields of engineering and computation such as robotics, nano-technology, and informatics; third, to provide a single forum within which commonalities and differences in the workings of complex systems may be discerned, hence leading to deeper insight and understanding.

UCS will publish monographs, lecture notes, and selected edited contributions aimed at communicating new findings to a large multidisciplinary audience.

More information about this series at http://www.springer.com/series/5394

Visarath In · Patrick Longhini ·
Antonio Palacios
Editors

Proceedings of the 5th International Conference on Applications in Nonlinear Dynamics

Springer

Editors
Visarath In
Space and Naval Warfare Systems Center
San Diego, CA, USA

Patrick Longhini
Space and Naval Warfare Systems Center
San Diego, CA, USA

Antonio Palacios
Department of Mathematics and Statistics
San Diego State University
San Diego, CA, USA

ISSN 1860-0832 ISSN 1860-0840 (electronic)
Understanding Complex Systems
ISBN 978-3-030-10891-5 ISBN 978-3-030-10892-2 (eBook)
https://doi.org/10.1007/978-3-030-10892-2

Library of Congress Control Number: 2018966115

This Springer imprint is published by the registered company Springer Nature Switzerland AG
The registered company address is: Gewerbestrasse 11, 6330 Cham, Switzerland

To Christine, Beredei and Reynard for their love and support throughout the years.

Visarath In

To Melanie, Paul, Caroline, Emma and Luke for their love and support.

Patrick Longhini

To Brenda and Daniel for their encouragement and support.

Antonio Palacios

Organizers

Adi Bulsara, SPAWAR Systems Center Pacific (SSC Pacific)
Antonio Palacios, San Diego State University
Michael F. Shlesinger, Office of Naval Research (ONR)
Ned J. Corron, U.S. Army RDECOM
Patrick Longhini, SPAWAR Systems Center Pacific (SSC Pacific)
Visarath In, SPAWAR Systems Center Pacific (SSC Pacific)

Sponsor

Nonlinear Dynamical Systems Group
Department of Mathematics
San Diego State University
San Diego, CA 92182-7720

Preface

The field of Nonlinear Science involves the study of phenomena that changes in both space and time. Examples include: the flocking of birds, laser systems, central pattern generators in biological systems, collective behavior of bubbles in fluidization systems, nano-oscillators in microelectronics, communication systems, and electronic nonlinear oscillators in antennas and radars. Regardless of the applications, nonlinear science provides methods to study the long-term behavior of how a system evolves in space and time. Yet, while there has been significant progress in developing theoretical ideas and methods to study nonlinear phenomena under an assortment of system boundary conditions, there exist comparatively fewer experiments and technological devices that actually take advantage of the rich behavior exhibited by theoretical models. Consider, for instance, the fact that a shark's sensitivity to electric fields is 400 times more powerful than the most sophisticated, currently available, electric-field sensor. In fact, in spite of significant advances in material properties, in many cases it remains a daunting task to duplicate the superior signal processing capabilities of most animals.

Bridging the gap between theory and biologically inspired devices can only be accomplished by bringing together researchers working in theoretical methods in nonlinear science with those performing experimental works. Other areas of strong interest among the research community, where theoretical findings can one day lead to novel technologies that exploit nonlinear behavior, include: chaos gates, social networks, communication, sensors, lasers, molecular motors, biomedical anomalies, and stochastic resonance. A common theme among these and many other related areas is the fact that nonlinear systems tend to be highly sensitive to perturbations when they appear near the onset of a bifurcation. This behavior is universal among many nonlinear phenomena and, if properly understood and manipulated, it can lead to significant enhancements in systems response. Representative examples have been observed in a large number of laboratory experiments on systems ranging from solid-state lasers to superconducting loops, and such behavior has been hypothesized to account for some of the more striking information-processing properties of biological neurons. Furthermore, background noise can precipitate this behavior, thereby playing a significant role in the optimization of the response of these systems to small external perturbations.

Since 2005, we have held a series of meetings to bring together researchers across various disciplines working on theory and experiments in nonlinear science. The first meeting was 2005 DANOLD (Device Applications of Nonlinear Dynamics) meeting, held in Catania, Italy. Then, in 2007 ICAND, the research community met again in Poipu Beach, Koloa (Kauai), Hawaii, USA. More recently, the 2010 ICAND meeting was held in Alberta, Canada, at the luxurious Fairmont Chateau in Lake Louise. The 2012 ICAND was held in Seattle, Washington and then in 2016 in Westminster, Colorado. This last meeting brought together researchers from physics, engineering, and biology who were involved in the analysis and development of applications that incorporate and, indeed, exploit the nonlinear behavior of certain dynamical systems. The focus for 2018 ICAND was equally divided between theory and implementation of theoretical ideas into actual devices and systems. Contemporary topics on complex systems, such as social networks, were also featured among selected lecturers.

The organizers extend their sincerest thanks to the principle sponsors of the meeting: Army Research Office (Washington, DC), Office of Naval Research (Washington, DC), Office of Naval Research-Global (Tokyo), San Diego State University (College of Sciences), and SPAWAR Systems Center Pacific. A special mention to Dr. Samuel Stanton from the Army Research Office and to Dr. Michael Shlesinger from the Office of Naval Research for their support and insight to hold such a diverse meeting. In addition, we extend our appreciation to Tania Gomez at SDSU for their hard work in preparation and financial duty, which enabled the conference to run smoothly. We would also like to thank our colleagues who chaired the session and to all the personal who spent many hours making this meeting a success. Finally, we thank Springer for their production of an elegant proceeding.

San Diego, USA
October 2018

Visarath In
Patrick Longhini
Antonio Palacios

Acknowledgements

The organizers extend their sincerest thanks to the principle sponsors of the meeting: Office of Naval Research (Washington, DC), Office of Naval Research-Global (London), San Diego State University (College of Sciences), and SPAWAR Systems Center Pacific. A special mention to Dr. Samuel Stanton from the Army Research Office and to Dr. Michael Shlesinger from the Office of Naval Research for their support and insight to hold such a diverse meeting. In addition, we extend our appreciation to SDSU students, Lourdes Coria, Horacio Lopez, and Brian Sturgis-Jensen, for their hard work in preparation and financial duty, which enabled the conference to run smoothly. We would also like to thank our colleagues who chaired the session and to all the personal who spent many hours making this meeting a success. Finally, we thank Springer for their production of an elegant proceeding.

San Diego, USA
October 2018

Visarath In
Patrick Longhini
Antonio Palacios

Contents

Chapter 1
The Cost of Remembering

Luca Gammaitoni[1(✉)], Igor Neri[1], Miquel López-Suárez[2], Davide Chiuchiù[3], and Maria Cristina Diamantini[4]

[1] Dipartimento di Fisica e Geologia, NiPS Laboratory, Università degli studi di Perugia, 06123 Perugia, Italy
luca.gammaitoni@nipslab.org, igor.neri@unipg.it
[2] Institut de Ciència de Materials de Barcelona (ICMAB-CSIC), Campus de Bellaterra, 08193 Bellaterra (Barcelona), Spain
mlopez@icmab.es
[3] Okinawa Institute for Science and Technology, Okinawa, Japan
davide.chiuchiu@oist.jp
[4] Dipartimento di Fisica e Geologia and INFN, NiPS Laboratory, Università degli studi di Perugia, Sezione di Perugia, 06123 Perugia, Italy
cristina.diamantini@pg.infn.it

Abstract. In 1961, Rolf Landauer pointed out that resetting a binary memory requires a minimum energy of $k_B T ln(2)$. However, once written, any memory is doomed to loose its content if no action is taken. To avoid memory losses, a refresh procedure is periodically performed. In this work we present a theoretical and experimental study of sub-$k_B T$ system to evaluate the minimum energy required to preserve one bit of information over time. Two main conclusions are drawn: (i) in principle the energetic cost to preserve information for a fixed time duration with a given error probability can be arbitrarily reduced if the refresh procedure is performed often enough; (ii) the Heisenberg uncertainty principle sets an upper bound on the memory lifetime, thus no memory can last forever.

1.1 Introduction

The act of remembering is of fundamental importance in human experience. While usually we refer as remembering as an act to preserve information the concept can be easily extended to any aspect of human life which is subject to deterioration as in objects and artifacts that tend to loose their original shape. In order to preserve the original shape, i.e. in order to keep the memory, we usually perform restoration work/memory reinforcement. Among others, memory degradation is a common problem also for computer memories that tend to lose their content over time. In order to counterbalance the memory degradation, a periodic refresh operation is performed, which consists in periodically reading and writing back the content of the memory.

V. In et al. (Eds.): *Proceedings of the 5th International Conference on Applications in Nonlinear Dynamics*, Understanding Complex Systems, https://doi.org/10.1007/978-3-030-10892-2_1

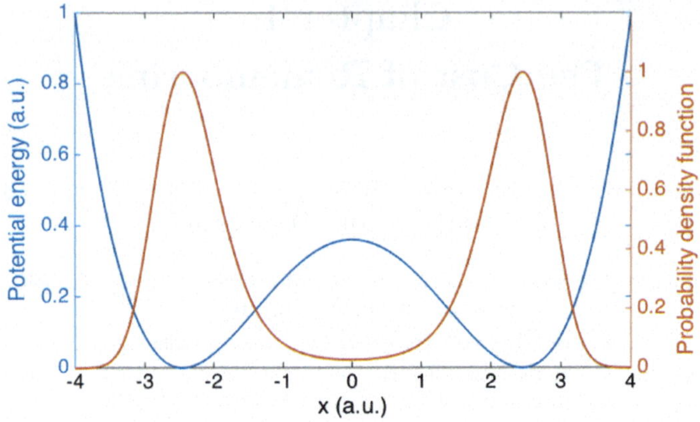

Fig. 1.1. Bistable energy potential and relative probability density function at thermal equilibrium of a single-degree-of-freedom system used to represent a memory device

The scope of this work is to study the fundamental energetic cost associated to preserve a given bit of information for a given time, $\bar{t}$, with a final probability of failure P_E, while executing a refresh procedure with periodicity t_R. To perform this study we first define a physical model for a 1-bit memory. Then, we compute the minimum required refresh time to satisfy the required probability of failure and retention time. Finally we experimentally evaluate the energetic cost of a single refresh operation and estimate the overall fundamental energetic cost associated to preserve the memory.

1.2 Physical Model for 1-Bit Memory

Information is encoded in a memory device by means of a physical property, like charge on a capacitor or orientation of the magnetic field on a magnetic dot. A single physical property can be used to represent an arbitrary number of bits in memory devices, however for sake of simplicity we will consider a single bit. The bit is encoded in the physical property of the device respect to a fixed threshold value. Without loss of generality we can consider as a memory device a particle trapped inside a bistable energy potential where the position, x, of the particle encodes the information [1–5]. Such a memory is represented as a single-degree-of-freedom system as pictured in Fig. 1.1.

We can now set the threshold value at $x = 0$ and thus we can define the logic state 0 if the particle is in the left well ($x < 0$) and the logic state 1 if the particle is in the right well ($x > 0$). The energy barrier, with a maximum at $x = 0$, separates the two stable states. To take into account a more realistic representation of the memory device dynamics we should assume that the single degree-of-freedom system is coupled to a thermal bath at temperature T. The effect of this coupling is that the dynamics of the system depend not only on its potential energy and initial conditions but also on the stochastic fluctuating force

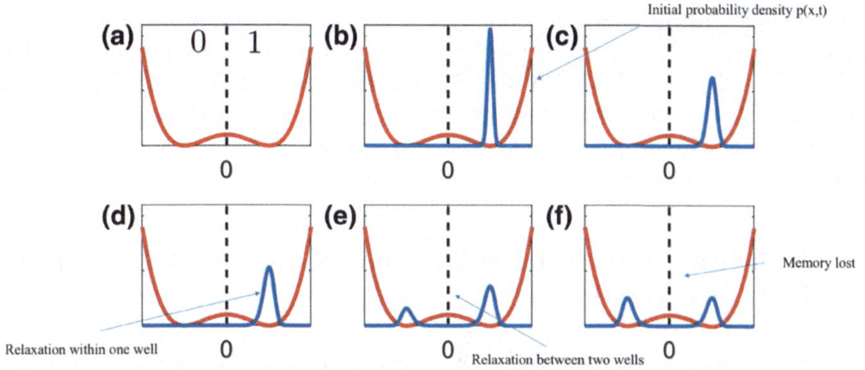

Fig. 1.2. Energy potential (in red) of a bistable system used to encode one bit of information and time evolution of the probability density function of the observable (in blue) once the bit is stored

and relative damping coefficient. According to this description the dynamics of the system can be described in terms of a Langevin equation in the form of:

$$m\ddot{x} = -\frac{dU(x)}{dx} - m\gamma\dot{x} + \xi + F \tag{1.1}$$

where $U(x)$ is a bistable potential, γ is the friction coefficient, ξ is the fluctuating force due to the contact with the thermal bath, and F is an arbitrary force used to modify the state of the memory. The equation of motion has now become a stochastic dynamical equation and its solution can be approached in statistical terms. One relevant quantity for describing the system dynamics is represented by the probability density function $P(x,t)$. Specifically, $P(x,t)dx$ represents the probability for the observable x (the position of the particle) to be at time t within the interval between x and $x + dx$. In particular $p_0(t) = \int_{-\infty}^{0} P(x,t)$ and $p_1(t) = \int_{0}^{\infty} P(x,t)$ are respectively the probability to find the bit 0 and 1 encoded on the system at a given time t.

Once written, the stored information is doomed to be lost due to thermal fluctuations. The average retention time depends on the physical parameters of the system. A schematic of the time evolution of the probability density function of the observable used to encode the information, relative to the energy potential represented in Fig. 1.2a, is represented with a blue curve in Fig. 1.2, panels from (b) to (f).

Initially the bit is stored with a given probability of error. Then, the probability density function relaxes mostly inside the well encoding the desired bit. Afterward the system relaxes between the two wells increasing the probability of finding the wrong bit encoded on the system. Finally, once the system is completely thermalized, the information is statistically lost.

The probability to find the bit in the wrong state increases over time, this is due to the fact that we start from an out of equilibrium condition and the system evolves towards equilibrium. To prolong the information life-span it is possible to periodically refresh the information stored, to counteract the effect of relaxation to thermal equilibrium. Each refresh-operation consists of reading and writing back the read information [6,7]. Once the operation of refresh is performed the potential error is not corrected but only the probability density function of the physical observable is modified.

1.3 Relation Between Probability of Error, Refresh and Retention Time

Assuming to start from having the bit 1 encoded in the memory, $p_0(t)$ defines the time evolution of the probability of finding the wrong value in the memory. Whit this assumption it is possible to define an overall probability of error of finding the wrong bit stored in the system at any time between the initial writing and any interrogation as function of the refresh time, t_R, as [5]:

$$P_E = [1 - p_0(t_R)]^{\lfloor \frac{\bar{t}}{t_R} \rfloor} \tag{1.2}$$

where $\lfloor \frac{t}{t_R} \rfloor$ represents the number of refresh operations performed in the time interval $[0 - \bar{t}]$.

A good model for the energy potential, able to capture the main characteristics of a bistable memory device, is the Duffing potential:

$$U(x) = 4\left(-\frac{x^2}{2} + \frac{x^4}{4}\right) \tag{1.3}$$

and its statistical time evolution can be obtained solving the relative Fokker–Plank equation [8,9]:

$$\frac{\partial}{\partial t}\, p(x,t) = \frac{\partial}{\partial x}\left(\frac{\partial U}{\partial x}\, p(x,t)\right) + T\frac{\partial^2}{\partial x^2}\, p(x,t), \tag{1.4}$$

where T is the temperature of the thermal bath.

The evolution of the system depends on its initial condition, more specifically it depends on the starting probability density function distribution. Considering to have the bit 1 stored on the system the initial probability density function can be approximated to a Gaussian distribution centered in $x = 1$:

$$p(x,0) = \frac{\exp\left(-\frac{(x-1)^2}{2\sigma_i^2}\right)}{\sqrt{2\pi}\sigma_i}. \tag{1.5}$$

where σ_i is the initial standard deviation of the Gaussian peak of the observable.

The solution of Eq. (1.4) permits to obtain the maximum refreshing interval t_R that satisfies the *a priori* requirements for $\bar{t}$ and P_E. The results agree with the common sense: large times $\bar{t}$ and small probabilities of error P_E yield short refresh times t_R.

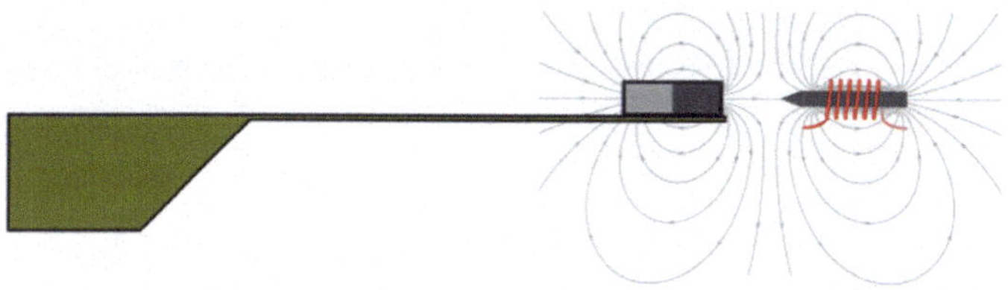

Fig. 1.3. Schematic of the experimental setup. The interaction between a magnet applied on the tip of the cantilever and an external electromagnet permits to control the effective stiffness of the beam

1.4 Experimental Evaluation of Energetic Costs for Refresh Operation

We now consider the energy cost of a single refresh operation. Based on our model, the refresh operation consists in bringing the $p(x,t)$ back to its initial condition:

$$p(x, t_R) \to p(x, 0) \tag{1.6}$$

We now assume that the motion of the system when it is trapped inside one well can be approximated by the dynamics of a harmonic oscillator. This is reasonable if the refresh time is much smaller than the system relaxation time ($t_R << \tau_k$). Considering the refresh protocol described above the probability density function of the system at any time is approximated to the sum of two Gaussian peaks centred around the minima of $U(x)$, each one with the same standard deviation. The refresh operation consists in applying an external force that shrinks the potential wells and thus change the standard deviation inside each well, from $\sigma_f = \sigma(t_R)$ to $\sigma_i = \sigma(0)$. The value of σ_f changes in time according to the physical parameter of the system, and initial distribution σ_i as [5]:

$$\sigma_f = \sqrt{\sigma_w^2 + \exp\left(-\frac{t_R}{\tau_w}\right)(\sigma_i^2 - \sigma_w^2)} \tag{1.7}$$

where τ_w is the relaxation time of the harmonic oscillator approximating the single well.

1.4.1 Experimental Measurement of Energy Required for a Single Refresh Operation

To measure the minimum energy required for the refresh, i.e. to "squeeze" the density function inside an harmonic well, we perform an experiment with a micro-mechanical V-shaped cantilever where the relevant observable, x, is the tip position. The interaction between a magnet applied on the tip of the cantilever and an external electromagnet permits to control the effective stiffness of the beam and thus the standard deviation of x at equilibrium. Figure 1.3 shows a schematic of the experimental setup.

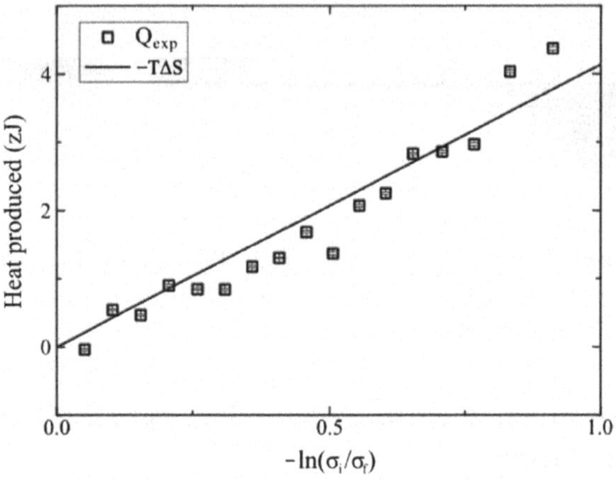

Fig. 1.4. Produced heat in the quasi-static regime during a single refresh operation for different entropy variations. Squares represent the estimated heat from experiments while the solid line is the theoretical prediction

The experimental realization of the refresh operation in this setup consists in changing the stiffens of the cantilever by means of the electromagnetic force and restoring the initial configuration, allowing the system to relax.

We expect each refresh operation to have a minimum energy cost related to the entropy variation of the system. This limit is met for quasi-static transformations, when frictional phenomena become negligible. The energy required to perform the refresh operation is estimated computing the work done on the system using the trajectory of the tip position and the variation of the potential energy of the system [5,10–12]. Without loss of generality, we can consider the approximation of harmonic potential inside each potential well. This assumption leads to the approximation of Gaussian distribution for the probability density function. The refresh procedure thus modifies the standard deviation of the probability density function of the system from its relaxed value, σ_f, to the initial one σ_i. These two quantities define the entropy variation during the refresh operation as:

$$\Delta S = k_B \ln \left(\frac{\sigma_i}{\sigma_f} \right) \tag{1.8}$$

In Fig. 1.4 we report the measured energy required for a single refresh operation (dots) as function of entropy variation, along with the theoretical prediction (continuous line).

Experimental data and theoretical prediction are in good agreement confirming that the harmonic model assumption for the mechanical system holds.

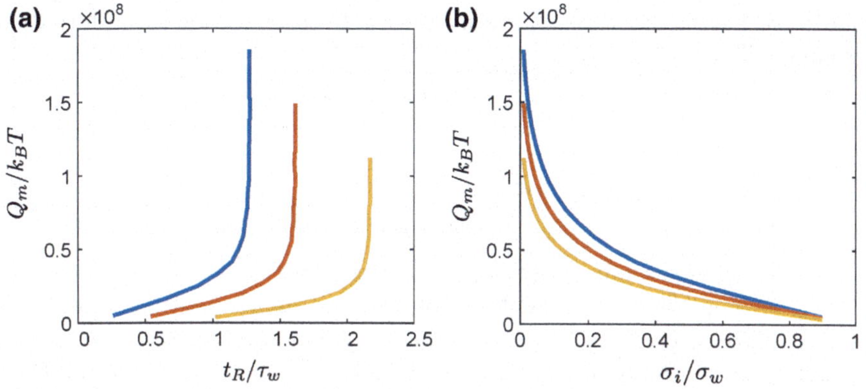

Fig. 1.5. Plots of Q_m to preserve the memory for $\bar{t} = 1\text{e}3\tau_k$ as a function of t_R **a** and σ_i **b**. Blue (dotted) lines are obtained with P_E=1e-6, red (dashed) lines with P_E=1e-4, and yellow (solid) lines with P_E=1e-2

1.4.2 Evaluation of Energy Requirement for Memory Preservation

Now that we have demonstrated that the theoretical limit can be achieved experimentally we can write the expression of the minimum energetic cost for preserving a memory for a given time with a finite probability of error as [5]:

$$Q_m = -NT\Delta S = \frac{\bar{t}}{t_R} k_B T \ln \left(\frac{\sqrt{\sigma_w^2 + e^{-\frac{t_R}{\tau_w}} (\sigma_i^2 - \sigma_w^2)}}{\sigma_i} \right) \tag{1.9}$$

In Fig. 1.5 we report the minimum energetic cost Q_m as function of the refresh time t_R (Fig. 1.5a) and σ_i (Fig. 1.5b) for different values of probability of error P_E.

From the results it is evident that we can preserve a memory for a given time with a given error probability while spending an arbitrarily little amount of energy. This is accomplished if the refresh procedure is performed arbitrarily often or arbitrarily close to thermal equilibrium.

1.5 Discussion

So far we have shown that preserving a memory for a given time and probability of error is possible expending an arbitrarily little amount of energy. However it should be noted that the initial standard deviation of the system σ_i, and thus the target standard deviation σ_f cannot be made arbitrarily small without spending an infinite amount of energy. This is also clear considering the Heisenberg indetermination principle that prevents the arbitrary confinement of the probability density, without spending an infinite amount of energy: the uncertainty on the impulse diverges when the uncertainty on the position shrinks [13]. In the best scenario we have $\sigma_x \sigma_p = \frac{\hbar}{2}$. If the memory refresh operation is performed at thermal equilibrium we have:

$$\sigma_p = m\sqrt{\langle v^2 \rangle - \langle v \rangle^2} = \sqrt{mk_B T} \tag{1.10}$$

and thus:

$$\sigma_x = \frac{\hbar}{2\sqrt{mk_B T}} \tag{1.11}$$

Since σ_i describes the uncertainty of the initial x value, we therefore have that $\sigma_i \geq \sigma_{iMin} = \frac{\hbar}{2\sqrt{mk_B T}}$. The existence of a σ_{iMin} implies that, even at $t = 0$, the probability of error p_0 is greater than zero. This probability of error then accumulates accordingly to Eq. 1.2 implying a minimum amount of error P_E.

It is thus possible to preserve your memory only for a limited amount of time for a fixed required probability of error. Within this limit, if the refresh procedure is done carefully enough, there is no need to spend any energy.

Acknowledgements. The authors gratefully acknowledge financial support from the European Commission (H2020, Grant agreement no: 732631, OPRECOMP, FPVII, Grant agreement no: 318287, LANDAUER and Grant agreement no: 611004, ICT-Energy) and ONRG grant N00014-11-1-0695.

References

1. R. Laudauer, IBM. J. Res. Dev. **5**, 183 (1961)
2. A. Berut, A. Arakelyan, A. Petrosyan, S. Ciliberto, R. Dillenschneider, E. Lutz, Nature (London) **483**, 187 (2012)
3. D. Chiuchiù, M.C. Diamantini, L. Gammaitoni, Europhys. Lett. **111**, 40004 (2015)
4. I. Neri, M. López-Suárez, Sci. Rep. **6**, 34039 (2016)
5. D. Chiuchiù et al., Phys. Rev. A **97**(5), 052108 (2018)
6. P.A. Laplante, *Comprehensive Dictionary of Electrical Engineering* (CRC Press, Boca Raton, 2005)
7. J. Bruce, D. Ng, D. Wang, *Memory Systems: Cache, DRAM, Disk*, 1st edn. (Morgan Kaufmann, Burlington, 2008)
8. *The Fokker-Planck Equation: Methods of Solution and Applications*, 2nd edn. (Springer, New York, 1989)
9. C. Gardiner, *Stochastic Methods: A Handbook for the Natural and Social Sciences*, 4th edn. (Springer, New York, 2009)
10. M. López-Suárez, I. Neri, L. Gammaitoni, Nat. Commun. **7**, 12086 EP (2016)
11. K. Sekimoto, *Stochastic Energetics*, 1st edn. (Springer, Berlin, 2010)
12. U. Seifert, Rep. Prog. Phys. **75**, 126001 (2012)
13. M.C. Diamantini, L. Gammaitoni, C.A. Trugenberger, Phys. Rev. E **94**, 012139 (2016)

Chapter 2
Modulation of NF Kinetics and Axonal Morphology Near the Excavation of the Mouse Optic Nerve

Yinyun Li[1(✉)], Tung Nguyen[2], and Peter Jung[2]

[1] School of Systems Science, Beijing Normal University, Beijing 100875, China
yinyun@bnu.edu.cn
[2] Department of Physics and Astronomy and Quantitative Biology Institute, Ohio University, Athens 45701, Ohio, USA
jungp@ohio.edu

Abstract. Neurofilaments (NFs) are the most abundant cytoskeletal structures in the axon and also cargo of axonal transport. Neurofilaments are synthesized in the neuronal cell body and transported bidirectionally along microtubule tracks in the axon with a net anterograde movement toward the nerve terminal. Based on this dual role of neurofilaments as space filling structures and cargo of axonal transport we hypothesize that neurofilament transport velocity regulates axon caliber. In this study, we combine results from a previous study of neurofilament kinetics in optic nerve with published morphometric features of the mouse optic nerve near the excavation to show that the sharp increase in the caliber of optic nerve is consistent with a slowing of neurofilament velocity.

2.1 Introduction

Neurons send electric signals, called action potentials, to other neurons along their axons, and the speed of this electric wave is proportional to their diameter (for a review see e.g. [1]). This linear relation renders axon caliber critical for neuron function and we are interested in how axons acquire their shape and caliber.

The cytoskeleton of the axon is composed of mainly three cytoskeletal proteins: neurofilaments, microtubules, and microfilaments (actin fibers) [2–5]. Among these cytoskeletal filaments, neurofilaments are the most abundant filaments and space filling structures determining the axonal caliber [2–4,6]. Besides being space-filling structures, neurofilaments are also cargo of slow transport. Neurofilaments are assembled in the cell body and transported bidirectionally along microtubule tracks with a net average velocity of 0.1–3 mm/day [2,7–9] towards the nerve terminal. This dual role has important implications for the

V. In et al. (Eds.): *Proceedings of the 5th International Conference on Applications in Nonlinear Dynamics*, Understanding Complex Systems, https://doi.org/10.1007/978-3-030-10892-2_2

formation of axon caliber and axon morphology [10] and potentially also for the understanding of neurodegenerative diseases such as ALS, associated with local swelling of axons [11–13]. For a constant neurofilament flux J, a decrease of the average transport velocity v, will give rise to an increase of neurofilament content and axon caliber controlled by the relation $J = \rho v$, where ρ denotes the linear density of neurofilaments, while an increase in velocity will give rise to a thinning of the axon.

Neurofilament transport kinetics has been modeled with a *stop and go model* (see Fig. 2.1) [14,15], where neurofilaments alternate stochastically between kinetic states a and r in which they move either anterogradly toward the nerve terminal, or retrogradely toward the cell body, states $a0$ and $r0$ in which they pause briefly for seconds to few minutes before they move again, and states ap and rp, in which they pause extensively for hours. We have termed the states, $a, a0, r, r0$, in which neurofilaments move or pause briefly, *on-track moving* and *on-track pausing* states, and the states, ap, rp, where they pause extensively *off-track states.* The underlying motivation for this nomenclature is that we envision that neurofilaments, which require a microtubule track for movement, move in a stop-and-go fashion along those tracks and pause extensively when they detach from the microtubule tracks and search diffusively for another one.

Transition rates of the neurofilaments between their 6 kinetic states (see Fig. 2.1), $\gamma_{10}, \gamma_{01}, \gamma_{\rm off}, \gamma_{\rm on}, \gamma_{\rm ar}, \gamma_{\rm ra}$, have been obtained by either direct observation of single neurofilaments in single small axons for minutes, analyzing in-vivo movement of ensembles of radio-labeled neurofilaments for weeks and months [14,15], and by using the novel fluorescent pulse-escape method [16] which reveals all except the reversal rates between anterograde and retrograde movement. Reversal rates $\gamma_{\rm ar}$ and $\gamma_{\rm ra}$ are small and therefore difficult to determine directly. Estimates of reversal rates of the order of $10^{-5}{\rm s}^{-1}$–$10^{-4}{\rm s}^{-1}$ have been reported based on the fraction of neurofilaments moving anterogradely and retrogradely [15], and an overall small reversal rate [17] of about $10^{-4}{\rm s}^{-1}$. A schematic of this model is shown in Fig. 2.1.

Neurofilament transport can be heterogeneous along the axon and this has implications for axon caliber. In myelinating cultures, for example, it was found that axons have a larger caliber A_m and exhibit a larger abundance of neurofilaments N_m along the myelinated segments than along the non-myelinated segments [10] with caliber A_n and abundance N_n. These differences in axon caliber correlated with a reduced neurofilament transport rate in myleinated segments v_m versus non-myelinated segments v_n. More specifically, the transport velocity of neurofilaments in the myelinated sections v_m was smaller than in the non-myelinated sections v_n, consistent with the prediction of the equation of continuity, i.e. $v_n N_n = v_m N_m$. This equation implies, that the number of neurofilaments moving into a segment of an axon per unit time must be the same as moving out. If that would not be the case, neurofilaments would accumulate and the axon would swell or neurofilaments would decline and the axon atrophy. It also implies, that in segments where neurofilaments move slower, the axon will be fatter, and vice verse. More recently, a study with ex-vivo extracted

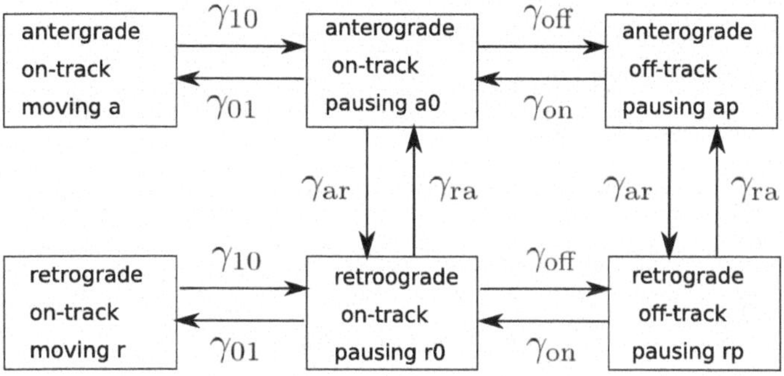

Fig. 2.1. 6-state model for neurofilament transport: Neurofilaments move anterograde and retrograde with velocities v_a and v_r in states a and r, respectively, are interrupted by brief pauses of seconds to few minutes in states $a0$ and $r0$, and pause extensively for hours in states ap and rp

myelinated axons of the mouse tibial nerve [18] revealed similar insights. In fully developed tibial nerve axons, with a regular repeated pattern of nodes of Ranvier and internodes, myelination at the internodes is far more substantial than the myelination seen in segments of axons in the myelinating cultures [10] and the axon exhibited a substantial reduction in caliber at the nodes of Ranvier (a factor of 10 in this particular experiment). It was found in that study, that the decrease in axon caliber at the nodes of Ranvier is accompanied with a proportionate increase of the transport velocity of neurofilaments, again, consistent with the equation of continuity $v_{\text{node}}N_{\text{node}} = v_{\text{inter}}N_{\text{inter}}$, where v_{node} and v_{inter} are the neurofilament transport velocities in the nodal and internodal segments of the axon, and N_{node} and N_{inter} the corresponding abundances of neurofilaments. Hence, local axon caliber correlates with neurofilament transport velocity. What the analysis in [10, 18] has also revealed is that the increase of axon caliber and neurofilament content along the myelinated sections of axons is facilitated by larger fractions of neurofilaments in the off-track states.

In this paper, we consider the morphologically heterogeneous structure of the mouse optic nerve near the retinal excavation [5, 19] (see Fig. 2.1). The cross sectional area (caliber) of the optic nerve undergoes a sharp, approximately 2.5-fold, expansion about 150 μm distal of the retinal excavation. The number of neurofilaments approximately doubles across this expansion and the number of microtubules increases approximately 1.5-fold. In this paper we explore the mechanism for the increased accumulation of neurofilaments in the optic nerve 150 μm distal from the retinal excavation and beyond.

2.2 Modeling Neurofilament Transport

Approximating the axon as a one-dimensional structure, we describe the transport of neurofilaments mathematically by the following set of partial differential equations [15],

$$
\begin{aligned}
\frac{\partial \rho_a}{\partial t} &= -v_a \frac{\partial \rho_a(x,t)}{\partial x} + \gamma_{01}\rho_{a0} - \gamma_{10}\rho_a + j_{\rm in}\delta(x-x_0) \\
\frac{\partial \rho_{a0}}{\partial t} &= -(\gamma_{01} + \gamma_{ar} + \gamma_{\rm off})\rho_{a0} + \gamma_{10}\rho_a + \gamma_{ra}\rho_{r0} + \gamma_{\rm on}\rho_{ap} \\
\frac{\partial \rho_{ap}}{\partial t} &= -(\gamma_{\rm on} + \gamma_{ar})\rho_{ap} + \gamma_{\rm off}\rho_{a0} + \gamma_{ra}\rho_{rp} \\
\frac{\partial \rho_r}{\partial t} &= -v_r \frac{\partial \rho_r(x,t)}{\partial x} + \gamma_{01}\rho_{r0} - \gamma_{10}\rho_r \\
\frac{\partial \rho_{r0}}{\partial t} &= -(\gamma_{01} + \gamma_{ra} + \gamma_{\rm off})\rho_{r0} + \gamma_{10}\rho_r + \gamma_{ar}\rho_{a0} + \gamma_{\rm on}\rho_{rp} \\
\frac{\partial \rho_{rp}}{\partial t} &= -(\gamma_{\rm on} + \gamma_{ra})\rho_{rp} + \gamma_{\rm off}\rho_{r0} + \gamma_{ar}\rho_{ap},
\end{aligned} \tag{2.1}
$$

where, $\rho_a(x,t)$ and $\rho_r(x,t)$ describe the linear densities of neurofilaments in the on-track moving states, $\rho_{a0}(x,t)$ and $\rho_{r0}(x,t)$ in the on-track pausing states, and $\rho_{ap}(x,t)$ and $\rho_{rp}(x,t)$ in the off-track pausing states. The partial derivatives in the equations for the motile neurofilaments indicate convective terms associated with their movement.

The transition rate γ_{10} represents the rate at which the NFs switch from the on-track running states, a and r, to the on-track pausing state, $a0$ and $r0$, (anterograde and retrograde), and γ_{01} represents the rate of the transition from the on-track pausing states, $a0$ and $r0$, back to the on-track running states, a and r. The rate constant $\gamma_{\rm on}$ represents the rate at which NFs switch from the off-track pausing states, ap and rp, to the on-track pausing states, $a0$ and $r0$. $\gamma_{\rm off}$ represents the transition from the on-track pausing state to the off-track pausing state. The reversal rate constants, $\gamma_{\rm ra}$, $\gamma_{\rm ar}$, represent the rates at which NFs switch from the retrograde pausing states, $r0$ and rp, to the anterograde pausing states, $a0$ and ap, and vice verse, respectively.

The last term on the right hand side of the first equation in Eq. 2.1 models injection of neurofilament at the proximal end x_0 of the axon at a rate of $j_{\rm in}$.

2.2.1 Detailed Balance and Transport Velocity

A system is said to obey *detailed balance* if the number of transitions between any two kinetic states, A and B, is the same as for the reverse transition. For our kinetic 6-state model these conditions result in the following equations

$$
\begin{aligned}
&\rho_a\gamma_{10} = \rho_{a0}\gamma_{01};\, \rho_{a0}\gamma_{\rm off} = \rho_{ap}\gamma_{\rm on};\, \rho_{a0}\gamma_{ar} = \rho_{r0}\gamma_{ra} \\
&\rho_r\gamma_{10} = \rho_{r0}\gamma_{01};\, \rho_{r0}\gamma_{\rm off} = \rho_{rp}\gamma_{\rm on};\, \rho_{ap}\gamma_{ar} = \rho_{rp}\gamma_{ra}\,,
\end{aligned} \tag{2.2}
$$

and consequently

$$
\frac{\partial \rho_a}{\partial x} = 0\,, \quad \frac{\partial \rho_r}{\partial x} = 0\,. \tag{2.3}
$$

If the rate constants do not vary along the axon, the distribution of neurofilaments in each state is uniform along the axon and are given by

$$\rho_a = \frac{j_{\rm in}}{v_a}, \rho_{a0} = \frac{j_{\rm in}}{v_a} q_1, \rho_{ap} = q_1 q_2 \frac{j_{\rm in}}{v_a}$$
$$\rho_r = q_3 \frac{j_{\rm in}}{v_a}, \rho_{r0} = \frac{j_{\rm in}}{v_a} q_3 q_1, \rho_{rp} = q_3 q_1 q_2 \frac{j_{\rm in}}{v_a}. \quad (2.4)$$

where $q_1 = \gamma_{10}/\gamma_{01}, q_2 = \gamma_{\rm off}/\gamma_{\rm on}, q_3 = \gamma_{\rm ar}/\gamma_{\rm ra}$. The transport velocity is determined by the fraction of neurofilaments in the mobile states a and r and their respective velocities v_a and v_r. i.e.

$$\bar{v} = v_a \frac{\rho_a}{\rho_{\rm all}} + v_r \frac{\rho_r}{\rho_{\rm all}}$$
$$= \frac{1}{((1 + q_1(1 + q_2))(\gamma_{ra} + \gamma_{ar}))} (\gamma_{ra} v_a + \gamma_{ar} v_r), \quad (2.5)$$

where $\rho_{\rm all} = \rho_a + \rho_{a0} + \rho_{ap} + \rho_r + \rho_{r0} + \rho_{rp}$.

Of importance in this paper are situations, where some rate constants vary along the axon. Modulation of the reversal rate constants γ_{ar} and γ_{ra}, as proposed in [20] to model a heterogeneous steady-state distribution of neurofilaments, is not consistent with detailed balance since the requirement of spatial constancy (see Eq. 2.3) of the distributions $\rho_{a0,r0}$ and $\rho_{a,r}$ is in conflict with $\rho_{a0}\gamma_{ar}(x) = \rho_{r0}\gamma_{ra}(x)$, which is obtained from Eq. 2.2. This means that the expression for the velocity in Eq. 2.5, which is based on detailed balance, is not an exact solution if the reversal rates vary along the axon. If the spatial modulation of the reversal rates, however, is weak, such as in the study in [20], Eq. 2.5 is still a good approximation for the transport velocity.

A more serious problem with using a spatially dependent reversal rate is that a sharp spatial modulation of the reversal rate does not result in a sharp change in neurofilament velocity because the reversal rates are small, i.e of the order of 10^{-5}–10^{-4}s^{-1}. For a reversal rate of 10^{-5}s^{-1} the average time between two reversals is 10^5s. For an average transport velocity of 1mm/day $\approx 0.01\,\mu$ms^{-1}, a neurofilament will travel a distance of 10^5s $\cdot 0.01\,\mu$m/s $= 1000\,\mu$m before it will respond to a changed reversal rate. The resulting velocity and hence axon-caliber profile, even for a sharp change in a reversal rate, will be too smooth to model the sharp transition of the caliber of the optic nerve we are considering.

A spatial change of the on-track rate $\gamma_{\rm on}(x)$ (or off-track rate $\gamma_{\rm off}$) does not destroy detailed balance. The conditions for detailed balance (see Eq. 2.2) where the on-track rate $\gamma_{\rm on}(x)$ is involved, i.e. $\rho_{a0,r0}\gamma_{\rm off} = \rho_{ap,rp}\gamma_{\rm on}(x)$ still allow spatially constant distributions ρ_{a0}, ρ_{r0} and hence ρ_a, ρ_r, as it is required for detailed balance (see Eq. 2.3). The densities of neurofilaments in the off-track states, however, will vary along the axon and are given by $\rho_{ap}(x,t) = q_1 q_2(x) j_{\rm in}$, and $\rho_{rp}(x,t) = q_1 q_2(x) q_3 j_{\rm in}$, with $q_2(x) = \gamma_{\rm off}/\gamma_{\rm on}$. Hence the equation for the transport velocity (see Eq. 2.5) where $q_2 = \gamma_{\rm off}/\gamma_{\rm on}$ is replaced by a variable $q_2(x) = \gamma_{\rm off}/\gamma_{\rm on}(x)$ is still valid for a heterogeneous rate $\gamma_{\rm on}(x)$.

2.3 Modeling Neurofilament Transport in the Optic Nerve

We have studied in earlier work [20] neurofilament transport in mouse optic nerve to address the controversial proposal in [21] that only a small fraction of neurofilaments are motile and the majority of neurofilaments are deposited in a stationary cross-linked cytoskeletal network. In that computational study we have been using a wealth of published kinetic and morphometric data to calibrate our mathematical model. The values of the transition rates in the above described 6-state model have been estimated by matching simulations of Eq. (2.1) to measured kinetic and morphometric features in mouse optic nerve [21]. These features included the propagation of a radioactively labeled pulse of neurofilaments over a time period of about 6 months, the decline of radioactivity in a 7mm window along the mouse optic nerve, and the observed linear increase of the abundance of neurofilaments along the nerve.

Of particular interest here is that the distribution of neurofilaments is not uniform along the optic nerve. It increases distally approximately linearly along the nerve [21]. To address this heterogeneity we followed the idea outlined earlier in this paper that if all neurofilaments are motile, neurofilament abundance $N(x)$ along the axon is controlled by their transport velocity $v(x)$, i.e. $J = N(x)v(x)$, where J is the net flux of neurofilaments. An increase in neurofilament abundance is therefore associated with a decrease of their transport velocity. Given the abundance profile $N(x)$ along the axon, we reconstructed the velocity profile $v(x)$ and then translated this into a profile of the reversal rate $\gamma_{ra}(x)$. A modulation of the on-rate γ_{on} accomplished the same goal. The required relative change of the on-rate, however, is larger than that of the reversal rate, because the velocity is more sensitive to changes in the reversal rates.

The study in [20] was focussed on the optic nerve, 1mm and further away from the retinal excavation, i.e. beyond the onset of myelination. In this study we focus on the first 1mm distal of the retinal excavation, where the nerve exhibits the above-described sharp increase in caliber. We follow the same modeling strategy as in [20]. We first extract neurofilament abundance from [5], i.e.

$$N(x) = 10^3 \begin{cases} 14.5 & \text{for } 0 < x < 50\,\mu\text{m} \\ 8.5 + 0.12x & \text{for } 50\,\mu\text{m} < x < 150\,\mu\text{m} \\ 29 + 2.7\cdot 10^{-3}x & \text{for } 150\,\mu\text{m} < x < 700\,\mu\text{m} \end{cases}, \tag{2.6}$$

where we connected the data points extracted from [5] linearly. The number of neurofilaments, $N(x)$, denotes the number of neurofilaments per thousand axons in the optic nerve. Given the expression for the velocity in Eq. 2.5, which is exactly valid even for heterogeneous on-rates $\gamma_{\text{on}}(x)$, the velocity profile then is $v(x) = J/N(x)$, with the neurofilament flux J. Resolving Eq. 2.5 for $q_2(x)$, i.e.

$$q_2(x) = \frac{1}{q_1}\left(\frac{\gamma_{ra}v_a + \gamma_{ar}v_r}{v(x)(\gamma_{ar} + \gamma_{ra})} - 1\right) - 1\,, \tag{2.7}$$

and $\gamma_{\text{on}}(x) = \gamma_{\text{off}}/q_2(x)$. As in the previous study of the optic nerve [20] we use the rate constants $\gamma_{10} = 0.093\text{s}^{-1}, \gamma_{01} = 0.041\text{s}^{-1}$ and velocities

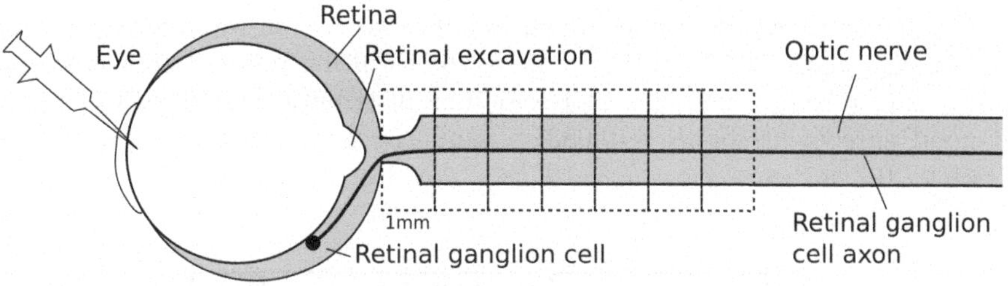

Fig. 2.2. Retina and optic nerve: This figure shows the eye, the retina, a representative retinal ganglion cell (full circle) with its axon (solid line), the retinal excavation, and the optic nerve with a sharp increase in caliber at about 150 μm distal from the retinal excavation. The sketched location of the sharp increase of the caliber of the optic nerve is not drawn to scale. Radioactive proteins are injected into the eye, are absorbed by retinal ganglion cells and incorporated into the neurofilaments which are assembled in their cell bodies for about 6 hours. The ensuing wave of radio-labeled neurofilaments travels through the axons of the retinal ganglion cells the average distance of about 1mm before it enters the optic nerve (see also [20]). The dotted rectangular regions of width of 1mm are the segments in which the abundance of radio-labeled neurofilaments are recorded in [21] (see Fig. 2.3)

$v_a = 0.53\,\mu\text{ms}^{-1}, v_r = -0.60\,\mu\text{ms}^{-1}$ which we obtained from in-vitro single-neurofilament tracking experiments [9,15]. The rates γ_{off} and γ_{on} have been obtained in the same neuronal culture using a fluorescent photoactivation pulse-escape method [16] resulting in $\gamma_{\text{off}} = 4.5 \cdot 10^{-3}\text{s}^{-1}$ and $\gamma_{\text{on}} = 2.75 \cdot 10^{-4}\text{s}^{-1}$. In this study, the on-rate γ_{on} is replaced by the heterogeneous on-rate Eq. (2.7) reconstructed from the morphometric profile of the optic nerve near the retinal excavation. Reversal rates γ_{ar} and γ_{ra} have been obtained in [15] from the fraction of retrogradely moving neurofilaments, given in terms of the model as γ_{ar}/γ_{ra}, and rough estimates of the number of rare reversals, given by γ_{ar} and γ_{ra}, resulting in $\gamma_{\text{ar}} = 3.1 \cdot 10^{-5}\text{s}^{-1}$ and $\gamma_{\text{ra}} = 6.9 \cdot 10^{-5}\text{s}^{-1}$. These values for the reversal rates served as a starting point in this study and have been adjusted as described below.

The last unknown parameter is the neurofilament flux J, which we obtain by simulating the fate of a pulse of radio-labeled neurofilaments injected into the retina for a duration of about 5 hours [21], i.e. $J(t) = j_0$ for $0 < t < 1.8 \cdot 10^4$s and otherwise $J(t) = 0$. The pulse of radio-labeled neurofilaments propagates into the optic nerve, spreads in width, and was recorded for about 6 months. We model the injection of neurofilaments into the retina with the source-term for neurofilaments in the anterograde moving state a, i.e.

$$\frac{\partial \rho_a}{\partial t} = -v_a \frac{\partial \rho_a}{\partial x} - \gamma_{10}\rho_a + \gamma_{01}\rho_{a0} + J(t)\delta(x+1)\,. \tag{2.8}$$

The source term $\delta(x+1)$ incorporates the average distance of about 1mm of the retinal excavation, the beginning of the optic nerve, to the retinal ganglion cells (see Fig. 2.2). In the domain $-1\text{mm} < x < 0$, we use a constant on-rate

$\gamma_{\text{on}} = \gamma_{\text{on}}(x = 0)$ which connects continuously to the on-rate extracted from morphometric data at $x = 0$. We compare the resulting pulse of neurofilaments 10 days after injection with the corresponding distribution reported in [21] and find good agreement for $J = 0.083\,\text{s}^{-1}$ and the fine tuned reversal rates of $\gamma_{ar} = 5.0 \cdot 10^{-5}\text{s}^{-1}$ and $\gamma_{ra} = 7.2 \cdot 10^{-5}\text{s}^{-1}$ (see Fig. 2.3).

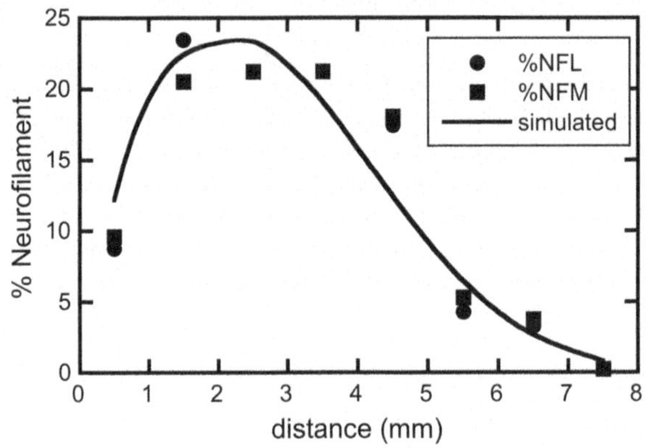

Fig. 2.3. Distribution of radio-labeled neurofilaments: The distribution of radio-labeled neurofilament subunits L and M along 8mm of the optic nerve is shown 10 days after injection [21] alongside the distribution of neurofilaments predicted by our model (solid line)

2.4 Results

The main goal of this paper is to elucidate the mechanism by which the optic nerve generates the sharp increase in caliber near the retinal excavation. We have constructed the rate constants for our kinetic model consistent with a number of key-experiments as described in the previous section. The solid line in Fig. 2.4a represents the reconstructed profile of the neurofilament distribution. The corresponding distributions of neurofilaments in their kinetic states is shown in Fig. 2.5a. The significance of the results shown in this figure is that the number of neurofilaments in on-track states (labeled with $a, a0, r$ and $r0$) is constant along the optic nerve, implying the validity of detailed balance. The numbers of neurofilaments in the off-track states ap and rp exhibit sharp increases at about $150\,\mu\text{m}$ distal from the retinal excavation ($x = 0$), providing the space-filling structures necessary for the structural integrity along the expanded caliber of the axon. As the axon expands in caliber distally, the average velocity of the neurofilaments decreases accordingly (Fig. 2.5b). The decrease of the velocity is facilitated by a decrease of the on-rate γ_{on} (Fig. 2.5c), as this causes more neurofilaments to accumulate off-track.

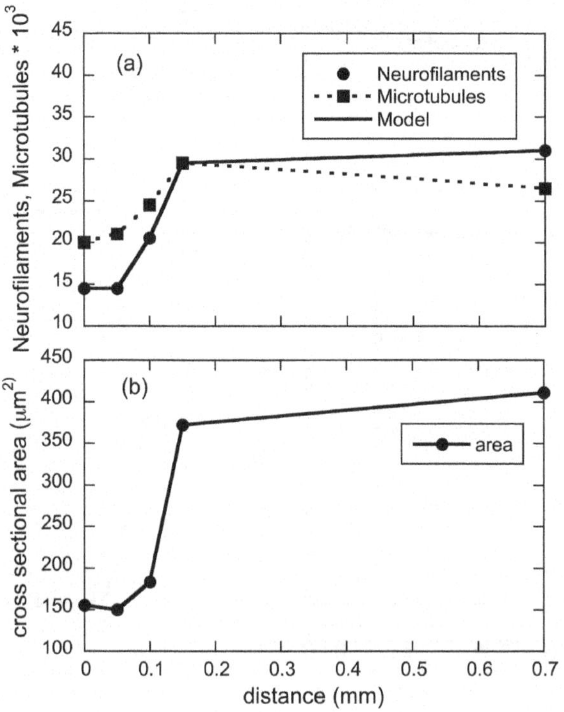

Fig. 2.4. Neurofilament, microtubules and axon caliber: In **a**, the numbers of neurofilaments (solid circles) and microtubules (solid squares) recorded in [5] are shown along the first mm of the optic nerve. The numbers of neurofilaments and microtubules shown are the average numbers recorded in 1000 axons at the respective locations in the optic nerve. The solid line shows the neurofilament content generated by our model. In **b**, we show the corresponding cross sectional areas (for thousand axons) along the optic nerve. Most importantly, the sharp increase in axon caliber of a factor of about 2.5 at about 150 μm distal from the retinal excavation nerve correlates with an increase of neurofilaments of about a factor of 2 and an increase of microtubules of a factor of 1.5. All data points are taken from [5] and redrawn

While our model doesn't provide the actual mechanism by which the average neurofilament velocity decreases, or equivalently, by which the on-rate decreases, it reveals that the on-rate, $\gamma_{\rm on}$, is reduced by about 60% (see Fig. 2.5b) where the caliber of the nerve (see Fig. 2.4b) and the abundance of neurofilaments (see Fig. 2.4a) exhibit sharp increases of about a factor of two. This behavior can be explained if we assume that the long-term pausing of the neurofilaments is associated with a diffusive search of neurofilaments for microtubules [22] after disengaging from another microtubule track to become motile again.

In the following we estimate a relation between the areal density of microtubules ρ_M and the expected on-rate $\gamma_{\rm on}$. 150 μm distal from the retinal excavation, the caliber of the optic nerve (per 1000 axons) exhibits an increase from

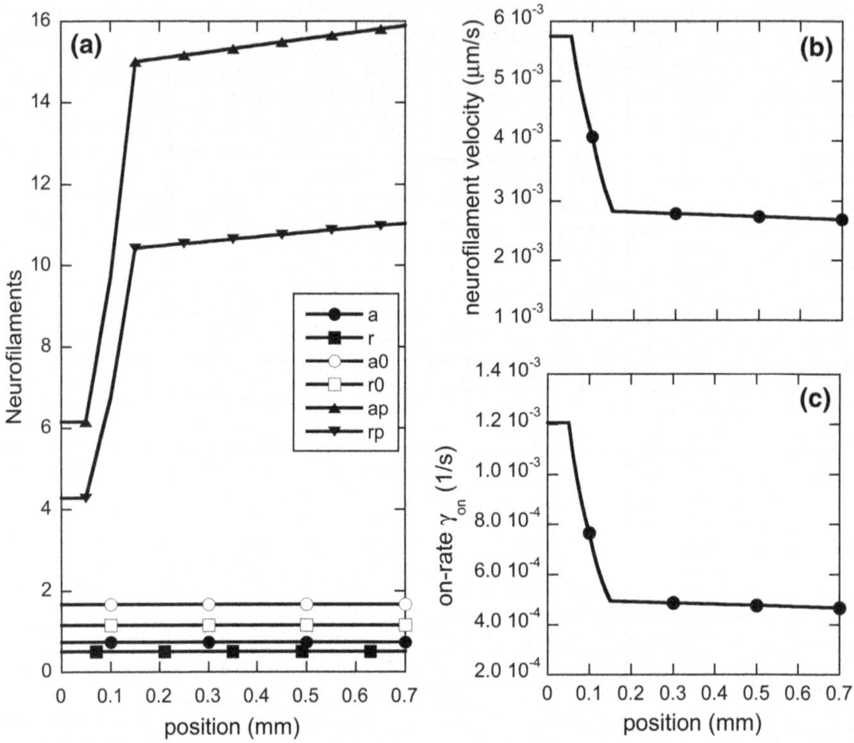

Fig. 2.5. Distribution of neurofilaments In panel **a**, we show the distributions of neurofilaments in the 6 kinetic states. The distributions of neurofilaments in the on-track states (moving and pausing),a,a0,r,r0, are constant along the axon. The numbers of neurofilaments in the off-track states, ap,rp, exhibit a sharp increase at 150 μm, where the optic nerve increases sharply in caliber (see Fig. 2.4b). In panel **b**, we show the reconstructed average velocity of the neurofilaments along the optic nerve. The decreasing velocity is associated with a decreasing on-rate $\gamma_{\rm on}$ as shown in the panel of **c**

about 150 μm^2 to about 400 μm^2 (see Fig. 2.4b) and the number of microtubules increases from 20 to about 30 (see Fig. 2.4a). This corresponds to a decrease of the microtubule areal density, ρ_M, of about 45%, from 0.13 μm^2 to 0.075 μm^2. Assuming a uniform distribution of microtubules in the cross section of each axon, we can calculate the average distance between two nearest-neighbor microtubules as $d_0 = 1/(2\sqrt{\rho_M})$ [23]. Assuming that neurofilaments detach from one microtubule, loose motility along the axon, but search by radial diffusion with diffusion constant D for another microtubule, the mean first passage time for neurofilaments to reach the distance of d_0 at which they, in the statistical average, will find another microtubule, is given by $T = d_0^2/(4D)$. The on-rate, accordingly, is then $g_{on} \approx 1/T \propto \rho_M$. Hence, the 45% decrease of microtubule density is expected to result in a 45% decrease of the on-rate $\gamma_{\rm on}$. Our predictions in

Fig. 2.5, based on our modeling is a decrease of the on-rate γ_{on} of about 60%. The discrepancy between the estimate and the model predictions is probably due to the fact that we lumped 1000 axons together into one fat axon to estimate the change of microtubule densities. In reality, axons of retinal ganglion cells are very thin with an area of about 0.2–0.4 μm^2 and the diffusion of neurofilaments is more constrained. Hence, we expect to over estimate diffusivity and on-rate.

2.5 Summary

We discuss the formation of a sharp increase in the caliber of the optic nerve near the retinal excavation within the paradigm that the dual role of neurofilaments as space-filling structures and cargo of slow axonal transport provides a mechanism to regulate axon caliber through changes of neurofilament kinetics. Our model predicts that the increase in the caliber of the optic nerve is generated by an increased fraction of off-track neurofilaments, reducing their average transport velocity. We further hypothesize that the increased fraction of off-track neurofilaments is related to the decreased density of microtobule tracks increasing the time it takes off-track neurofilaments to find a microtubule track necessary for motility by radial diffusion.

Acknowledgements. YL acknowledges financial support from the Chinese Natural Science foundation for Young Scientist No. 31601145 and support from Fundamental Research Funds for Central Universities, China (Y. Li). P. Jung is supported by the US National Science Foundation by grant IOS-1656765. We are grateful for extensive discussion with Anthony Brown from the Department of Neuroscience of Ohio State University, Columbus, Ohio.

References

1. C. Hildebrand, S. Remahl, H. Persson, C. Bjartmar, Myelinated nerve fibres in the CNS. Prog. Neurobiol. **40**, 319–384 (1993)
2. A. Brown, Slow axonal transport. Encycl. Neurosci. **9**, 1–9 (2009)
3. A. Brown, P. Jung, A critical reevaluation of the stationary axonal cytoskeleton hypothesis. Cytoskeleton (Hoboken, N.J.) **70**, 1–11 (2013)
4. R.A. Nixon, K.B. Logvinenko, Multiple fates of newly synthesized neurofilament proteins: evidence for a stationary neurofilament network distributed nonuniformly along axons of retinal ganglion cell neurons. J. Cell Biol. **102**, 647–659 (1986)
5. R.A. Nixon, P.A. Paskevich, R.K. Sihag, C.Y. Thayer, Phosphorylation on carboxyl terminus domains of neurofilament proteins in retinal ganglion cell neurons in vivo: influences on regional neurofilament accumulation, interneurofilament spacing, and axon caliber. J. Cell Biol. **126**, 1031–1046 (1994)
6. P.N. Hoffman, J.W. Griffin, B.G. Gold, D.L. Price, Slowing of neurofilament transport and the radial growth of developing nerve fibers. J. Neurosci. **5**, 2920–2929 (1985)
7. A. Brown, Slow axonal transport: stop and go traffic in the axon. Nat. Rev. Mol. Cell Biol. **1**, 153–156 (2000)

8. L. Wang, C.L. Ho, D. Sun, R.K. Liem, A. Brown, Rapid movement of axonal neurofilaments interrupted by prolonged pauses. Nat. Cell Biol. **2**, 137–141 (2000)
9. L. Wang, A. Brown, Rapid intermittent movement of axonal neurofilaments observed by fluorescence photobleaching. Mol. Biol. Cell **12**, 3257–3267 (2001)
10. P.C. Monsma, Y. Li, J.D. Fenn, P. Jung, A. Brown, Local regulation of neurofilament transport by myelinating cells. J. Neurosci. **34**, 2979–2988 (2014)
11. J.Q. Trojanowski, V.M.Y. Lee, Aggregation of neurofilament and alpha-synuclein proteins in Lewy bodies: implications for the pathogenesis of Parkinson disease and Lewy body dementia. Arch. Neurol. **55**, 151–152 (1998)
12. C.C.J. Miller, S. Ackerley, J. Brownlees, A.J. Grierson, N.J.O. Jacobsen, P. Thornhill, Axonal transport of neurofilaments in normal and disease states. Cell Mol. Life Sci. **59**, 323–330 (2002)
13. R.E. Schmidt, L.N. Beaudet, S.B. Plurad, D.A. Dorsey, Axonal cytoskeletal pathology in aged and diabetic human sympathetic autonomic ganglia. Brain Res. **769**, 375–83 (1997)
14. A. Brown, L. Wang, P. Jung, Stochastic simulation of neurofilament transport in axons: the ?stop-and-go? hypothesis. Mol. Biol. Cell **16**, 4243–4255 (2005)
15. P. Jung, A. Brown, Modeling the slowing of neurofilament transport along the mouse sciatic nerve. Phys. Biol. **6**, 046002 (2009)
16. N. Trivedi, P. Jung, A. Brown, Neurofilaments switch between distinct mobile and stationary states during their transport along axons. J. Neurosci. **27**, 507–516 (2007)
17. A. Uchida, A. Brown, Arrival, reversal, and departure of neurofilaments at the tips of growing axons. Mol. Biol. Cell **15**, 4215–4225 (2004)
18. C.L. Walker, A. Uchida, Y. Li, N. Trivedi, J.D. Fenn, P.C. Monsma, R.C. Larivićre, J.-P. Julien, P. Jung, A. Brown, Local acceleration of neurofilament transport at nodes of Ranvier [submitted]
19. R.A. Nixon, Dynamic behavior and organization of cytoskeletal proteins in neurons: reconciling old and new findings. Bioessays **20**, 798–807 (1998)
20. Y. Li, P. Jung, A. Brown, Axonal transport of neurofilaments: a single population of intermittently moving polymers. J. Neurosci. **32(**, 746–758 (2012)
21. A. Yuan, T. Sasaki, M.V. Rao, A. Kumar, V. Kanumuri, D.S. Dunlop, R.K. Liem, R.A. Nixon, Neurofilaments form a highly stable stationary cytoskeleton after reaching a critical level in axons. J. Neurosci. **29**, 11316–11329 (2009)
22. T. Nguyen, P. Jung, Neurofilament proximity to microtubules determines their motility [in preparation]
23. P. Hertz, Über den gegenseitigen durchschnittlichen Abstand von Punkten, die mit bekannter mittlerer Dichte im Raume angeordnet sind. Math. Ann. **67**, 387–398 (1909)

Chapter 3
Coupled Crystal Oscillator System and Timing Device

Antonio Palacios[1(✉)], Pietro-Luciano Buono[2], Visarath In[3], and Patrick Longhini[3]

[1] Nonlinear Dynamical Systems Group, Department of Mathematics, San Diego State University, San Diego, CA 92182, USA
apalacios@sdsu.ed
[2] Faculty of Science, University of Ontario Institute of Technology, 2000 Simcoe St. N, Oshawa, ON L1H 7K4, Canada
Pietro-Luciano.Buono@uoit.ca
[3] Space and Naval Warfare Systems Center, Code 71740, 53560 Hull St, San Diego, CA 92152-5001, USA
{visarath,patrick.longhini}@spawar.navy.mil

Abstract. At the National Observatory in Washington D.C., time is measured by averaging the times of an uncoupled ensemble. The measurements show a scaling law for phase-error reduction as, where is the number of crystals in the ensemble. Analytical and computational works show that certain patterns of collective behavior produced by a network of nonlinear oscillators leads to optimal phase-error that scales down as. In this talk we use symmetry-based methods to classify all possible patterns of oscillations, and their stability properties. Then we show why, among all possible patterns, a traveling wave, in which consecutive oscillators are out of phase by, yields the best phase-error reduction. Finally, we prove, analytically, that is the fundamental limit of of phase-error reduction that can be obtained with a network of nonlinear oscillators of any type, not just crystals.

3.1 Introduction

We present a computational and analytical study of a network-based model of a high-precision, inexpensive, Coupled Crystal Oscillator System and Timing (CCOST) device. A bifurcation analysis of the network dynamics shows a wide variety of collective patterns, mainly various forms of discrete rotating waves and synchronization patterns. Results from computer simulations seem to indicate that, among all patterns, the *standard* traveling wave pattern in which consecutive crystals oscillate out of phase by $2\pi/N$, where N is the network size, leads to phase drift error that decreases as $1/N$ as opposed to $1/\sqrt{N}$ for an uncoupled ensemble. The results should provide guidelines for future experiments, design and fabrication tasks.

V. In et al. (Eds.): *Proceedings of the 5th International Conference on Applications in Nonlinear Dynamics*, Understanding Complex Systems, https://doi.org/10.1007/978-3-030-10892-2_3

3.2 Modeling

A crystal oscillator circuit sustains oscillation by taking a voltage signal from the quartz resonator, amplifying it, and feeding it back to the resonator. The frequency of the crystal is slightly adjustable by modifying the attached capacitances. A varactor, a diode with capacitance depending on applied voltage, is often used in voltage-controlled crystal oscillators, VCO. The analog port of the VCO chip is modeled by a nonlinear resistor R^-, which obeys the voltage-current relationship

$$v = -ai + bi^3,$$

where a and b are constant parameters. In addition, parasitic elements can be represented by a series resonator (L_2, C_2, R_2) connected in parallel with the nonlinear resistor. The resulting circuit, depicted in Fig. 3.1(left), forms a two-mode resonator model. See Fig. 3.1.

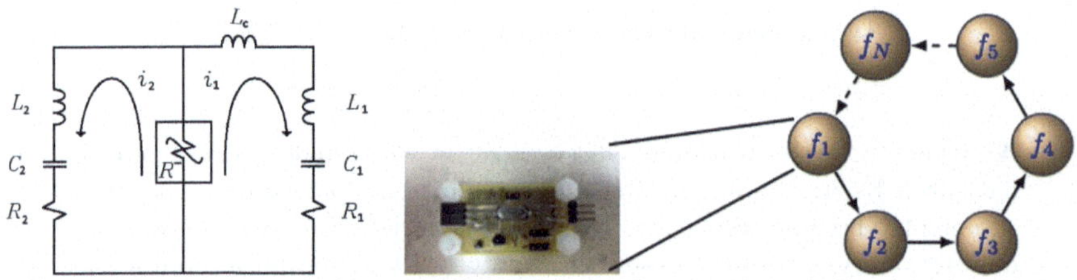

Fig. 3.1. (Left) Two-mode crystal oscillator circuit. A second set of spurious RLC components (R_2, L_2, C_2) are introduced by parasitic elements. (Right) CCOST Concept

Applying Kirchhoff's voltage law yields the following governing equations

$$L_j \frac{d^2 i_j}{dt^2} + R_j \frac{di_j}{dt} + \frac{1}{C_j} i_j = [a - 3b(i_1 + i_2)^2] \left[\frac{di_1}{dt} + \frac{di_2}{dt} \right], \qquad (3.1)$$

where $j = 1, 2$ and L_c has been included in L_1.

3.3 Governing Equations for Coupled System

In this section we consider a Coupled Crystal Oscillator System (CCOST) made up of N, assumed to be identical, crystal oscillators. Typical coupling topologies include unidirectional and bidirectional coupling in a ring fashion. Figure 3.1(right) shows the former case. The spatial symmetry of the unidirectionally coupled ring is described by the group $\mathbf{Z}_N$ of cyclic permutations of N objects. In the bidirectionally coupled case the symmetry group is d_N, which describes the symmetries of an N-gon.

Applying Kirchhoff's law to the CCOST network with unidirectional coupling yields the following (dimensionless version) governing equations

$$
\begin{aligned}
&\frac{d^2 i_{k,1}}{dt^2} + \Omega_1^2 i_{k,1} = \\
&\varepsilon \Bigg\{ -R_1 \frac{di_{k,1}}{dt} + \left[a - 3b\left(i_{k,1} + i_{k,2} - \lambda\left[i_{k+1,1} + i_{k+1,2}\right]\right)^2\right] \\
&\left[\frac{di_{k,1}}{dt} + \frac{di_{k,2}}{dt} - \lambda\left(\frac{di_{k+1,1}}{dt} + \frac{di_{k+1,2}}{dt}\right)\right]\Bigg\} \\
&\frac{d^2 i_{k,2}}{dt^2} + \Omega_2^2 i_{k,2} = \\
&\varepsilon L_r \Bigg\{ -R_2 \frac{di_{k,2}}{dt} + \left[a - 3b\left(i_{k,1} + i_{k,2} - \lambda\left[i_{k+1,1} + i_{k+1,2}\right]\right)^2\right] \\
&\left[\frac{di_{k,1}}{dt} + \frac{di_{k,2}}{dt} - \lambda\left(\frac{di_{k+1,1}}{dt} + \frac{di_{k+1,2}}{dt}\right)\right]\Bigg\},
\end{aligned}
\tag{3.2}
$$

where $L_{k,1} = L_1$, $L_{k,2} = L_2$, $R_{k,1} = R_1$, $R_{k,2} = R_2$, $C_{k,1} = C_1$ and $C_{k,2} = C_2$. Letting $t = \sqrt{L_1 C_1}\tau$, $\Omega_1^2 = 1$, $\Omega_2^2 = \frac{L_1}{L_2}\frac{C_1}{C_2}$, $L_r = \frac{L_1}{L_2}$, $\varepsilon = \sqrt{\frac{C_1}{L_1}}$. The new time variable τ has been relabeled as t.

3.4 Averaging

After applying the following set of invertible coordinates transformations

$$
\begin{aligned}
i_{kj} &= x_{kj} \cos\phi_{kj}; \\
i'_{kj} &= -\Omega_j x_{kj} \sin\phi_{kj}; \\
i''_{kj} &= -\Omega_j x'_{kj} \sin\phi_{kj} - \Omega_j^2 x_{kj} \cos\phi_{kj} - \Omega_j x_{kj} \psi'_{kj} \cos\phi_{kj}; \\
\phi_{kj} &= \Omega_j t + \psi_{kj};
\end{aligned}
\tag{3.3}
$$

for $j = 1, 2$ we arrive at the following set of equations, written symbolically as:

$$
\begin{bmatrix} \mathbf{x}'_k \\ \phi'_k \\ \phi'_s \end{bmatrix} = \begin{bmatrix} 0 \\ 0 \\ \mathbf{\Omega}^0 \end{bmatrix} + \varepsilon \begin{bmatrix} \mathbf{X}^{[1]}(\mathbf{x}_k, \phi_k + \phi_s, \phi_{k+1} + \phi_s, \varepsilon) \\ \mathbf{\Omega}^{[1]}(\mathbf{x}_k, \phi_k + \phi_s, \phi_{k+1} + \phi_s, \varepsilon) \\ 0, \end{bmatrix}. \tag{3.4}
$$

where $\mathbf{x}_k = (x_{k1}, x_{k2})$, $\phi_k = (\phi_{k1}, \phi_{k2})$ and $\mathbf{\Omega}^0 = (\Omega_1, \Omega_2)$. These equations include the shift $\phi_k \mapsto \phi_k + \phi_s$ and $\phi_{k+1} \mapsto \phi_{k+1} + \phi_s$, where $\phi_s = (\phi_{s1}, \phi_{s2})$.

After applying the averaging method, we arrive at a new set of equations, which can be written in complex form to facilitate analysis. The equations are of the form

$$
\begin{aligned}
\dot{z}_{k1} &= f_1(z_{k1}, z_{k2}, z_{k+1,1}, z_{k+1,2}, \mu) \\
\dot{z}_{k2} &= f_2(z_{k1}, z_{k2}, z_{k+1,1}, z_{k+1,2}, \mu),
\end{aligned}
\tag{3.5}
$$

where μ is a vector of parameters. A similar set of equations are obtained for the bidirectional case. The complete equations can be found in [1,2]. The symmetry of these averaged amplitude-phase equations is captured by the groups

$\mathbf{Z}_N \times \mathbf{O(2)} \times \mathbf{O(2)}$ and $\mathrm{d}_N \times \mathbf{O(2)} \times \mathbf{O(2)}$ for the unidirectional and bidirectional coupling cases, respectively. A complete analysis of the equations can be found in [1,2]. We summarize the main results. Steady-states of the averaged system with symmetry group $\Sigma \subset \Gamma \times \mathbf{SO(2)}$, with $\Gamma = \mathbf{Z}_N$ and $\Gamma = \mathrm{d}_N$ lead to periodic solutions with spatio-temporal symmetry $\Sigma \subset \Gamma \times \mathbf{S}^1$. Then, the tangent space to the trivial steady-state can be decomposed along irreducible representations of the $\mathbf{Z}_N$ and d_N actions and thus we obtain a block diagonalization of the linearization of the complexified governing equations. Symmetry-preserving and symmetry-breaking bifurcations are then determined by examining the eigenvalues computed directly from the block diagonalization. Criticality computations are also performed to determine the direction of bifurcations.

3.5 Phase Drift

Phase error is defined as the drift of the period of oscillation of an oscillating system away from the expected period length. To study phase drift, the governing equations are rewritten in Langevin form

$$\begin{aligned} d_t X_k &= F(X_k) - \lambda \sum_{j \to k}^{N} h(X_j, X_k) + \eta_k \\ d_t \eta_k &= -\frac{\eta_k}{\tau_c} + \frac{\sqrt{2D}}{\tau_c} \xi_k, \end{aligned} \tag{3.6}$$

where the noise function η_k is assumed to be Gaussian, band-limited, having a zero mean, a variance σ^2, and have a specific correlation time, τ_c. The noise is assumed to not drive the dynamics of the system, this corresponds to $\tau_f \ll \tau_c$, where τ_f is the time-constant of each oscillator [3,4]. $X_k = [i_{k1}, i'_{k1}, i_{k2}, i'_{k2}]$ is the state variable of each crystal oscillator, τ_c, D are correlation time and noise intensity respectively, F represents the internal dynamics of each oscillating unit, i.e., each crystal oscillator, h is the coupling function between two oscillators, in which the summation is taken over those cells j that are coupled to each cell k, λ is the coupling strength, ξ_k is a Gaussian distributed random variable with zero mean, and standard deviation σ.

Figure 3.2(top-left) illustrates the performance with respect to the scaling exponent, i.e., this figure is a log plot phase error, $Err(N, \lambda) = N^{m(\lambda)}$. Samples are taken for 100 values of λ. For each value of λ, the mean phase error for 50 repeated simulations is calculated for $N = 3, 5, \ldots, 21$. Then a least squares regression is performed on the log of these values, producing the scaling exponents depicted in Fig. 3.2. This analysis suggests that strong coupling is preferable to weak coupling to produce optimal scaling. From Fig. 3.2, the optimal scaling is found at $\lambda = 0.99$ with $m = -0.8947$. Figure 3.2(top-right) illustrates the design and network response captured by an oscilloscope. The white box in the figure contains appropriate potentiometers to control the gain of the operational amplifiers, which in turn, are used to manipulate coupling strength, and thus, control the network response to the desired pattern of oscillation.

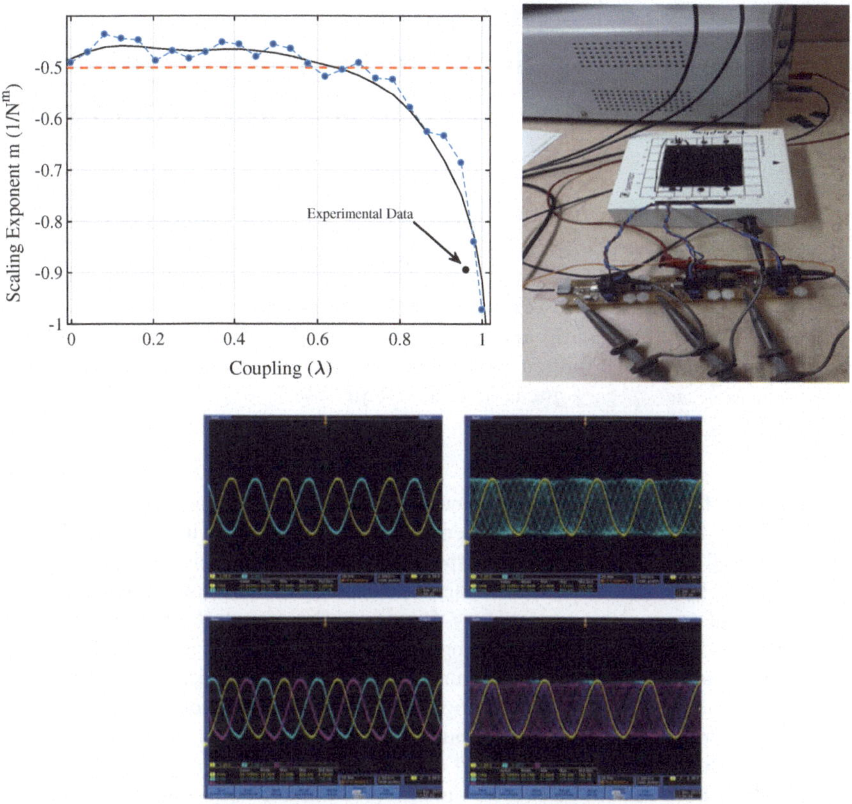

Fig. 3.2. (Top) Experimental realization of a network of coupled crystal oscillators implemented via PIC boards. (Bottom-left) Experimental measurements for $N = 2$ and $N = 3$ reveal, as expected, a traveling wave pattern among the oscillations. (Bottom-right) When the oscillators are uncoupled the pattern disappears

References

1. P.-L. Buono, B. Chan, J. Ferreira, A. Palacios, S. Reeves, P. Longhini, V. In, Symmetry-breaking bifurcations and patterns of oscillations in rings of crystal oscillators. SIAM J. Appl. Dyn. Syst. **17**(2), 1310–1352 (2018a)
2. P.-L. Buono, V. In, P. Longhini, L. Olender, A. Palacios, S. Reeves, Phase drift on networks of coupled of crystal oscillators for precision timing. Phys. Rev. E **98**, 012203 (2018b)
3. C. Gardiner, Complexity, *Handbook of Stachastic Methods*, 3rd edn. (Springer, Berlin, 2003)
4. S. Wio, M. L., R. Deza, *An Introduction to Stochastic Processes and Nonequilibrium Statistical Physics* (World Scientific Publishing, Singapore, 2012)

Chapter 4
Engineering Scalable Digital Circuits From Non-digital Genetic Components

Alexander P. Nikitin[1(✉)], Jordi Garcia–Ojalvo[2], and Nigel G. Stocks[1]

[1] School of Engineering, University of Warwick, Coventry CV47AL, UK
a.p.niktin@warwick.ac.uk, n.g.stocks@warwick.ac.uk

[2] Department of Experimental and Health Sciences, Parc de Recerca Biomedica de Barcelona, Universitat Pompeu Fabra, Dr. Aiguader 88, 08003 Barcelona, Spain
jordi.g.ojalvo@upf.edu

Abstract. Synthetically engineered single-cellular biological systems could be designed to classify patterns of chemical signals with high specificity and invoke appropriate responses. This requires cells to produce accurate logical computation over their multiple inputs and then trigger cellular response in a binary form like the signals YES and NO. However, current engineered biological systems, as a rule, are built from components like combinatorial promoters that, although displaying 'logic like' capabilities, fall short of supporting true binary (Boolean) computation. Consequently misclassification of inputs or errors in processing commonly occur that in turn lead to an incorrect cellular response. Here we show how that increased nonlinearity combined with noise suppression leads to genetic circuits capable of true Boolean logic operation able to support scalable logic circuit design.

4.1 Introduction

Potential applications of engineered single-cellular biosensory systems (biosensors) could be very broad, examples include identification of specific cancer cells [11, 14] or the detection of heavy metals like lead and mercury in the environment [8, 10, 13]. Furthermore, multi-input biosensors could produce more complex functions, for example they could classify patterns of chemical signals with high specificity [11, 14]. To do this the biosensors must produce logical computations over their multiple inputs and trigger cellular responses in a form of high and low levels of chemical signals that is similar to 1 and 0 of the Boolean algebra [7]. For example, in cancer therapy biosensors can trigger apoptosis in the presence of a specific set of cancer-specific biomarkers only [14]. However, modern biosensors generally are not robust, i.e. frequently their response to inputs holds significant errors. In cancer therapy such errors mean that some healthy

V. In et al. (Eds.): *Proceedings of the 5th International Conference on Applications in Nonlinear Dynamics*, Understanding Complex Systems, https://doi.org/10.1007/978-3-030-10892-2_4

cells are killed by mistake and some cancer cells survive [11,14]. Error reduction in biosensors is an open problem [11,14] and hence mainstream therapeutic interventions or applications in environmental sensing are still to be realised.

There are lots of sources of errors that include not only fluctuations of different sorts [12] but also stem from the inherent analogue response of genetic components [6]. Indeed, transfer functions of genetic circuits are similar to the Hill function [9] rather than the ideal Heaviside step function. Here we show that even if a genetic gate is able to produce digital-like computations with an acceptable small error (a deviation from the true binary levels), the non-Heaviside nonlinearity leads to the non-scaleability in large genetic circuits due to an amplification and propagation of this error.

Here we report the design of genetic gates capable of true logic function that suppress propagation error. These gates can therefore be combined in complex circuits that are scalable and operate with high accuracy. We also present a theoretical framework that can briefly be formulated as *a synthesis of scalable digital circuits from non-digital genetic components.*

4.2 Non-scalability of Digital-Like Genetic Circuits

According to the thermodynamic models of gene expression it is assumed that the level of gene expression is proportional to the equilibrium probability that RNA polymerase is bound to the promoter of interest [2]. Statistical mechanics provides a framework for computing this probability as a function of concentrations of regulatory proteins and molecular complexes in the cell [2]. For example, if the regulatory protein is an activator then the level of gene expression has the following form,

$$R = \mu \frac{C + \Gamma}{C + M}, \tag{4.1}$$

where μ, Γ and M are positive constants, and C is the concentration of the protein that plays a role of an activator when $\Gamma \ll M$. It is easy to show that there are two saturation levels corresponding to two limits $\lim_{C \to 0} R = \mu\Gamma/M$ and $\lim_{C \to \infty} R = \mu$. i.e. the digital-like inputs in the form of low and high concentrations of the activator induces the digital-like response in the form of the low and high levels of gene expression correspondingly.

It is easy to see that the thermodynamic model Eq. (4.1) is similar to the Hill function [9] a phenomenological model commonly used in the mathematical description of gene expression.

Because gene expression could support Boolean-like logic there have been significant efforts to create digital circuits in cells [1,3–5]. For example the genetic AND gate can be created by manipulating gene expression using two molecules A and B that, due to mutual interactions, create a complex AB. In turn complex AB is an activator for a gene whose expression leads to a synthesis of a protein D. If the concentration of A and B represent logical inputs then the concentration of D approximates the AND operation.

Equilibrium mechanics can be used to undertake a more detailed analysis. The stationary value of the concentration C_{AB} of the complex AB is proportional to the concentrations C_A and C_B of the molecules A and B respectively. Correspondingly, $C_{AB} \propto C_A C_B$, and the expression rate of D takes on the following form,

$$R_D = \mu_D \frac{C_A C_B + \Gamma_D}{C_A C_B + M_D}, \tag{4.2}$$

where Γ_D and μ_D are M_D are positive constants. According to the rate equation

$$\frac{dC_D}{dt} = R_D - k_D C_D, \tag{4.3}$$

the stationary concentration of the protein D is

$$C_D = F(C_A, C_B) \equiv \frac{\mu_D}{k_D} \frac{C_A C_B + \Gamma_D}{C_A C_B + M_D}. \tag{4.4}$$

In Eqs. (4.3) and (4.4) the coefficient k_D describes the degradation of the protein D.

Let C_A and C_B be binary quantities, i.e. they are able to take on the low and high values. In practice, the low chemical value is rarely zero and hence the concentration of the complex AB can deviate significantly from binary values. Consequently the output of the gate will take on four values corresponding to four cases: (i) both C_A and C_B are low, (ii) C_A is low and C_B is high, (iii) C_A is high and C_B is low, and (iv) both C_A and C_B are high. We can introduce the normalised output of the gate $Q = C_D/F(C_{A,h}, C_{B,h})$ where $C_{A,h}$ and $C_{B,h}$ are the high levels of the inputs C_A and C_B, and Boolean inputs of the gate I_A and I_B take on 0 and 1 when C_A and C_B take on their low and high values respectively. Because the AND gate is not ideal, the truth Table 4.1 shows deviations of Q from the expected Boolean quantity Q_e.

But the deviation of the output Q from the expected ideal output Q_e is not significant. Therefore we may suppose that the AND logic gate is acceptable for integration in large-scale circuits.

Is this assumption correct? The answer depends on the architecture of the circuit but we can use some common types of circuits as a test-bed to investigate

Table 4.1. Two-input AND gate. The nondimentional low value of C_A and C_B equals 1.0 and the high value equals 80.0. The parameters are $\mu_D = 5$, $\Gamma_D = 32$, $M_D = 1600$ and $k_D = 0.05$

Inputs		Expected output	Output
I_A	I_B	Q_e	Q
0	0	0	0.03
0	1	0	0.08
1	0	0	0.08
1	1	1	1.00

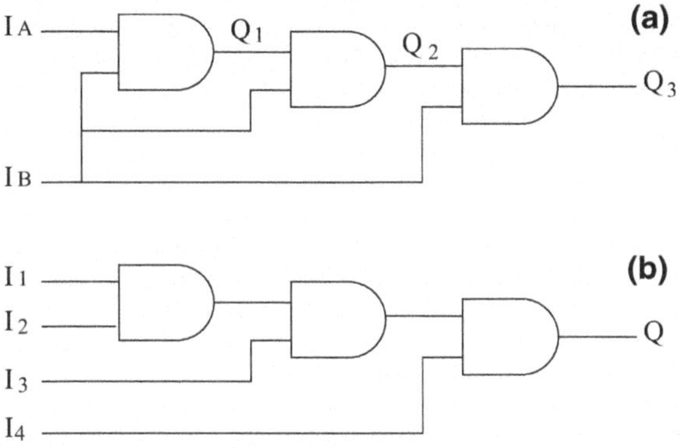

Fig. 4.1. The digital like circuits. **a** The chain of AND logic-like gates. **b** The circuit that functions like a multi-input AND logic gate

Table 4.2. Two-input AND gate. The parameters are identical to the parameters in Table 4.1

Inputs		Expected outputs			Outputs		
I_A	I_B	Q_{1e}	Q_{2e}	Q_{3e}	Q_1	Q_2	Q_3
0	0	0	0	0	0.03	0.03	0.03
0	1	0	0	0	0.08	0.33	0.72
1	0	0	0	0	0.08	0.03	0.03
1	1	1	1	1	1.00	1.00	1.00

their robustness. Here our choice is a simple chain of similar AND gates with one common input (see Fig. 4.1a). The chain allows investigation of important problems such as error propagation and parametric stability.

In the circuit shown in Fig. 4.1a the outputs of the gates are indexed. The truth table of the logical circuit is drawn in Table 4.2.

Indeed, for some inputs the output Q takes on a value of 0.72 instead of the correct value 0. This error occurs due to error propogation and would clearly lead to incorrect cellular response if implemented as designed. Finding a solution to the problem of the nonscaleability of the AND gate is the goal of this study.

4.3 Circuit with the Correction Module

The AND gate can be modified by adding a correction module. The main function of the correction module is to keep the output in the right range of values and prevent error propagation in the circuit. The correction module should modify the common nonlinearity of the system so that the attractor must lose its uniqueness, and the output must gain a dependence on the input I_A (Fig. 4.2).

The role of the correction module can equivalently be performed by a gene activated by the AND gate or a double inversion module (see Appendix). This

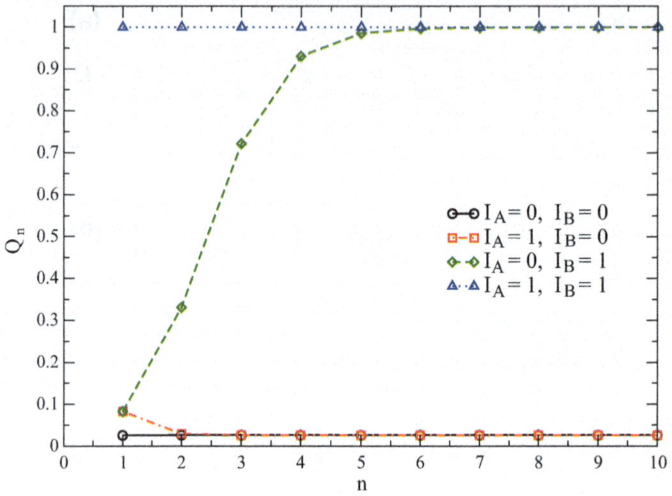

Fig. 4.2. The normalised outputs Q_n of the chain $A \to D_1 \to \cdots \to D_{10}$ with the common input B. The circuit for the first three gates is shown in Fig. 4.1a. The parameters are identical to Table 4.1

statement gives us some flexibility in a choice of interpretations for the model of the correction module,

$$C_{Z_n} = \Psi(C_{D_n}, \varkappa_Z) \equiv \nu_Z \, \frac{C_{D_n}^m + \Gamma_Z^m}{C_{D_n}^m + M_Z^m}, \tag{4.5}$$

where $\varkappa_Z$ denotes the set of parameters ν_Z, Γ_Z and M_Z. Here $\nu_Z = \mu_Z/k_Z$, μ_Z describe an expression rate, and k_Z is the degradation of the protein Z.

With the correction module Eq. (4.5) the new circuit is organized as a chain $A \to D_1 \to Z_1 \to D_2 \to Z_2 \to \cdots \to D_n \to Z_n \to D_{n+1} \to Z_{n+1} \to \cdots$, where the same signal B is applied to all AND gates. In this circuits, D_n is the result of the nth AND gate and Z_n is the output of its correction module.

By substitution of the equation

$$C_{D_{n+1}} = \Phi(C_{Z_n}, \varkappa_D) \tag{4.6}$$

into Eq. (4.5) it is easy to obtain a combined transfer function of the computation module and the correction module,

$$\begin{aligned} C_{Z_{n+1}} &= \Psi(\Phi(C_{Z_n}, \varkappa_D), \varkappa_Z) \\ &= \nu_Z \, \frac{\nu_D^m (C_{Z_n} C_B + \Gamma_D)^m + \Gamma_Z^m (C_{Z_n} C_B + M_D)^m}{\nu_D^m (C_{Z_n} C_B + \Gamma_D)^m + M_Z^m (C_{Z_n} C_B + M_D)^m} \\ &= \nu_Z \, \frac{\sum_{k=0}^m C_{Z_n}^k C_B^k \binom{m}{k} \left[\nu_D^m \Gamma_D^{m-k} + \Gamma_Z^m M_D^{m-k}\right]}{\sum_{k=0}^m C_{Z_n}^k C_B^k \binom{m}{k} \left[\nu_D^m \Gamma_D^{m-k} + M_Z^m M_D^{m-k}\right]} . \end{aligned} \tag{4.7}$$

In limit $n \to \infty$, we can expect $C_{Z_{n+1}} = C_{Z_n}$. Such solution to Eq. (4.7) we denote ϕ. In this case, Eq. (4.7) can be transformed into the following,

$$\sum_{k=0}^{m+1} a_k \phi^k = 0, \tag{4.8}$$

where the coefficients

$$\begin{aligned} a_0 &= -\nu_Z(\nu_D^m \Gamma_D^m + \Gamma_Z^m M_D^m), \quad a_{m+1} = C_B^m(\nu_D^m + M_Z^m), \\ a_k &= C_B^{k-1}\binom{m}{k-1}\left[\nu_D^m \Gamma_D^{m-k-1} + M_Z^m M_D^{m-k-1}\right] \\ &- \nu_Z C_B^k \binom{m}{k}\left[\nu_D^m \Gamma_D^{m-k} + \Gamma_Z^m M_D^{m-k}\right], \quad 1 \le k \le m, \end{aligned} \tag{4.9}$$

were introduced.

The case $m = 1$ corresponds to a simple activator or two simple repressors in the correction module. If $m = 1$, Eq. (4.8) becomes the parabolic equation $a_2\phi^2 + a_1\phi + a_0 = 0$ with the coefficients

$$\begin{aligned} a_2 &= C_B(\nu_D + M_Z), \\ a_1 &= \nu_D \Gamma_D + M_Z M_D - \nu_Z C_B(\nu_D + \Gamma_Z), \\ a_0 &= -\nu_Z(\nu_D \Gamma_D + \Gamma_Z M_D) \end{aligned} \tag{4.10}$$

The solution to the parabolic equation is well known,

$$\phi_\pm = \frac{-a_1 \pm \sqrt{a_1^2 - 4a_2 a_0}}{2a_2}. \tag{4.11}$$

According to Eq. (4.10) the inequality $a_2 a_0 < 0$ holds, therefore one root of the equation always is positive, $\phi_+ > 0$, and the other one always is negative, $\phi_- < 0$. Because the concentration C_Z cannot be negative, the positive solution ϕ_+ only is observed. It is easy to show the positive solution ϕ_+ always has a stable point, i.e. it is an attractor.

Moreover, the value ϕ_+ is a monotonic function of C_B, i.e. it increases with increasing C_B. In addition the dependence of ϕ_+ on the input C_A is completely lost like in the previously observed case of the chain with the AND gates without the correction modules. This means that the module characterized by $m = 1$ cannot be exploited for corrections of the output levels in the AND gate.

The case $m = 2$ corresponds to a dimer activator in the correction module or one dimer repressor and one simple repressor in the double inversion module. If $m = 2$, Eq. (4.8) becomes the cubic equation $a_3\phi^3 + a_2\phi^2 + a_1\phi + a_0 = 0$. It is easy to show that there is one real solution to the cubic equation for small (and very large) values of C_B, and there are three real solutions for large values of C_B, All solutions are positive, $\phi_1 > 0$, $\phi_2 > 0$ and $\phi_3 > 0$. They are fixed points of Eq. (4.7). The stability analysis shows that ϕ_1 and ϕ_3 are stable, and ϕ_2 is the unstable fixed point. I.e., if the input C_B is low then the system is

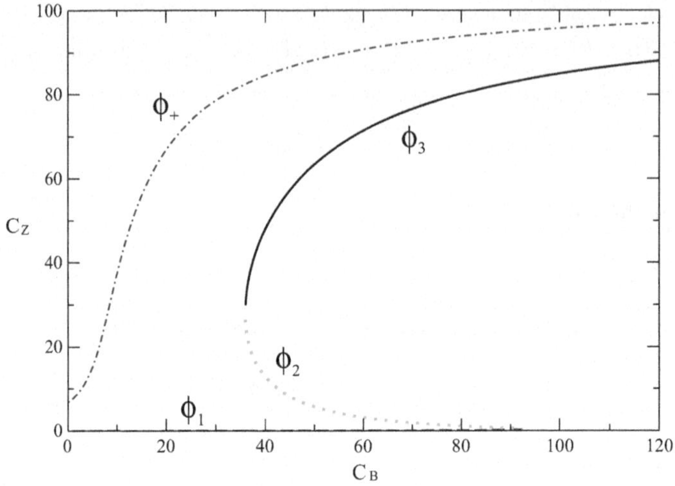

Fig. 4.3. Case $m = 1$: the solution ϕ_+ (Eq. (4.11)) is always monotonic. Case $m = 2$: the stable solutions ϕ_1 and ϕ_3, and the unstable solution ϕ_2. If the input C_B is low ($C_B < 35$) then the system is monostable and its output C_Z always corresponds the low level. If the input C_B is high ($35 < C_B < 92$) then the system is bistable. The case of very large input C_B when the system is monostable is not observed in this paper. Parameters: $\mu_D = 5$, $k_D = k_Z = 0.05$, $\Gamma_D = 32$, $M_D = 1600$, $\mu_Z = 10.2$, $\Gamma_Z = 1.41$, and $M_Z = 100$

monostable and its output always corresponds the low level (see Fig. 4.3); if the input C_B is high then the system is bistable and its output could be in the low or high levels that are dependent on the input C_A. The unstable solution ϕ_2 separates the basins of two attractors ϕ_1 and ϕ_3 (see Fig. 4.3). The bistability is able to suppress possible small deviations near the low and high levels, and prevent transitions between them.

4.4 Conclusion

Here we report the requirements and theoretical background for the design of digital genetic AND gates that are suitable for integration into large scale genetic circuits. The genetic circuits constructed from such gates can be characterized by their improved robustness and predictable function. Indeed, the correction modules are able to tune the output levels of the digital-like processing units to the right values so that small deviations of digital signal levels and random fluctuations will be supressed and not propagated along the large-scale genetic circuits. We hope that such circuits will pave the way for the development of real world applications.

Acknowledgements. We thank Alfonso Jaramillo for fruitful discussions. This work was funded by the BBSRC/EPSRC grant to WISB (BB/M017982/1).

Appendix

Let the circuit $D \dashv E \dashv Z$ represents a double inversion module, where the protein D is repressing the synthesis of the protein E, and the protein E is repressing the synthesis of the protein Z. According to the thermodynamic model of transcription [2], the stationary concentration of the protein E is dependent on the concentration of D as following,

$$C_E = \omega \frac{\Omega^m}{C_D^m + \Omega^m}, \tag{4.12}$$

where the coefficients ω and Ω are some constants, m is integer, $m = 1, 2, 3, 4, \ldots$. It is assumed that $\omega > 0$ and $\Omega > 0$. Here $m = 1$ corresponds to a case of the protein D that is a simple repressor of E. In contrast to $m = 1$, the case $m > 1$ means the protein D is assembled into a complex to be the repressor of E. For examples, $m = 2$ corresponds to a dimer, and $m = 4$ means the complex is a tetramer, e.t.c. We can write a similar equation for a relationship between the concentrations of the proteins Z and E,

$$C_Z = \lambda \frac{\Lambda^i}{C_E^i + \Lambda^i}, \tag{4.13}$$

where the coefficients λ and Λ are some positive constants, $\lambda > 0$ and $\Lambda > 0$, the index i is integer, $i = 1, 2, 3, 4, \ldots$.

By substitution Eq. (4.12) into Eq. (4.13), we obtain the transfer function of the double inversion module,

$$\begin{aligned} C_Z &= \lambda \frac{\Lambda^i \left(C_D^m + \Omega^m\right)^i}{\omega^i \Omega^{im} + \Lambda^i \left(C_D^m + \Omega^m\right)^i} \\ &= \lambda \frac{\Lambda^i \sum_{k=0}^{i} \binom{i}{k} C_D^{mk} \Omega^{(i-k)m}}{\omega^i \Omega^{im} + \Lambda^i \sum_{k=0}^{i} \binom{i}{k} C_D^{mk} \Omega^{(i-k)m}}, \end{aligned} \tag{4.14}$$

where $\binom{i}{k}$ are the binomial coefficients.

In case $i = 1$, Eq. (4.14) can be simplified,

$$C_Z = \frac{\lambda}{\Lambda} \frac{C_D^m + \Omega^m}{C_D^m + \left(\frac{\omega}{\Lambda} + 1\right) \Omega^m}. \tag{4.15}$$

It is easy to find a similarity between Eq. (4.15) and the thermodynamic model of transcription with a simple activator [2], $D \rightarrow Z$,

$$C_Z = \nu \frac{C_D^n + \Gamma^m}{C_D^m + M^m}. \tag{4.16}$$

Indeed, Eqs. (4.15) and (4.16) are identical when $M^m = (\omega/\Lambda + 1)\Omega^m$, $\Gamma = \Omega$ and $\nu = \frac{\lambda}{\Lambda}$. Therefore, the double inversion module can be replaced by the single activator module with the same order m of the complex, i.e. a dimer activator can be used instead of a dimer repressor and a single represser together. On one hand, the simplification of the genetic circuit could be an advantage. On the other hand, the double inversion module has an advantage over the activation module. The number of free parameters in the double inversion module is greater than in the activation module therefore the circuit with double inversion module could easily be tuned to parameter levels of interest. For example, we need a module with a very low saturation level in limit $C_D \to 0$. Then, Eq. (4.14) is transformed into the following,

$$\lim_{C_D \to 0} C_Z = \lambda \frac{\Lambda^i \Omega^{im}}{\omega^i \Omega^{im} + \Lambda^i \Omega^{im}} = \lambda \frac{1}{\left(\frac{\omega}{\Lambda}\right)^i + 1}. \tag{4.17}$$

According to Eq. (4.17), the low saturation level is independent from the parameters Ω and m. If the ratio $\omega/\Lambda > 1$ then $\lim_{C_D \to 0} C_Z$ rapidly approaches to zero with growing i. Therefore the low level limit can be reduced by increasing both the ratio ω/Λ and i.

In contrast to the low saturation level, the high saturation level is only dependent on one parameter, $\lim_{C_D \to \infty} C_Z = \lambda$.

References

1. Y. Benenson, Biomolecular computing systems: principles, progress and potential. Nat. Rev. Genet. **13**, 455–468 (2012)
2. L. Bintu, N.E. Buchler, H.G. Garcia, U. Gerland, T. Hwa, J. Kondev, R. Phillips, Transcriptional regulation by the numbers: models. Curr. Opin. Genet. Dev. **15**, 116–124 (2005)
3. R.W. Bradley, B. Wang, Designer cell signal processing circuits for biotechnology. New Biotechnol. **32**, 635–643 (2015)
4. R.W. Bradley, M. Buck, B. Wang, Recognizing and engineering digital-like logic gates and switches in gene regulatory networks. Curr. Opin. Microbiol. **33**, 74–82 (2016)
5. J.A.N. Brophy, C.A. Voigt, Principles of genetic circuit design. Nat. Methods **11**, 508–520 (2014)
6. R.S. Cox-III, M.G. Surette, M.B. Elowitz, Programming gene expression with combinatorial promoters. Mol. Syst. Biol. **3**, 145 (2007)
7. R.L. Goodstein, *Boolean Algebra* (Pergamon Press, Oxford, 1963)
8. M.B. Gu, Environmental biosensors using bioluminescent bacteria, in *Environmental Chemistry*, ed. by E. Lichtfouse, J. Schwarzbauer, D. Robert (Springer, Berlin, 2005), Chap 63, pp. 691–698
9. A.V. Hill, The possible effects of the aggregation of the molecules of hæmoglobin on its dissociation curves. Proc. Physiol. Soc. **40**, iv–vii (1910)
10. Y. Lei, W. Chen, A. Mulchandani, Microbial biosensors. Anal. Chim. Acta **568**, 200–210 (2006)
11. M. Morel, R. Shtrahman, V. Rotter, L. Nissim, R.H. Bar-Ziv, Cellular heterogeneity mediates inherent sensitivity-specificity tradeoff in cancer targeting by synthetic circuits. PNAS **113**, 8133–8138 (2016)

12. J.M. Raser, E.K. O'Shea, Noise in gene expression: origins, consequences, and control. Science **309**, 2010–2013 (2005)
13. B. Wang, M. Barahona, M. Buck, A modular cell-based biosensor using engineered genetic logic circuits to detect and integrate multiple environmental signals. Biosens. Bioelectron. **40**, 368–376 (2013)
14. Z. Xie, L. Wroblewska, L. Prochazka, R. Weiss, Y. Benenson, Multi-input RNAi-based logic circuit for identification of specific cancer cells. Science **333**, 1307–1311 (2011)

Chapter 5
A Brainmorphic Computing Hardware Paradigm Through Complex Nonlinear Dynamics

Yoshihiko Horio(✉)

Research Institute of Electrical Communication, Tohoku University,
2-1-1, Katahira, Sendai, Aoba-ku 980-8577, Japan
horio@riec.tohoku.ac.jp

Abstract. In a brainmorphic computing paradigm, a hardware system should process information imitating the anatomical and physiological mechanisms of the brain by naturally using physical and dynamical characteristics of the constituent devices, especially through nonlinear analog circuits and devices. The latest knowledge from brain science, especially, on high-order brain functions emerged from high-dimensional complex neuro-dynamics, are reflected in the design of brainmorphic hardware. In addition, the bodily and environmental constraints are considered and utilized as embodiment in this hardware paradigm. In this paper, we propose a brain/body whole organism computation paradigm where brain-intrinsic efficient and distinct information-processing styles and functions are expected to emerge through high-dimensional complex nonlinear dynamics and the embodiment. In particular, we employ a chaotic neuron in a reservoir neural network to emerge the reference-self in the brain/body whole organism computing framework. Chaotic behavior is usually avoided in the reservoir computing because it will violate the echo state property. However, we deliberately introduce high-dimensional chaotic dynamics through the chaotic neurons, but preserving the echo state property. The high-dimensional chaotic dynamics create a rich variety of neural patterns, and at the same time, integrate information in the neural patterns into a unique dynamical state as a high-dimensional attractor. We show preliminary results for chaotic time-series predictions through the chaotic reservoir neural network to demonstrate feasibility of the chaotic dynamics introduced in the reservoir.

5.1 Introduction

Although the current brain-inspired VLSI hardware systems employ some brain-like architecture such as fine-grained local memories and in-situ learning, they are far from the real brain. For example, information is not really distributed and integrated for representation, processing, and storage. In addition, they

V. In et al. (Eds.): *Proceedings of the 5th International Conference on Applications in Nonlinear Dynamics*, Understanding Complex Systems, https://doi.org/10.1007/978-3-030-10892-2_5

largely ignore high-complexity and complex dynamics of the brain, which are particularly important for high-order brain functions [1–3]. Moreover, they do not consider enough the bodily and environmental constraints and interactions (embodiment), which may lead to a unique and efficient information-processing paradigm of the brain [4].

In order to get one step closer to the brain, we propose a "brainmorphic" hardware paradigm [5,6], which is a natural extension of the neuromorphic paradigm [7]. In this paradigm, the brainmorphic hardware

1. processes information imitating the anatomical and physiological structure and mechanisms of the brain;
2. naturally mimics physicochemical biophysics directly using physics and dynamics of the circuits and devices, especially through analog nonlinear circuits and devices;
3. reflects the latest knowledge from brain science, especially, on high-order brain functions including emotion and consciousness; and
4. considers and utilizes the bodily and environmental constraints with a complex dynamical internal state (reference-self).

In this paper, we propose a novel "Brain/Body Whole Organism Computing" framework [5,6] focusing on the bodily and environmental constraints as an embodiment [4], which is one of the key elements for emergence of unique and efficient brain-like information processing. In particular, we employ a reservoir network [8] consisting of chaotic neurons to generate the reference-self in this framework. Although chaotic behavior is usually avoided in the reservoir computing, we deliberately introduce high-dimensional chaotic dynamics to obtain a rich variety of neural patterns to represent information, and at the same time, to integrate the information as a unique state for implementing the dynamical internal state. We illustrate preliminary simulation results for chaotic time-series predictions with the chaotic reservoir neural network showing feasibility of the chaotic dynamics in the reservoir network.

5.2 Brain/Body Whole Organism Computing Framework

We are advocating the brain/body whole organism computing framework [5,6] in order to overcome the problems in recent brain-inspired computers. Our initial target in this framework is a small and low-power integrated circuit implementation of brainmorphic hardware, especially for intelligent edge computational devices.

Possible required elements of brainmorphic hardware for the whole organism computing would be:

A Generation of stable and rich neural patterns that dynamically represent "reference-self," which consistently keeps an internal state of the system itself.

B Dynamical generation of sensitive neural patterns that represent the corresponding external objects.

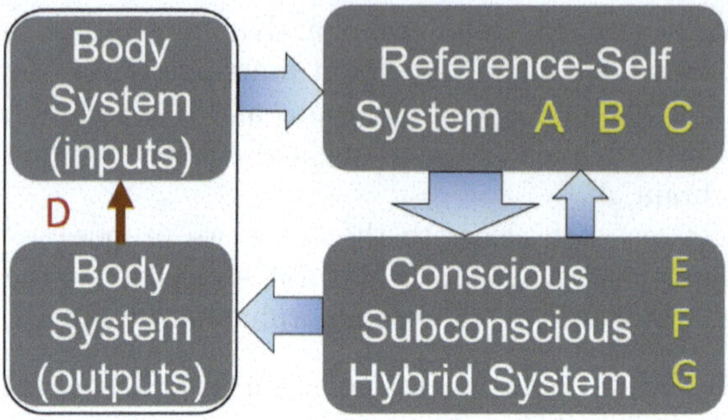

Fig. 5.1. A possible architecture of the brain/body whole organism computing hardware

- C A quick change in the internal state by mutual interaction between A and B above.
- D Mechanisms for the embodiment, and interaction with external objects and environment.
- E Conscious and sub-conscious processes, and high-order functions arisen from the mutual interaction of them.
- F Memory creations and retrievals through a macroscopic learning mechanism.
- G Global regulations to rapidly control and modify processing modes.

High-dimensional complex nonlinear dynamics play an important role, especially, in A–C and E [9].

Figure 5.1 shows a construction example of the brain/body whole organism computing hardware [5,6], which consists of "reference-self subsystem," "body subsystem," and "conscious/subconscious hybrid subsystem."

5.2.1 Hardware Architecture

Although we have shown possible hardware architecture for each subsystem in Fig. 5.1 [5,6], we will concentrate, in this paper, on the reference-self subsystem, in which we employ the reservoir computing framework [8].

The reference-self subsystem in Fig. 5.1 consists of three elements, that is, ① robust but dynamic retention of a neural pattern that represents the internal state as "reference-self," ② rich neural pattern generation in response to the external objects, and ③ mutual interaction between ① and ②, resulting in a novel neural pattern.

Possible hardware architecture for the reference-self subsystem is shown in Fig. 5.2 [5,6]. As shown in the figure, this subsystem is constructed with three neural networks (NNs), each of which corresponds to ① to ③ above; that is, ① an internal state NN, ② an object representation NN, and ③ a state-change detection NN. High-dimensional complex dynamics, especially chaotic dynamics, are deeply involved in these neural networks as follows.

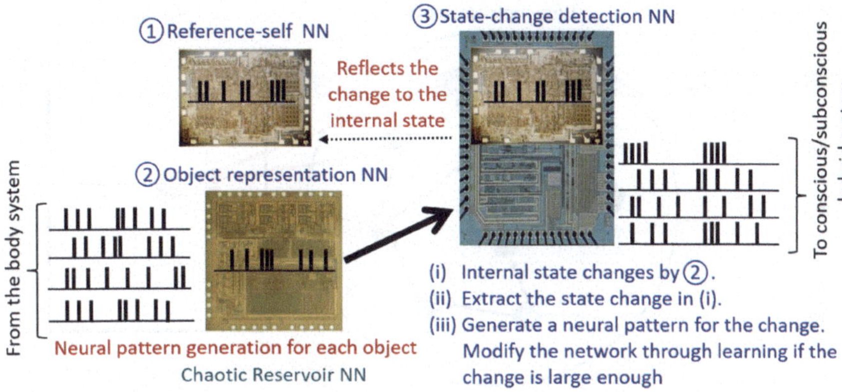

Fig. 5.2. A possible hardware architecture for the reference-self subsystem

The NN for ① should robustly maintains a high-dimensional attractor as its internal state even if the environmental parameters are changed. Therefore, we utilize a robust property of high-dimensional chaotic systems such as consistency [10,11] in this NN.

In contrast, the NN for ② should rapidly respond to the external input by changing its neural pattern. In addition, this NN should be able to produce a rich variety of neural patterns (attractors). Therefore, a possible candidate of this NN is a reservoir neural network [8], but with chaotic dynamics. For example, the default state of this NN would be chaotic itinerant dynamics [12], and an infinite number of low-dimensional quasi-attractors represent external objects.

Finally, ③ will have a triple NN structure with (i) a NN that retains a copy of the reference state of ①, and whose internal state is altered by ②, (ii) a NN that extracts the change in the NN of (i), and (iii) a NN that produces a neural pattern according to (ii). The NN in (i) uses the same neuron circuit as that in ①, while a simple integrated-and-fire based spiking neuron circuit would be used in (ii). The reservoir network would also be suitable for the NN in (iii) for a variety of complex spatio-temporal spiking patterns.

5.3 Chaotic Reservoir with Chaotic Neurons

A general structure of the reservoir neural network is shown in Fig. 5.3 [8]. As shown in the figure, the network consists of the input layer, the reservoir layer (recurrent neural network), and the output layer. One of the distinct features of the reservoir network is that only connection weights from the reservoir to the output layer are updated during the learning process, so that even simple learning algorithm can be used [8]. In addition, other fixed connection weights can be randomly chosen.

Since the reservoir network with chaotic dynamics is a strong candidate for the NNs in the reference-self subsystem shown in Sect. 5.2, we propose a chaotic reservoir network using the chaotic neural network model [9,13].

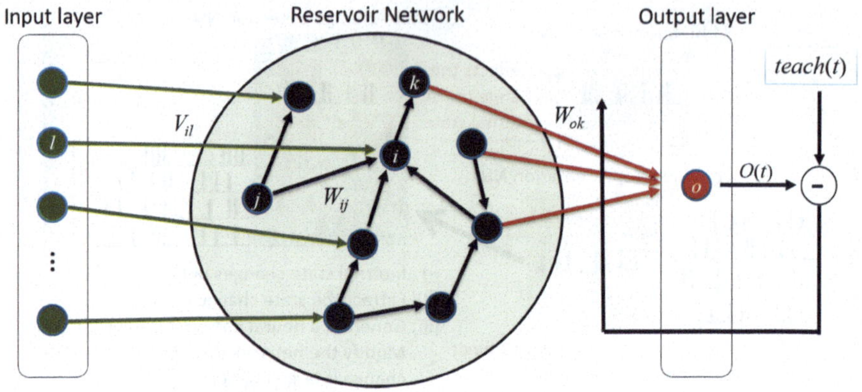

Fig. 5.3. A general structure of the reservoir neural network

Typical approach to destabilize the reservoir network into a chaotic state is to make the spectral radius $\rho(\boldsymbol{W})$ be greater than 1 by properly choosing the connection weight matrix $\boldsymbol{W}$ [8]. This may violate the echo state property, so that the chaotic behavior was avoided in the conventional reservoir networks [8]. However, we deliberately introduce chaotic dynamics into the reservoir network without completely destroying the echo state property by replacing ordinary neurons with chaotic neurons [13] (Eqs. (5.1) and (5.2)) while keeping $\rho(\boldsymbol{W}) < 1$.

$$y_i(t+1) = ky_i(t) + \sum_{j=1}^{M} W_{ij} f(y_j(t)) + \sum_{l=1}^{P} V_{il} I_l(t) - \alpha f(y_i(t)) - \theta_i(1-k), \tag{5.1}$$

$$x_i(t+1) = f(y_i(t+1)), \tag{5.2}$$

where $y_i(t)$ and $x_i(t)$ are the internal state and output of the neuron i at time t, respectively, W_{ij} and V_{il} are the connection weights from neuron j to neuron i in the reservoir, and that from the lth-input I_l to neuron i as shown in Fig. 5.3, $\alpha = 0.01$ and $k = 0.01$ are the parameters for refractoriness, M is the number of neurons in the reservoir, P is the number of inputs, and $f(\cdot)$ is a sigmoidal function with $\varepsilon = 0.02$.

While the chaotic neurons are used in the reservoir network, one standard sigmoidal neuron o is used in the output layer. The connection strength to the output neuron o from the chaotic neuron k is W_{ok} as shown in Fig. 5.3. For learning, we employ a standard mean square method where only connection weights to the output neuron from the reservoir neurons are changed, while other weight values are kept.

In order to verify feasibility of the reservoir network with chaotic neurons, preliminary simulations for one-step predictions of chaotic time-series from the logistic map (Eq. (5.3)), and Hénon map (Eqs. (5.4) and (5.5)) are used.

$$r(t+1) = pr(t)(1 - r(t)), \tag{5.3}$$

Table 5.1. Network parameters for simulations

The number of neurons in the reservoir, M	100
The percentage of connections among neurons inside the reservoir	10%
The percentage of the reservoir neurons connected to the output layer	20%
Distributions of the random values of W_{ij} in the reservoir	Uniform $[-0.01; 0.01]$
Spectrum radius $\rho(\boldsymbol{W})$	<0.01
Distributions of the random values of θ_i in the reservoir	Uniform $[-0.01; 0.01]$
Training length	1900 steps
Testing length T	600

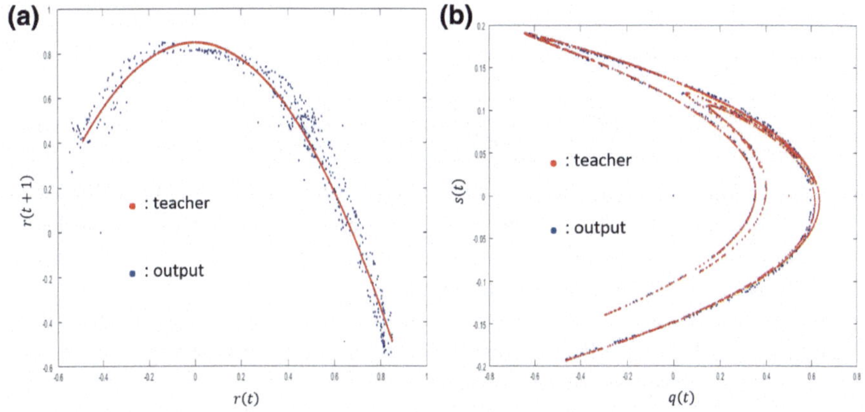

Fig. 5.4. Attractors obtained from the original and predicted time-series for (a) logistic map ($r(t)$), and (b) Hénon map ($q(t)$ and $s(t)$)

and

$$q(t+1) = 1 - aq^2(t) + s(t), \tag{5.4}$$
$$s(t+1) = bq(t). \tag{5.5}$$

In the simulations, $p = 3.7$, $a = 1.3$, and $b = 0.4$ are used for the above maps.

The network parameters for the simulations are summarized in Table 5.1. In the case of logistic map, we used one input neuron, that is, $P = 1$, while for Hénon map, $P = 2$.

The 1-step prediction results after the learning are shown in Fig. 5.4. In the figure, the blue dots show the predicted points, while red dots are correct points. The average errors AE defined in Eq. (5.6) were of the order of $<10^{-2}$ for both

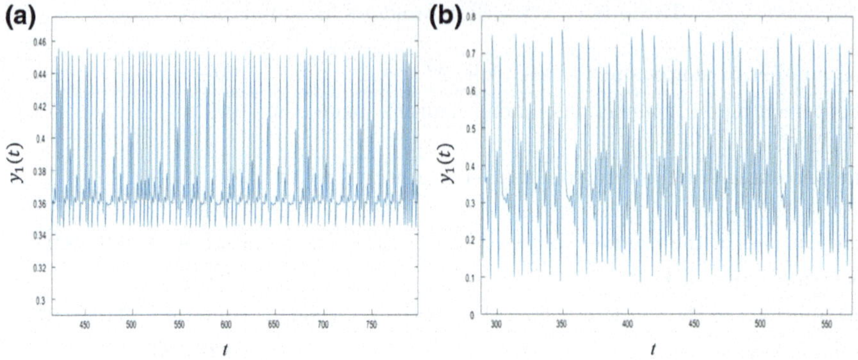

Fig. 5.5. Time waveforms of the internal state $y_1(t)$ of the chaotic neuron 1 in the reservoir for (a) logistic map, and (b) Hénon map

maps.

$$AE = \frac{1}{T}\sum_{t=1}^{T}\sqrt{(O(t) - teach(t))^2}, \tag{5.6}$$

where $T = 600$ is the length of testing phase, $O(t)$ is an output value of the output neuron o at time t, and $teach(t)$ is a correct value of the time series.

In addition, we confirmed the chaotic dynamics in the reservoir network during the testing phase as shown in Fig. 5.5.

The obtained results confirm that the proposed chaotic reservoir network has an ability to work as an efficient reservoir. Therefore, we will apply the chaotic reservoir in the reference-self subsystem.

5.4 Conclusions

We proposed a brainmorphic hardware paradigm in which complex nonlinear dynamics play an important role. As one of the important ingredients of this paradigm, the brain/body whole organism computing framework with its possible hardware architecture was proposed. We also proposed the reservoir network with chaotic neurons to generate a variety of neural patterns, and at the same time, to regulate these patterns into an integrated state, for the reference-self system in the brain/body whole organism computing framework. We confirmed feasibility of the chaotic dynamics introduced in the reservoir via chaotic neurons, instead of destabilizing the reservoir with $\rho(\boldsymbol{W}) > 1$, by preliminary simulations for chaotic time-series predictions. We will further study on the properties of the chaotic reservoir. At the same time, we are currently working on the chaotic reservoir with spiking neurons for efficient hardware implementation.

Acknowledgements. This work is supported by JSPS KAKENHI Grant Numbers 16K00340 and 17H0693, and the Cooperative Research Project Program of the Research Institute of Electrical Communication, Tohoku University.

References

1. A. Damasio, *Feeling of What Happens* (Harcourt Brace, 1999). ISBN 0156010755
2. G. Edelman, G. Tononi, *A Universe of Consciousness* (Basic Books, 2000). ISBN 0465013775
3. A. Damasio, *Self Comes to Mind –Constructing the Conscious Brain–* (Pantheon, 2010). ISBN: 13: 9780307378750
4. J. Hawkins, What intelligent machines need to learn from the neocortex. IEEE Spectr. **54**(6), 33–37; 68–69 (2017)
5. Y. Horio, Towards a neuromorphic computing hardware system, in *Proceedings of International Symposium on Nonlinear Theory and Its Applications* (2017), pp. 189–192
6. Y. Horio, Towards a brainmorphic computing paradigm and a brain/body whole organism computation system, in *Proceedings of RISP International Workshop on Nonlinear Circuits, Communicaltion and Signal Processing* (2017), pp. 703–706
7. C. Mead, Neuromorphic electronic systems. Proc. IEEE **78**(10), 1629–1636 (1990)
8. M. Lukoševičius, H. Jaeger, Reservior computing approaches to recurrent neural network training. Comput. Sci. Rev. **3**, 127–149 (2009)
9. Y. Horio, K. Aihara, Analog computation through high-dimensional physical chaotic neuro-dynamics. Physica-D **237**(9), 1215–1225 (2008)
10. Z.F. Mainen, T.J. Sejnowski, Reliability of spike timing in neocortical neurons. Science **268**, 1503–1506 (1995)
11. A. Uchida, R. McAllister, R. Roy, Consistency of nonlinear system response to complex drive signals. Phy. Rev. Lett. **93**, 244102 (2004)
12. K. Kaneko, I. Tsuda, Chaotic itinerancy. AIP Chaos **13**(3), 926–936 (2003)
13. K. Aihara, T. Takabe, M. Toyoda, Chaotic neural networks. Phy. Rev. Lett. A **144**, 333–340 (1990)

Chapter 6
Nonlinear Computing and Nonlinear Artificial Intelligence

Behnam Kia and William Ditto(✉)

Nonlinear Artificial Intelligence Lab, North Carolina State University, Raleigh, NC, USA
bkia@ncsu.edu, wditto@ncsu.edu

Abstract. The importance and the necessity of nonlinearity in Artificial Intelligence, AI, and deep learning are very well understood. A multi-layer neural network with linear activation function is equivalent to a single layer of neurons. It is nonlinearity of activation functions that adds complexity to each layer, transforming the network to a universal computing machine that can approximate any continuous function. However, nonlinearity and the complexity that it creates have not been investigated enough in AI and modern deep learning systems. NC State University's Nonlinear Artificial Intelligence Lab focuses on nonlinearity and the complexity that comes with it, and investigates how this can be an engine of artificial intelligence. We peruse our research at different levels with different goals. In this article we explain our approach, and present an overview of our results.

6.1 Introduction

We live in a nondeterministic, noisy, and stochastic world. Furthermore, it is believed that noise, stochasticity, and chaos play a crucial role in our brain and the way it processes information [1–3]

Transistors are the basic computer systems. The main approach to improve the performance of the computers has been following the Moore's law -scaling down the size of transistors and integrating more transistors into a computer chip [4]. The Moore's law has provided us with a roadmap to improve the performance of the computers for decades. But the challenge is that after decades of scaling the transistors, we have reached to a point that as we further scale down the size of transistors, we are reaching fundamental physical limitations of these devices, and we are losing the determinism of these binary switches. For example, electrons can tunnel through an open switch (quantum tunneling) [5]. And it is becoming exponentially harder and more expensive to design and fabricate fully deterministic systems that perform deterministic computing. On top of it, we are moving towards stochastic processing and computing,

V. In et al. (Eds.): *Proceedings of the 5th International Conference on Applications in Nonlinear Dynamics*, Understanding Complex Systems, https://doi.org/10.1007/978-3-030-10892-2_6

and the most notable example is AI. So why not utilize and embrace nonlinear, chaos-based hardware, and use it to perform computing methods that are inherently robust to noise? This is the approach that we have picked, and we design and fabricate nonlinear, chaotic hardware, and we utilize this platform to implement nondeterministic computation and AI. However, there is a lot of challenges facing adoption and utilization of nonlinear dynamics and chaos in artificial intelligence.

Adopting and engineering chaos and nonlinear dynamics into an engineering application is a two-edged sword. From one perspective, we can enjoy the great amount of processing power that chaos can deliver. For example, it is shown that a simple nonlinear circuit can represent an infinite number of different functions. On the other hand, chaos comes at a great cost too. Designing a robust, stable nonlinear, chaotic circuit, and manually or adaptively programing it to implement desired tasks is not a simple job, and furthermore, noise and fabrication nonidealities can degenerate the performance of the circuit.

There is a lot to learn from the story of deep learning. Deep neural networks—neural networks with multiple hidden layers—were very well known to researchers and machine learning practitioners, and their great performance as universal function approximators was very well understood. But they were deemed unpractical because when the nonlinear operations of multiple layers of neurons are composed together, the training of resulting function is mathematically intractable. In other words, training a multilayer deep neural network is a non-convex optimization problem to solve [6]. As a result, many abounded the idea of deep neural network in favor of simpler, but less powerful, machine learning methods such as Support Vector Machines (SVM) that are mathematically tractable [7]. But expressing the learning mechanism as a non-convex optimization problem brings immense representation, modeling, and learning power. In 2012, with the help of GPUs and large data sets for training, finally a practical method was introduced to optimize these non-convex learning problems, and after that AI never became the same [8]. The main take-home note from deep learning story is that if we manage to tame very complex nonlinear systems, we can unshackle the unprecedented high-performance that these complex systems can provide. This has been our mission in our research group from day one. Take a chaotic system that brings the maximum possible amount of diversity in behavior and complexity, tame it and utilize the performance that it can offer. In [9] we demonstrated that a simple nonlinear circuit contains an infinite number of different functions. In [10] we introduced nonlinear dynamics as an engine of computing.

In Sect. 6.2 we explain the main idea behind how we can utilize chaos and nonlinear dynamics in computation. In Sect. 6.3 we will overview our recent nonlinear hardware designs. And describe what type of processing we can perform on top of this hardware. In Sect. 6.4 we review sample applications that we have implemented. In Sect. 6.5 we discuss where our designs fit in the industry, how much compatible they are with exiting technology, and we conclude the article.

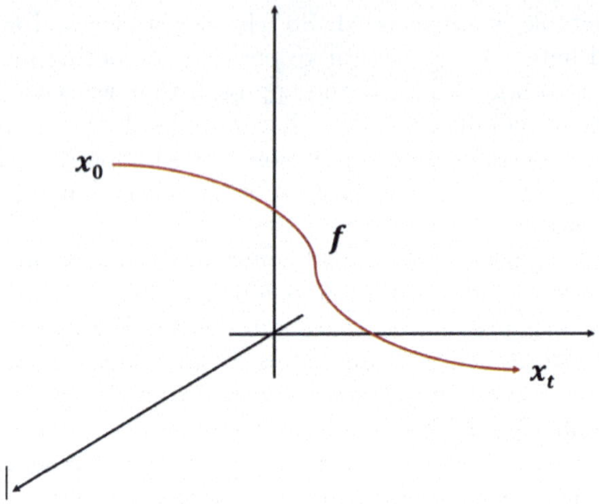

Fig. 6.1. A 3-dimensional dynamical system that maps an initial state x_0 to a future state x_t. Obviously, the dynamical system can be considered as a function

6.2 The Main Idea

A dynamical system is a system that evolves over time and maps states in its state space to some other future states. Let f be a dynamical equation, mapping initial states to future states:

$$f : \Re^n \rightarrow \Re^n \tag{6.1}$$

Figure 6.1 shows an example visualization of a dynamical system in 3-dimensional state space, $n = 3$.

It is clear from the definition and visualization of dynamical system that a dynamical system embodies a function, it implements a function.

A dynamical system can be linear or nonlinear. A linear dynamical system tends to build a simple, basic function, whereas a nonlinear dynamical system can implement much more complex functions. Much more importantly, a nonlinear dynamical system usually happens to be sensitive to its parameters. This provides us with a parametric function builder that given different parameters can implement different functions. See Fig. 6.2 where a parametric nonlinear dynamical system f_p is implementing two different functions for two different p values.

It is shown that indeed a nonlinear dynamical system contains an infinite number of functions [9], and nonlinear dynamics can be considered as an engine of computation [10]. In [10] it is shown that the number of distinguishable functions

$$N_f \propto e^{\lambda_C n} \tag{6.2}$$

increases exponentially with evolution time, where λ_C is the computing exponent, and n is the number of iterations the iterative dynamical system makes

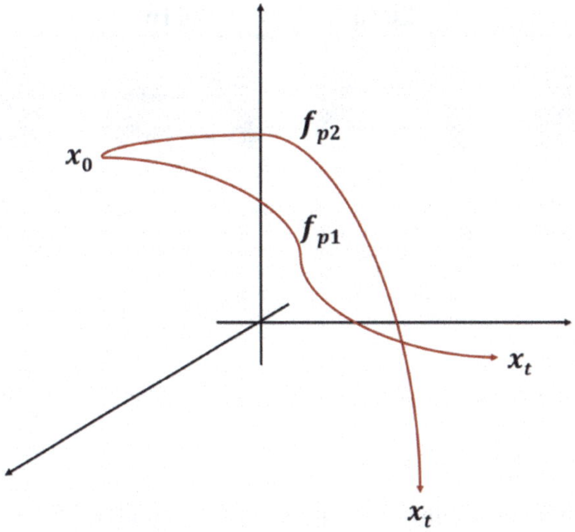

Fig. 6.2. A 3-dimensional parametric dynamical system that maps an initial state x_0 to two different future state x_t. With different parameter values one can potentially implement different functions

before producing the final state (or in continuous-time dynamical systems we will have evolution time t instead of iteration number n). The computing exponent λ_C was defined in parallel to Lyapunov exponent, with this difference that computing exponent measures and captures the number of different functions that a dynamical system can implement. Nonlinear dynamical system can have positive computing exponent, therefore the number of functions that they can implement exponentially increases as the iteration number n (or evolution time t) linearly increases. This demonstrate the capacity of the nonlinear systems in approximating and implementing different functions.

Our research has bifurcated into two avenues: first, manually finding and setting the parameters in order to program the nonlinear dynamical system to implement a desired function, and second, letting the nonlinear dynamical system itself learns which parameters it needs to select in order to implement the desired function. In the next sections we explain these two avenues, and what type of applications we can implement.

6.3 Hardware Design

We have designed and developed multiple generations of hardware for nonlinear computing. We have followed a similar path to design and develop nonlinear dynamics-based hardware that is simple in design, while complex in behavior. Such nonlinear hardware can implement complex and diverse tasks and functions using fewer transistors and less energy [11]. And they create an ideal hardware

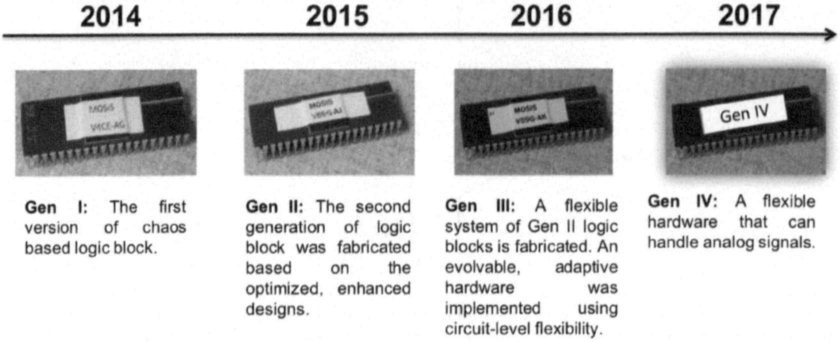

Fig. 6.3. Four generations of hardware developed by Nonlinear Artificial Intelligence Lab

platform to implement nonlinear computation. Currently we have designed and developed four generations of nonlinear dynamics-based hardware, and with each generation we have advanced both the hardware as well as the applications that it can enable and implement (Fig. 6.3).

It is important to note that this is a *technology platform* in the sense that many different applications can be designed and deployed. Figure 6.4 shows a model for our technology platform.

Device Level: We use conventional CMOS devices to design our circuits and we use conventional CMOS technology to fabricate our circuits and chips. Our hardware technology is a new design method that makes use of current devices in order to design nonlinear circuits that exhibit very complex behaviors.

It is worth noting that *beyond CMOS* devices can also be used to design nonlinear circuits. As an example, memristors can be suitable nonlinear devices to implement nonlinearity and complex behavior at the circuit level. However, for practical reasons at this point, we are mostly focused on conventional CMOS devices as the building blocks of our circuits.

Circuit Level: At the circuit layer, we design circuits that have nonlinear, complex behavior. This circuit design is nothing more that connecting a series of basic CMOS devices together, but with the crucial difference that we purposefully create nonlinearity and complexity in behavior, and thus derive complex processing out of this complex behavior. This is a philosophical and engineering departure from the conventional norm. In conventional design methods, designers make sure that all of their circuits have simple, fully predictable, stable dynamics. And then they put together many of these simple circuits in order to implement complex systems. In other words, complexity is achieved through a complex design with many devices and circuits. But in our approach, we develop simple-in-design, but complex-in-behavior, circuits and systems. Therefore, complex processing emerges from the complex dynamics of simple circuits that have fewer transistors and lower energy requirements.

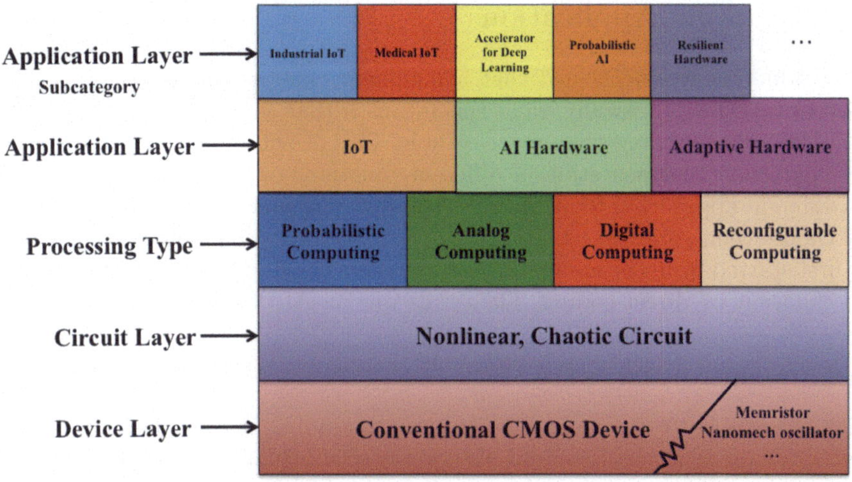

Fig. 6.4. Model of Nonlinear Artificial Intelligence's technology platform, showing its different layers of design

Processing Type: The main processing capabilities of this hardware emerge from its complex dynamics, and since complex dynamics is flexible in behavior, the nonlinear circuit can implement many different functions and tasks. More specifically, we have shown that the hardware can implement all of the following types of processing:

- Digital Computing: The circuits can emulate operations of different digital functions.
- Reconfigurable computing: Complex dynamics is flexible and contains many different behaviors; therefore it can emulate many different functions. And reconfiguration is instant since they all coexist within the same circuit as opposed to FPGAs, which require halting the processing and loading new control bits.
- Probabilistic Computing: The complex dynamics of the nonlinear circuits can operate as a probabilistic system and therefore can perform probabilistic computing.
- Analog computing: These nonlinear circuits are analog in nature, and they can receive and process both analog and digital inputs.

Application Layer: This hardware is a platform with all of the unique processing capabilities listed above, so many different applications can be designed and developed based on it. The Fig. 6.2 model shows some of these applications. These applications are enabled by one or more processing capabilities in the processing layer. We have designed different proof-of-concept examples to demonstrate the processing capabilities and possible applications the hardware can perform. Some of these examples are listed below.

6.4 Example Applications

In introduction we mentioned that a nonlinear chaotic system contains many different functions. Basically, what this means is that a chaotic system embodies many different functions that are selectable. This provides us with a platform for representation; representation of different functions or behaviors. We can take two different approaches to utilize this rich library of functions, (1) manually pick and choose them, and (2) let the system learn to pick and choose automatically. We first started from the manual selection, where the designer/programmer picks and choose it by direct coding. The result was an ALU unit.

6.4.1 Adaptive Hardware

Since our new hardware is flexible and programmable, it can adapt to different internal or external changes, and also adapt to its changing environment. This adaptation can be manually administrated, or it can be autonomous. For example, we purposefully overheated one of our fabricated hardware to a level (82 °C) well beyond its specification and tolerance level. As a result, it eventually failed to do what it is was programmed to do. However, because the hardware was flexible, we reprogrammed with a new set of control inputs to perform the same task, albeit using different control inputs [12] (Fig. 6.5).

6.4.2 Learning and Artificial Intelligence

By utilizing nonlinear dynamics, living systems exhibit diverse and complex behaviors while conserving their energy. And they can explore many different behaviors or reactions that their nonlinearity provides to them in order to (adaptively) pick and choose the ones that best meet their needs and conditions at the time. We explore such connections, and design and build intelligent hardware based on this concept. Our main hypotheses toward achieving artificial intelligence with morphable nonlinear systems are that: (1) nonlinear dynamics provides flexibility and morphability, and therefore it creates a suitable platform for

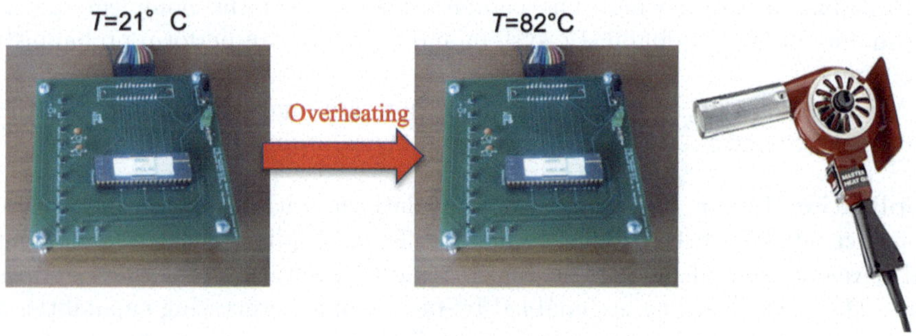

Fig. 6.5. An adaptive hardware maintaining operational capability despite external and internal changes (overheating in this specific experiment)

plasticity and learning (or intelligence in general); and (2) machine intelligence should be hardware based, as opposed to being software based. In nature there is no separate software; it is the physical organism itself that shows intelligence, and that intelligence is intertwined with the inherited genetics and physical make up of the organism. Combining these two hypotheses, we propose that to achieve nature-like intelligence, we need a nonlinear dynamics-based hardware that provides flexibility and plasticity at the hardware level. We have trained one of our fabricated hardware chips to evolve and learn different tasks, such as summation or subtraction, with no need for direct programing. The problem of automatically training a chaotic system to implement a given function can be formulated as an optimization problem below:

$$p_f = \underset{p}{argmin} \sum_i cost(x_i, y_i, \hat{y_i}) \tag{6.3}$$

where p is parameter of the chaotic system, x_i,y_i is a pair of input-output that the chaotic system is supposed to learn how to map (such pairs of given inputs-outputs are called training data in the context of AI; the data drawn from a desired function that maps x to y, and we use this training data to tune the parameters of chaotic system to implement the desired function), $\hat{y_i}$ is what chaotic system produces as the output to x_i, cost function can be defined as squared error if the outputs are continuous valued, or as binary hit/miss if the outputs are binary, i.e. $cost(x_i, y_i, \hat{y_i}) = 0 \quad \text{if} \quad y_i = \hat{y_i}, \text{otherwise} \quad 1$, and we calculate cost function over the entire training data (all i values). Now the problem of learning a desired function using a chaotic system is transformed to an optimization problem where we reduce the distance between y_i,$\hat{y_i}$ for all i values, and different optimization techniques can be used to minimize this cost function. The results of this experiment are under review to be published as a separate research article.

6.4.3 IoT Hardware

This application is a mixture from the examples above. We are introducing hardware for IoT nodes, where there is a massive influx of sensor data, and this data is filtered and processed to extract information to be sent to the higher layers of an IoT network. Figure 6.6 below shows the conventional general data acquisition and processing signal chain for IoT nodes and edge computing.

Our new hardware can implement the IoT node and computing at the node (edge computing) with a much more efficient chain shown in Fig. 6.7 below:

Our nonlinear dynamics-based hardware can:

- Directly receive analog inputs from sensors;
- Filter noise from analog signals;
- Convert analog signals to digital;
- Digitally process these digital inputs;
- Morph into new configurations at any cycle, and therefore digital processing can be reconfigurable, adaptive, and evolvable;

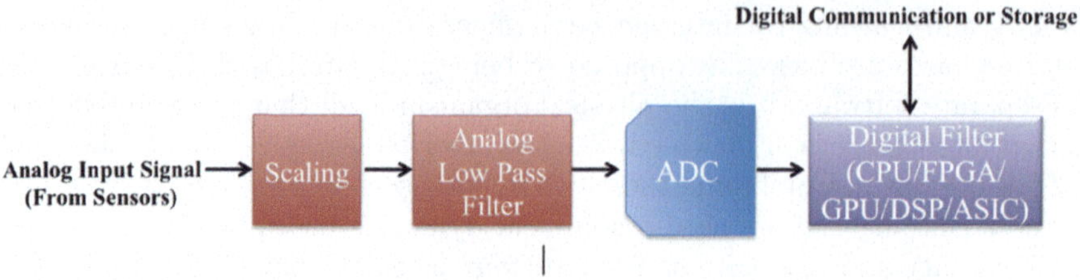

Fig. 6.6. Conventional data acquisition and processing signal chain

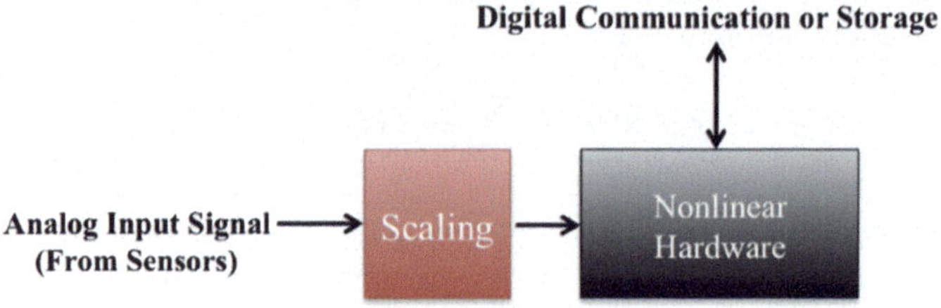

Fig. 6.7. Alternative chain, enabled by our hardware

- Implement many different operations including multiplications efficiently, which means it can implement multiplication-intensive applications such as deep learning with minimal power and silicon area requirements.

6.5 Conclusion

A chaotic system is hard to work with, it scares the engineers away, it is unstable, hard to design, fabricate, and utilize. But if all is done correctly, a chaotic system provides an unprecedented amount of performance, unmatched by any conventional linear system. The AI community has fully experienced this transformation of moving from tractable, elegant methods and mathematics to intractable, hard to optimize models, and this move resulted in huge leap in AI. We believe chaos is another uncharted territory that despite the challenges that come with it, can provide huge rewards.

Here we discussed our fabrications, sample applications, and the results. The main conclusion is that chaos can provide extremely fascinating features and capabilities with unique applications, however, there are challenges to overcome. NAIL has been following multiple different tracks to AI. On one extreme, we teach and practice the conventional AI and deep learning and team with government, research and technology companies to apply conventional AI to their needs. On the other extreme, NAIL is pioneering a novel approach to AI based on nonlinear dynamics and chaos to develop AI systems that demonstrate awareness, cognition and deeper intelligence and interactions.

References

1. T.M. McKenna, T.A. McMullen, M.F. Shlesinger, The brain as a dynamic physical system. Neuroscience **60**(3), 587–605 (1994)
2. M.D. Fox, A.Z. Snyder, J.L. Vincent, M. Corbetta, D.C. Van Essen, M.E. Raichle, The human brain is intrinsically organized into dynamic, anticorrelated functional networks. Proc. Natl. Acad. Sci. U.S.A. **102**, 9673-9678 (2005)
3. R.T. Canolty, M. Soltani, S.S. Dalal, E. Edwards, N.F. Dronkers, S.S. Nagarajan et al., Spatiotemporal dynamics of word processing in the human brain. Front. Neurosci. **1**, 185-196 (2007)
4. Chris A. Mack, Fifty years of Moore's law. IEEE Trans. Semicond. Manuf. **24**(2), 202–207 (2011)
5. Thomas N. Theis, H.-S. Philip Wong, The end of Moore's law: a new beginning for information technology. Comput. Sci. Eng. **19**(2), 41–50 (2017)
6. A. Blum, R.L. Rivest, Training a 3-node neural network is NP-complete. Advances in neural information processing systems (1989)
7. Marti A. Hearst et al., Support vector machines. IEEE Intell. Syst. Appl. **13**(4), 18–28 (1998)
8. A. Krizhevsky, I. Sutskever, G.E. Hinton, Imagenet classification with deep convolutional neural networks. Advances in neural information processing systems (2012)
9. B. Kia, J.F. Lindner, W.L. Ditto, A simple nonlinear circuit contains an infinite number of functions. IEEE Trans. Circuits Syst. II: Express Briefs **63**(10), 944–948 (2016)
10. Behnam Kia, John F. Lindner, William L. Ditto, Nonlinear dynamics as an engine of computation. Philos. Trans. R. Soc. A **375**(2088), 20160222 (2017)
11. B. Kia, K. Mobley, W.L. Ditto, An integrated circuit design for a dynamics-based reconfigurable logic block. IEEE Trans. Circuits Syst. II: Express Briefs (2017)
12. B. Kia et al., Nonlinear dynamics-based adaptive hardware, in *2017 NASA/ESA Conference on Adaptive Hardware and Systems (AHS)* (IEEE, 2017)

Chapter 7
Linear Chaos in a Tape Recorder

Ned J. Corron(✉)

Charles M. Bowden Laboratory, U.S. Army AMRDEC, Redstone Arsenal, Huntsville, AL, USA
ned.j.corron.civ@mail.mil

Abstract. A mathematical model of an analog tape recorder is developed and shown to exhibit linear chaos. The playback dynamics act as a wave and are modeled by a linear partial differential equation with a simple analytic solution. This linear dynamical system is shown to exhibit three properties commonly used to define chaotic dynamics: the solution set is dense with periodic orbits, contains transitive orbits, and exhibits extreme sensitivity to initial conditions. Thus, a tape recorder provides a common physical example of linear chaos.

7.1 Introduction

It is lore in the study of dynamical systems that chaos is an inherently nonlinear phenomenon [1]. However, examples of chaos in linear [2–5] and quasi-linear [6–12] systems have been known for some time now—a situation that many researchers still find surprising and even disturbing. This lore persists despite the fact that a positive Lyapunov exponent, a common indicator of chaos, indicates linear instability and can also be displayed by linear systems [13]. Analytic intractability is another closely held belief of chaotic dynamics, yet there are counterexamples here, too. Recent research has identified chaotic piecewise-linear oscillators that admit exact analytic solutions, which can be written as a linear convolution of a discrete information sequence and a fixed basis function, similar to a modern communication waveform [14–16]. Altogether, these counterexamples suggest a larger view of chaotic phenomena that may have practical implications. In this paper, we expand the sphere of linear, tractable chaos to include a common physical system, namely, an analog tape recorder.

The recognition and development of linear chaotic dynamics may be technologically important, as it enables the intriguing aspects of chaotic dynamics to be accessible to standard engineering practice [17]. For example, chaotic oscillators have been proposed as low-cost, high-speed physical random number generators to support encryption and Monte Carlo simulations [18]. Also, the wide bandwidth and non-repeating nature of chaotic waveforms suggest benefits for random-signal radar [19] and spread-spectrum communications [20,21]. Using

V. In et al. (Eds.): *Proceedings of the 5th International Conference on Applications in Nonlinear Dynamics*, Understanding Complex Systems, https://doi.org/10.1007/978-3-030-10892-2_7

Fig. 7.1. Audio tape recorder

chaos with linear characteristics may enable these benefits to be engineered into these and other technologies without paying the price for using highly nonlinear devices.

7.2 Model

Figure 7.1 shows a reel-to-reel tape recorder, which was a common analog technology for capturing and playing back audio signals prior to the advent of digital technologies. This electromechanical device uses a motor to move a magnetic tape across fixed read and write heads at a constant linear speed. The tape stores time-varying signals as magnetic spatial variations along the tape. In record mode, an input signal is written on the moving tape using the record head. In playback mode, the audio signal is reproduced as the tape moves across the read head.

We wish to develop a mathematical model of a tape recorder operating in playback mode. We define the state of the tape recorder at time t as $u(x, t)$, where u is the signal stored on the tape at the position x relative to the read head at $x = 0$. See Fig. 7.2. The time dependence of the state reflects that the tape is moving. We have an initial condition

$$u(x, 0) = f(x) \tag{7.1}$$

where $f(x)$ is a previously recorded waveform stored on the tape. During playback, the tape moves across the tape head at a constant speed, so the state of the machine evolves as

$$u(x, t) = f(x + t) \tag{7.2}$$

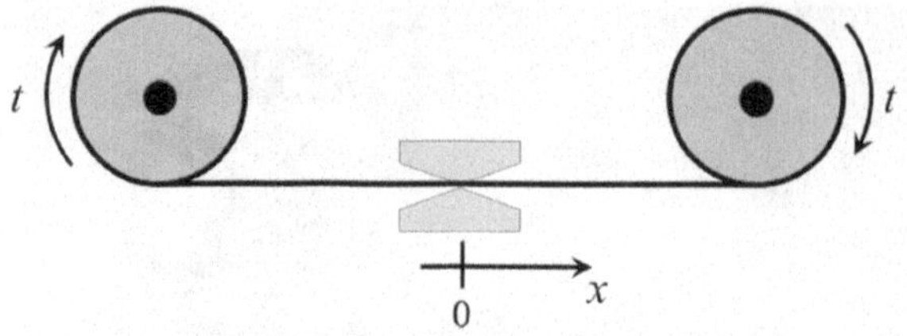

Fig. 7.2. Tape recorder model

where we assume unit velocity without loss of generality. For times $t > 0$, the read head detects the signal at position $x = 0$, so that playback provides $u(0,t) = f(t)$ which effectively converts the spatially stored waveform to a time signal.

Examining Eq. (7.2), we recognize the time-evolving state as a leftward-propagating wave. For sufficiently smooth signals $f(x)$, the tape recorder state formally satisfies the partial differential equation

$$\frac{\partial u}{\partial t} = \frac{\partial u}{\partial x} \tag{7.3}$$

which is a one-way wave equation. Established methods extend this equation for non-smooth and discontinuous functions and are consistent with a model of a physical tape recorder. For a mathematical model, we assume the domain $t \geq 0$ and $0 \leq x < \infty$, which represents a never-ending, infinitely long-playing tape. Here, we do not include states for $x < 0$, since such states do not affect future playback. In this idealized model, Eq. (7.2) provides the solution to the partial differential equation (7.3) subject to the initial condition given in Eq. (7.1).

7.3 Devaney's Chaos

The most widely accepted definition of chaos is due to Devaney [22]. This definition was originally developed in the context of an iterated, one-dimensional map function. However, the spirit of that definition has been extended and is often applied to a larger class of dynamical systems, including higher-dimensional maps and differential equations [5].

A partial generalization of Devaney's original definition formally considers a metric space Y and a mapping function $\phi : Y \to Y$. The function ϕ defines a dynamical system by the repeated iteration from an initial condition. For a continuous function ϕ, the associated dynamical system is chaotic on Y if it satisfies three requirements. First, periodic points are dense in Y, which can be rigorously written as

$$\forall U \subset Y \Rightarrow \exists y \in U, n > 0 : \phi^n(y) = y \tag{7.4}$$

where $\subset$ implies an open subset. Second, the iterated function is topologically transitive, or

$$\forall U, V \subset Y \Rightarrow \exists y \in U, n > 0 : \phi^n(y) \in V \tag{7.5}$$

which implies that the set Y cannot be decomposed into smaller, disconnected open sets. Third, the iterated function exhibits sensitive dependence, or

$$\exists\delta > 0 : \{\forall y \in U \subset Y \Rightarrow \exists z \in U, n > 0 : \|\phi^n(y) - \phi^n(z)\| \geq \delta\}. \tag{7.6}$$

Of these three conditions for chaos, only the last explicitly requires a metric to define distance. This definition can be further generalized for continuous time dynamics by applying analogous requirements to dynamical systems comprising ordinary and partial differential equations.

7.4 Linear Chaos

Despite common lore, it is well known that certain linear systems can formally satisfy the requirements in Devaney's definition of chaos [2–5]. Such systems are collectively referred to as linear chaos. Here, we claim the tape recorder model satisfies Devaney's definition for chaos, thereby providing a physically realizable example of linear chaos. In making this claim, we assume an extension of Devaney's definition that accommodates a partial differential equation for the dynamical system. For the tape recorder model, we explicitly show there exists a metric space in which the general solution $u(x,t)$ exhibits dense periodic orbits, a transitive orbit, and sensitive dependence on initial conditions.

First, we identify a metric space for the state of the dynamical system. Since the system is a partial differential equation, the time evolving state is a function of the spatial coordinate x. Thus, we use the notation $u_t(x) = u(x,t)$ to emphasize this function state at a fixed time t. Using this notation, we define a norm to measure the size of a function state using

$$\|u_t\| = \sup_{x\geq 0}\left\{|u_t(x)|\, e^{-\lambda x}\right\} \tag{7.7}$$

where $\lambda > 0$ is a fixed parameter and sup is the supremum of the function over the indicated range. We then identify a metric space using the set

$$W = \left\{u_t : \sup_{x\geq 0}\{|u_t(x)|\} < \infty\right\} \tag{7.8}$$

and the metric induced by the norm in Eq. (7.8).

The first of Devaney's conditions for chaos is that periodic orbits are dense. That is, for any initial condition $f \in U \subset W$, where $\subset$ implies an open subset, there exists an initial condition $\tilde{f} \in U$ such that the resulting solution trajectory $\tilde{u}(x,t) = \tilde{f}(x+t)$ is periodic with some period T. In this definition, U is any neighborhood containing f, and it is most demanding to consider a small neighborhood so that f and $\tilde{f}$ must be nearby (in the sense of the metric). We show this requirement by considering the particular initial condition

$$\tilde{f}(x) = f(x - nT), \quad nT \leq x < (n+1)T, \quad n \in Z \tag{7.9}$$

where $\tilde{f} \in W$ by construction, T is an arbitrary period, and Z is the set of integers. By design we have $\tilde{f}(x) = \tilde{f}(x+T)$, so that $\tilde{u}(x,t) = \tilde{u}(x,t+T)$ and the corresponding solution is periodic. Furthermore, we have that

$$\lim_{T\to\infty} \|\tilde{f}(x) - f(x)\| = 0 \tag{7.10}$$

which implies that the initial condition for the periodic orbit can be made arbitrarily close to $f(x)$ by increasing the period T. Recalling U is an open set, we are then assured that $\tilde{f} \in U$ for sufficiently large T, thereby showing periodic orbits are dense in W.

The second of Devaney's conditions is topological transitivity. We consider two arbitrary functions $f \in U \subset W$ and $g \in V \subset W$. We then construct the initial condition

$$\tilde{f}(x) = \begin{cases} f(x), & x < T \\ g(x-T), & x \geq T \end{cases} \tag{7.11}$$

where $\tilde{f} \in W$ and T is an arbitrary interval. The corresponding solution is $\tilde{u}(x,t) = \tilde{f}(x+t)$, so that

$$\lim_{T\to\infty} \|\tilde{u}(x,0) - f(x)\| = 0 \tag{7.12}$$

and we are assured that $\tilde{u}(x,0) \in U$ for sufficiently large T. Also, we have $\tilde{u}(x,T) = g(x) \in V$, which implies there exists a transitive orbit that connects U to V. Thus, the tape recorder model is topologically transitive on W.

The third of Devaney's conditions is sensitive dependence. Rigorous mathematical results have shown that dense periodic orbits and a transitive orbit are usually sufficient for a topological definition of chaos [23]. However, it is useful to also explicitly show sensitive dependence as implied by a positive Lyapunov exponent, since it is the famous hallmark of chaotic dynamics. To this end, we consider a function $f(x) \in W$ and choose $\tilde{f}(x) \in W$ such that

$$\sup_{0\leq x<T} \left\{ |f(x) - \tilde{f}(x)|\, e^{-\lambda x} \right\} = 0, \quad \sup_{x\geq T} \left\{ |f(x) - \tilde{f}(x)|\, e^{-\lambda x} \right\} = \varepsilon \tag{7.13}$$

which implies that $f(x)$ and $\tilde{f}(x)$ are identical on the interval $0 \leq x < T$. We then consider the solution trajectories $u(x,t) = f(x+t)$ and $\tilde{u}(x,t) = \tilde{f}(x+t)$ resulting from these initial conditions. For $t \leq T$, we find

$$\|u(x,t) - \tilde{u}(x,t)\| = \varepsilon\, e^{\lambda t} \tag{7.14}$$

which follows from the requirements in Eq. (7.13). Thus, this result explicitly shows an exponential growth in the initial separation of solution trajectories starting from $f(x)$ and $\tilde{f}(x)$. For an arbitrary $\delta > 0$, we may always choose T large enough such that $\varepsilon\, e^{\lambda T} > \delta$, which meets the requirement that arbitrarily small perturbations grow to a significant size. Thus, the wave equation exhibits sensitive dependence on W, thereby completing the claim that the wave equation model of a tape recorder is chaotic.

In establishing sensitive dependence, we find that Eq. (7.14) reveals that λ quantifies the growth rate of the exponential separation and, thus, acts like a positive Lyapunov exponent for the system. We note that the parameter λ does not appear in the original wave equation. Instead, this parameter was defined for the norm in Eq. (7.7), so that it is a characteristic of the metric used to measure the system. To define a norm using Eq. (7.7), it is only required that $\lambda > 0$. As such, only the existence of a positive Lyapunov exponent is fundamental to the nature of the physical system, while its magnitude depends on the measurement system and relays nothing about the physical system.

7.5 Discussion

We presented a linear wave-equation model for an analog tape recorder and showed that it meets the requirements for the most commonly cited definition of chaos. As such, this system provides a physical example that realizes linear chaos [5]. However, the chaos of this simple system is certainly not the same intriguing complex behavior that inspires the lore of conventional nonlinear chaos. Indeed, dynamical chaos was originally coined to describe the complex intractable oscillations observed in simple nonlinear systems, such as the iterated logistic map or the famous Lorenz oscillator [1, 13]. Quite different are the dynamics of a linear wave equation, for which an analytic solution is straightforward and its behavior is completely transparent. It may be argued that such trivial dynamics are not what chaos was intended to describe. Thus, one might reasonably conclude that the common definition of chaos has limitations, that it does not correctly identify what we know should and should not be chaos, and that it cannot be mechanically applied without potentially devaluing what we mean by chaos.

However, such a conclusion may be precarious and potentially dangerous. We note that the most fundamental chaotic dynamical system is a Bernoulli shift, with the mapping function $\phi(x) = 2x \bmod 1$ on the unit interval. Indeed, establishing conjugacy to a shift is considered conclusive evidence for the fold and shift dynamics that are essential for low-dimensional chaos [22]. However, it is also reasonable to argue that the dynamics of the Bernoulli shift are obvious and transparent, since this system exhibits an exact analytic solution and simple behavior [6, 7, 10]. If we accept the reasons for denying linear dynamics as chaotic, we might also rule out the trivial shift dynamics, which would be a conundrum. Perhaps we can only conclude that formally recognizing and defining chaos is a complex matter.

Acknowledgements. The author recognizes Dr. Daniel Hahs, Dr. Shawn Pethel, and Dr. Shangbing Ai for helpful discussions regarding the interpretation and presentation of the research results.

References

1. J.M.T. Thompson, H.B. Stewart, *Nonlinear Dynamics and Chaos* (Wiley, New York, 1986)

2. C.R. MacCluer, Chaos in linear distributed systems. J. Dyn. Syst. Meas. Control **114**, 322 (1992)
3. A. Gulisashvili, C.R. MacCluer, Linear chaos in the unforced quantum harmonic oscillator. J. Dyn. Syst. Meas. Control **118**, 337 (1996)
4. R. deLaubenfels, H. Emamirad, V. Protopopescu, Linear, chaos and approximation. J. Approx. Theory **105**, 176 (2000)
5. K.-G. Grosse-Erdmann, A.P. Manguillot, *Linear Chaos* (Springer, Berlin, 2011)
6. D.F. Drake, Information's Role in the Estimation of Chaotic Signals, Ph.D. dissertation (Georgia Institute of Technology, 1998)
7. S.T. Hayes, Chaos from linear systems: implications for communicating with chaos, and the nature of determinism and randomness. J. Phys. Conf. Ser. **23**, 215 (2005)
8. Y. Hirata, K. Judd, Constructing dynamical systems with specified symbolic dynamics. Chaos **15**, 033102 (2005)
9. N.J. Corron, S.T. Hayes, S.D. Pethel, J.N. Blakely, Chaos without nonlinear dynamics. Phys. Rev. Lett. **97**, 024101 (2006)
10. D.F. Drake, D.B. Williams, Linear, random representations of chaos. IEEE Trans. Signal Process. **55**, 1379 (2007)
11. N.J. Corron, S.T. Hayes, S.D. Pethel, J.N. Blakely, Synthesizing folded band chaos. Phys. Rev. E **75**, 045201R (2007)
12. D.W. Hahs, N.J. Corron, J.N. Blakely, Synthesizing antipodal chaotic waveforms. J. Franklin Inst. **351**, 2562 (2014)
13. E. Ott, *Chaos in Dynamical Systems* (Cambridge University Press, Cambridge, 1993)
14. N.J. Corron, An exactly solvable chaotic differential equation. Dyn. Contin. Discret. Impuls. Syst. A **16**, 777 (2009)
15. N.J. Corron, J.N. Blakely, M.T. Stahl, A matched filter for chaos. Chaos **20**, 023123 (2010)
16. N.J. Corron, J.N. Blakely, Exact folded-band chaotic oscillator. Chaos **22**, 023113 (2012)
17. N.J. Corron, J.N. Blakely, Chaos in optimal communication waveforms. P. Roy. Soc. Lond. A **471**, 20150222 (2015)
18. A. Uchida, K. Amano, M. Inoue, K. Hirano, S. Naito, H. Someya, I. Oowada, T. Kurashige, M. Shiki, S. Yoshimori, K. Yoshimura, P. Davis, Fast physical random bit generation with chaotic semiconductor lasers. Nat. Photon. **2**, 728 (2008)
19. M.S. Willsey, K.M. Cuomo, A.V. Oppenheim, Selecting the Lorenz parameters for wideband radar waveform generation. Int. J. Bifurc. Chaos **21**, 2539 (2011)
20. H. Leung (ed.), *Chaotic Signal Processing* (SIAM, 2014)
21. M. Eisencraft, R. Attux, R. Suyama (eds.), *Chaotic Signals in Digital Communications* (CRC Press, Boca Raton, 2014)
22. R.L. Devaney, *Introduction to Chaotic Dynamical Systems* (Addison-Wesley, Boston, 1989)
23. J. Banks, J. Brooks, G. Cairns, G. Davis, P. Stacy, On Devaney's definition of chaos. Am. Math. Mon. **99**, 332–334 (1992)

Chapter 8
Piezoelectric Cantilevers, Magnets and Stoppers as Building Blocks for a Family of Devices Performing in Vibrationally Noisy Environments

Salvatore Baglio[1(✉)], Carlo Trigona[1], Bruno Andò[1], and Adi R. Bulsara[2]

[1] DIEEI, University of Catania, Viale A. Doria 6, 95125 Catania, Italy
[2] SPAWAR Pacific Code 71000, San Diego, CA 92152, USA

Abstract. Vibration Energy Harvesting has received a lot of attention in recent years, because of the ubiquitous existence of vibrations in a variety of environments. In real-world device implementation, however, several problems are encountered particularly when the harvesters are intended to power miniaturized systems at micro and/or nano scale; in these cases, to store the harvested energy can pose significant problems due to the very low level of voltages involved, thereby conflicting with the threshold of blocking diodes. Investigations on this specific subject have lead us to the development of a family of devices which exploits the synergetic use of piezoelectric materials, flexible beams, magnets and mechanical stoppers together with some concepts of nonlinear dynamics used to accurately model and understand the device behaviors. Here we present an excursion that begins with the genesis of these ideas and leads to a family of devices able to capture mechanical energy, convert it into electrical energy, and store this energy regardless of the voltage level. The switching mechanism with the mechanical stopper is used to overcome the diode threshold. Few building blocks (Piezoelectric cantilevers, magnets and stoppers) have been identified that, once suitably arranged and used, can lead to novel devices operating as detectors and/or energy harvesters. Beyond energy harvesting, devices able to multiply voltages and rectify signals will be presented, these devices can perform, even at very low voltages because do not use diode. A review of these devices together with working principles, models and experimental characterization results is reported in this review paper.

8.1 Introduction

Environmental kinetic energy represents one of the richest sources for energy harvesting and has been, in recent years, frequently targeted by a number of research efforts aimed at providing an autonomous solution to power up small-scale and low-power electronic devices. In fact several applications exist [1] in which energy harvesting plays a crucial role e.g. self-powered sensors [2], implanted sensor nodes

V. In et al. (Eds.): *Proceedings of the 5th International Conference on Applications in Nonlinear Dynamics*, Understanding Complex Systems, https://doi.org/10.1007/978-3-030-10892-2_8

[3], and in general autonomous microsystems and smart systems [4] wherein batteries need to be replaced or recharged [5]. Kinetic energy comes in a large variety of forms, and, more generally, as noisy environmental vibrations [6]. While sometimes energy appears at specific frequencies, as in the case of rotating machinery [7], it is more common that it is distributed over a wide spectrum of frequencies [8]. Most devices for vibration energy harvesting were originally based on linear resonant systems that show optimal operating conditions when they are excited at resonance [9, 10].

Our work has been focused on the problem of collecting, most efficiently, electrical energy from noisy mechanical environmental vibrations whose energy often appears with a wide frequency spectrum at low frequencies. To tackle this issue we have switched from the traditional harmonic oscillator approach to a more complex, but richer in performance, strategy that exploits nonlinear dynamics and in particular bistable behaviors.

The conversion from kinetic to electrical energy is accomplished by using piezoelectric materials embedded into flexible cantilever beams with an inertial mass that deform in response to the inertial forces acting on the mass. Bistability has been obtained by adding two magnets to the original cantilever beam [11]. A better use of the magnets has led us to exploit other features in piezoelectric cantilever beam. In fact antiphase bistable systems [12], tri-stable [13] and 2D vibration harvesters [14] have been developed and characterized.

The need for miniaturized devices results in smaller amplitudes of the signals that convey the harvested energy to be stored. This scenario is not compatible with the use of diode, or other threshold current blocking components, which however are necessary to accumulate the energy harvested. The addition of mechanical stoppers, operated also as electrical contacts, to the piezo electric cantilever beam has led to a device that operates in such a way to harvest kinetic energy and transfer the electric energy into magnetic first, and then back to electric when it has to be stored into the capacitor [15]. This Random Mechanical Switching Harvester on Inductor (RMSHI) device lets us overcome the threshold of the blocking diodes at any input signal amplitude. Adding magnets to this device has led to the bi-RMSHI [16] which efficiently responds to noisy incoming vibrations.

By looking beyond the limit of the energy harvesting problem the above mentioned building blocks have been mixed and matched so that the result is some other devices and working principles that have been exploited for signal processing e.g. rectification [12] or amplification [17] of signals whose amplitude is smaller than diode threshold. Beyond this, the above strategies have been extended to exotic solutions for switched capacitor systems [18] wherein environmental vibrations and piezoelectric materials are still present to supply the power needed for basic functions and to convert mechanical to electric energy.

8.2 Cantilevers and Magnets for Energy Harvesting

The main components in a vibration energy harvesting system are shown in Fig. 8.1.

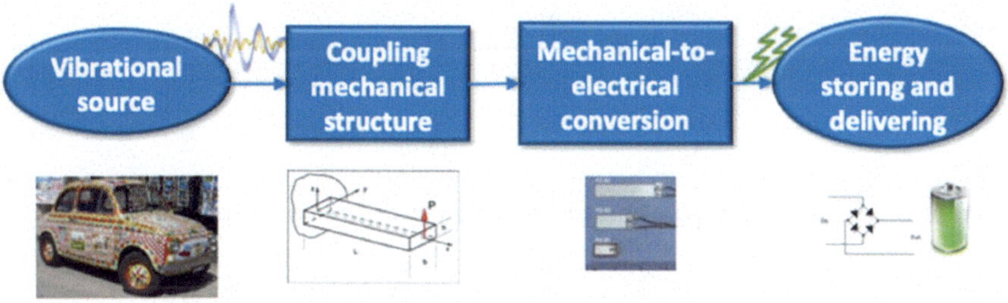

Fig. 8.1. Main functional blocks in a vibration energy harvesting system

The coupling is implemented through an inertial mass and therefore a force that deforms the flexible beam. The deformations are then picked by the piezoelectric material and converted into electrical signals whose energy can finally be stored.

8.2.1 Bistable Systems

Figure 8.2 shows the bistable setup using two magnets with opposing magnetization placed at the cantilever tip and the fixed frame respectively. The bistable potential energy function is also shown together with the conceptual drawing and the images of a MEMS scale prototype [19].

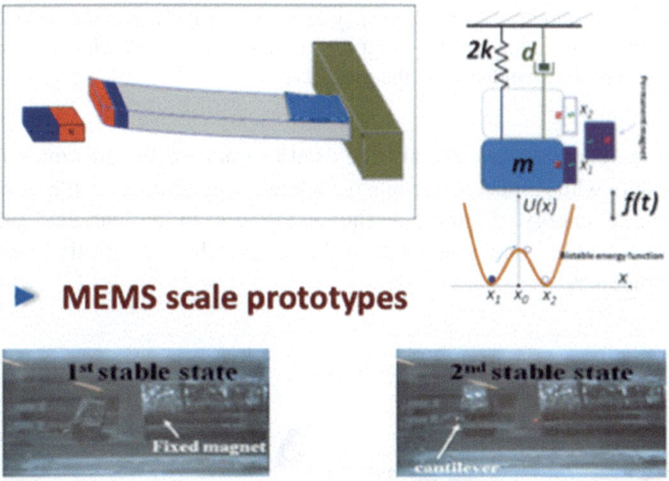

Fig. 8.2. Conceptual drawing of the bistable cantilever for efficient environmental vibration energy harvesting [19] shown together with the working principle and a MEMS prototype

This system can be modeled as [20]:

$$m\ddot{x} + d\dot{x} + \Psi = f(t) \tag{8.1}$$

$$\Psi \triangleq \frac{\partial U(x)}{\partial x} = U'^{(x)} \tag{8.2}$$

$$U(x) = kx^2 + (ax^2 + b\Delta^2)^{-\frac{3}{2}} + c\Delta^2 \tag{8.3}$$

Figure 8.3 shows the typical benefit gained via the bistable approach.

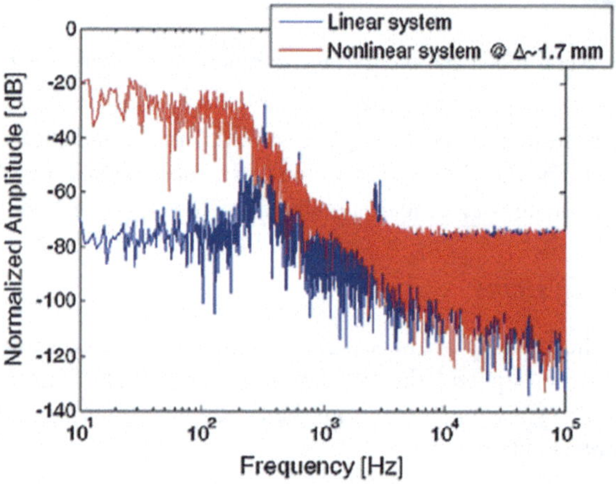

Fig. 8.3. Effect of bistability on the frequency spectrum of the harvesting device [19]. Blue signal refers to the intrinsic linear behavior of the cantilever in Fig. 8.2 while the red plot shows the effects of the bistability induced by the magnets

Opposing magnetic forces can also be used to improve the efficiency in vibration energy harvesters with respect to various parameters e.g. volume, or the direction of the incoming kinetic energy. Figure 8.4 shows the use in a "parallel" arrangement, opposed to the "inline" one, leading to a double bistable system that behaves in an "anti-phase" manner.

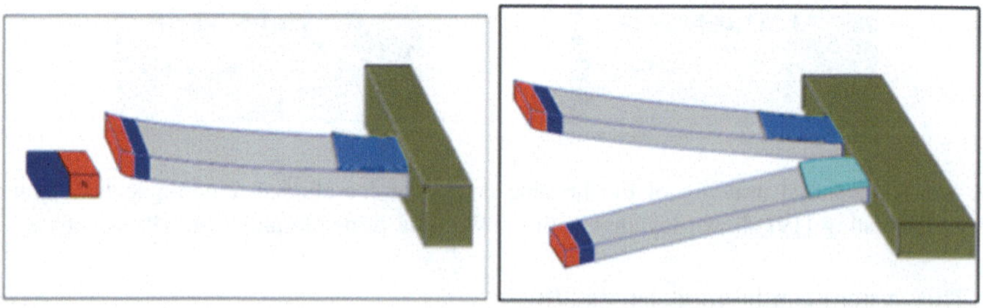

Fig. 8.4. "Parallel" use of the magnets to obtain bistable and "anti-phase" vibrating beams

A simple mathematical model can be written down [12]:

$$
\begin{aligned}
m\ddot{x}_1 &= -d\dot{x}_1 - kx_1 + k_{nl1}x_1 + k_{nl\_acc}(x_2 - x_1) + d_m(\dot{x}_2 - \dot{x}_1) + F(t) \\
m\ddot{x}_2 &= -d\dot{x}_2 - kx_2 + k_{nl2}x_2 + k_{nl\_acc}(x_2 - x_1) - d_m(\dot{x}_2 - \dot{x}_1) + F(t)
\end{aligned}
\tag{8.4}
$$

with reference to the symbols defined in Fig. 8.5.

$$
\begin{aligned}
k_{nl1} &= \alpha_1 - \beta_1 x_1^{\,2} \\
k_{nl2} &= \alpha_2 - \beta_2 x_2^{\,2} \\
k_{nl\_acc} &= \gamma - \delta(x_2 - x_1)^2
\end{aligned}
$$

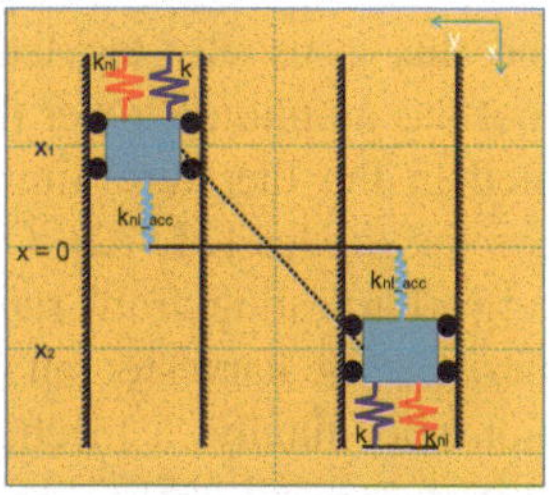

Fig. 8.5. Conceptual scheme used to develop the mathematical model of bistable and "anti-phase" vibrating beams and definition of the elastic constants (left) [12]

8.2.2 *Beyond Bistable Systems*

The approach outlined above can be extended to tri-stable or multistable systems [13], that improve the efficiency of the system, as well as 2D bistable devices [14] that respond to vibrations arriving from different direction. In Fig. 8.6 some of the prototypes developed are shown.

Fig. 8.6. From top left in clockwise direction: Bistable anti-phase device [12], multistable device, tri-stable [13] and 2D bistable [14] vibration energy harvester device protoypes

8.3 Adding Mechanical Stoppers to the Beam

As discussed above, our interest is focused on piezoelectric transducers that produce an AC output voltage in response to mechanical deformations induced into an elastic beam by the external vibrations whose energy is to be harvested. If dimension shrinkage is taken into account as a logical consequence of a possible MEMS scale realization, a significant reduction of the output voltage amplitude has to be faced.

In order to store the energy (the rightmost block in Fig. 8.1) a current rectification is necessary and this is usually tackled using diode bridge circuits that, however, fail when the input has amplitude lower than the diode threshold. Several approaches have been presented in the literature aiming to overcome this drawback by boosting the voltage across the diodes [21, 22]. Our approach focuses on the development of systems for energy harvesting from random, low amplitude, broadband vibrations that includes a piezoelectric harvester, an inductor, the current rectifying section, the charge storage section and, finally, a mechanical switch driven by the same environmental vibrations to be harvested [15].

8.4 RMSHI and Bi-RMSHI for Low Level Vibration Energy Harvesting

Figure 8.7 shows the functional block scheme of the system Random Mechanical Switching Harvesting on Inductor (RMSHI).

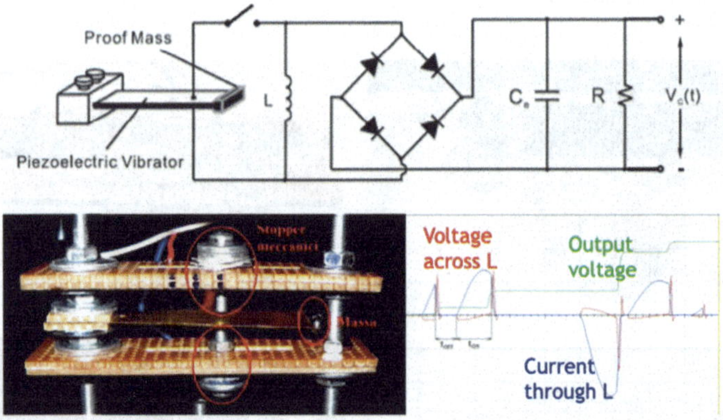

Fig. 8.7. Top. Schematic of the RMSHI system. The switch must be mechanical here. Bottom left. The experimental prototype. Bottom right. Signals. It is possible to observe as the inductor voltage (red) spikes every time the beam leaves the stopper thus allowing the magnetic energy stored into the inductor to be transferred to the capacitor [15]

One experimental prototype is also shown in Fig. 8.7 together with a screenshot of the signals in the system. When vibrations drive the beam to one of the stoppers (Fig. 8.7 bottom left) electric contact is made, the piezoelectric beam is connected to the circuit and current flows into the inductors because the voltage is smaller than the

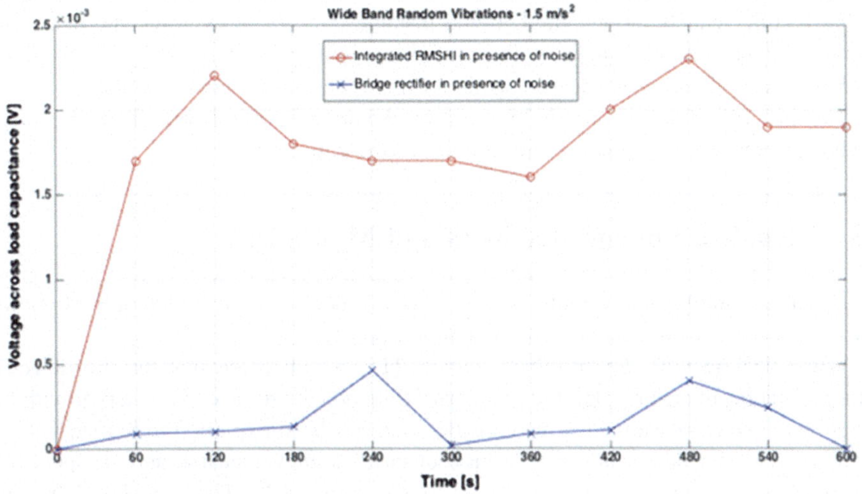

Fig. 8.8. Comparison between the output voltage in the RMSHI (red) and traditional diode bridge rectifier (blue) vibration energy harvester [23]

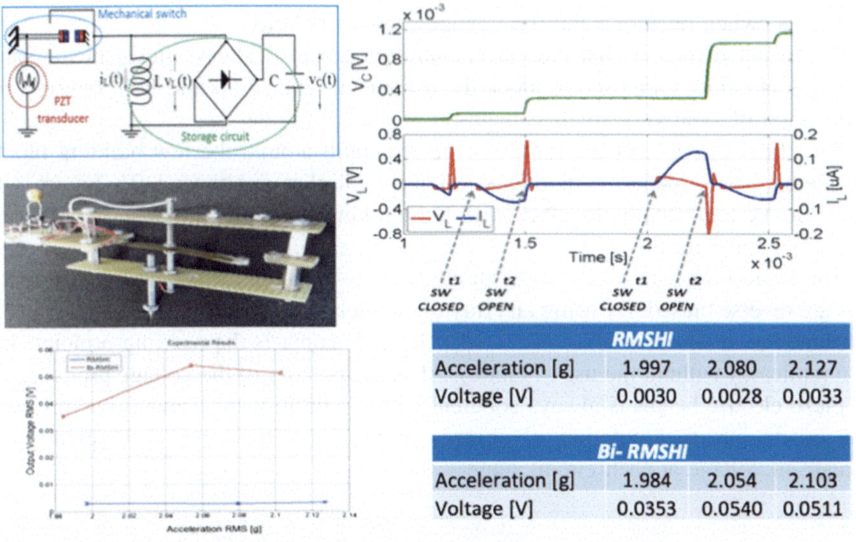

RMSHI			
Acceleration [g]	1.997	2.080	2.127
Voltage [V]	0.0030	0.0028	0.0033

Bi- RMSHI			
Acceleration [g]	1.984	2.054	2.103
Voltage [V]	0.0353	0.0540	0.0511

Fig. 8.9. Bistable RMSHI. From top left clockwise: the system schematic, relevant waveforms simulated, experimental test conditions, comparison with the RMSHI, experimental prototype [23]

diode thresholds. As soon as the beam leaves the stopper an over-voltage appears across the inductor that allows the magnetic energy stored in the inductance to be transferred to the capacitor through the diode bridge. Some experimental results are shown in Fig. 8.8 [23] where the voltage across the storage capacitor is shown in the case of a random vibration (bandwith 100 Hz) with acceleration 1.5 m/s^2 rms. The large difference with respect to the case of the same signal arriving into a regular diode bridge rectifier can be appreciated. If the bistable concept is exploited here we will obtain, as expected, advantages with random incoming vibrations. In Fig. 8.9 the bi-RMSHI device drawing is shown together with some experimental results.

The bistable strategy thus helps enforce the transfer of energy from the piezo electric element to the inductor and, finally, to the capacitor.

8.5 Diodeless Voltage Rectifiers and Multipliers

As a final section of this overview, we show some other non typical uses of the building blocks that have been introduced earlier in the paper.

While dealing with the antiphase devices [12] we observed that the piezoelectric nature of the beam ensures that the polarity of the voltage produced is coherent with the direction of the displacement. This consideration has been exploited to realize voltage rectifiers that, instead of diodes, use piezoelectric cantilever beams and stoppers that operate as electric contacts. This device is shown in Fig. 8.10. The conceptual drawing (upper left) is reported together with the working principle (upper right), the experimental prototype (bottom left) and some sample signals (bottom right). Here the contacts “B”&”D” are closed when the voltage is positive while, due to the antiphase behavior, the contacts “A”&”C” are open. The opposite configuration is obtained when the beams switch positions and the voltage changes polarity.

The result of this is that the current through the load always flows in the same direction. No diodes are used to block the reverse current, so no threshold opposes the rectification of even very small voltages.

As a final example of the family of devices built around the few building blocks considered in this paper, a diodeless voltage multiplier is shown [17]. Figure 8.11 shows the system scheme together with the working principle, the experimental prototype and some results.

The basic idea here is that even diodes are always forward biased when the odd ones are reverse biased. By using an array of parallel piezoelectric beams and a suitably distributed array of upper and lower stopper/electric contacts the working principle has been replicated without the use of diodes. Also, the polarity of the voltage produced by the piezo electric beams is always coherent with the beam displacement thus resulting in a continuous accumulation of charges in the capacitors and, consequently, an increase in voltage at each circuit stage.

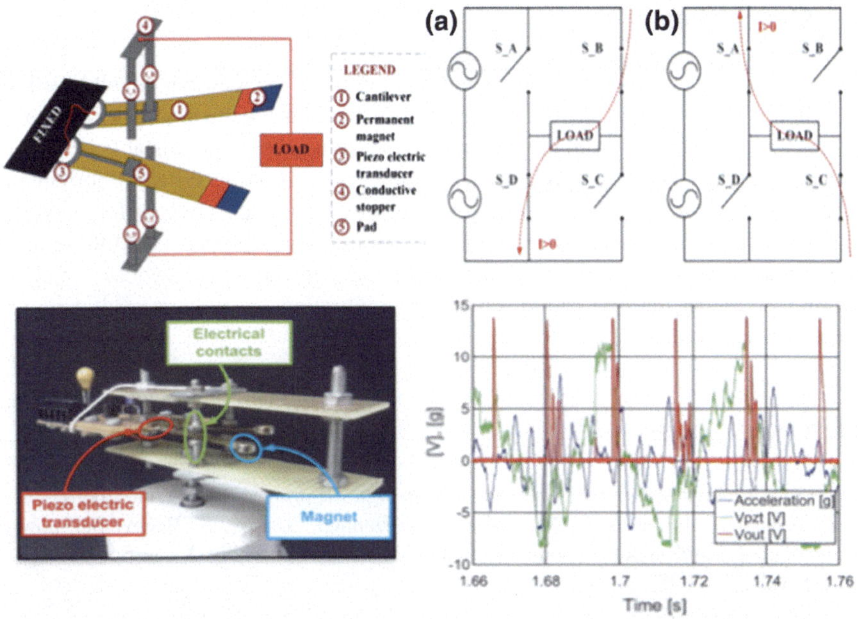

Fig. 8.10. Diodeless piezo electromechanical voltage rectifier [12]. From upper left in clockwise order: the conceptual scheme of the system, the working principle, the experimental prototype and experimental validation measures

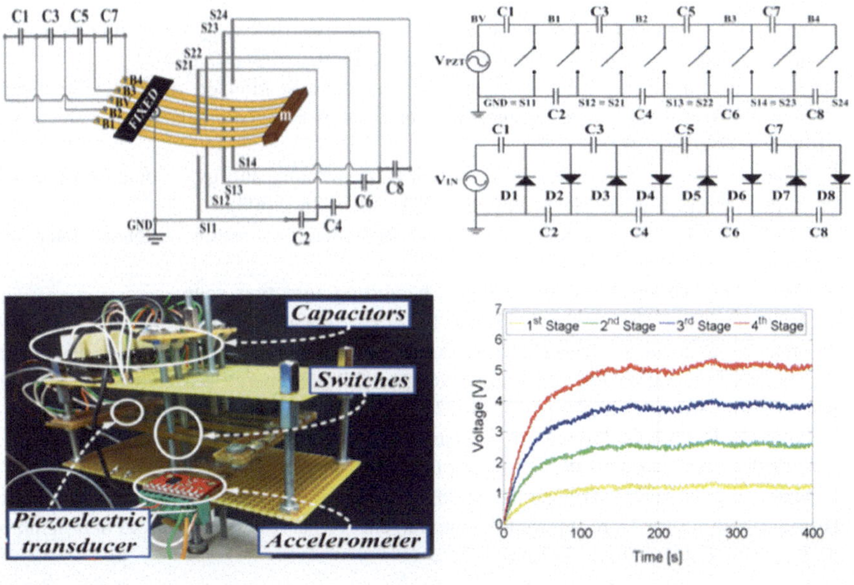

Fig. 8.11. Diodeless piezo electromechanical voltage multiplier [17]. From upper left in clockwise order: the conceptual scheme of the system, the working principle, the experimental prototype and experimental validation measures

8.6 Conclusions

This review paper reports the development of a family of devices that are all based on a few building blocks: piezo electric beams, magnets, and mechanical stoppers that act as electric contacts. Starting from the application of magnets to obtain bistable behavior in vibration energy harvesting applications, we have moved along different combination of the proposed building blocks that result in more complex devices for harvesting kinetic energy from weak and wide spectrum sources. But a proper use of these blocks has been also exploited here to demonstrate the realization of devices which can perform well, under the sole stimulus of external vibrations, as voltage rectification and multiplication systems.

References

1. S. Priya, D. Inman, *Energy Harvesting Technologies* (Springer, New York, 2008)
2. R. Torah, P. Glynne-Jones, M. Tudor, T. O'Donnell, S. Roy, S. Beeby, Self-powered autonomous wireless sensor node using vibration energy harvesting. Meas. Sci. Technol. **19** (12), 125202 (2008)
3. V. Raghunathan, A. Kansal, J. Hsu, J. Friedman, M. Srivastava, Design considerations for solar energy harvesting wireless embedded systems, in *Proceedings of the 4th International Symposium on Information Processing in Sensor Networks* (2005), pp. 459–462
4. M. Lallart, D. Guyomar, Y. Jayet, L. Petit, E. Lefeuvre, T. Monnier, P. Guy, C. Richard, Synchronized switch harvesting applied to self-powered smart systems: Piezoactive microgenerators for autonomous wireless receivers. Sens. Actuators A Phys. **147**(1), 263–272 (2008)
5. H. Sodano, D. Inman, G. Park, Comparison of piezoelectric energy harvesting devices for recharging batteries. J. Intell. Mater. Syst. Struct. **16**(10), 799–807 (2005)
6. B. Ando; S. Baglio; G. L'Episcopo; C. Trigona, Investigation on mechanically bistable MEMS devices for energy harvesting from vibrations. IEEE J. Microelectromechanical Syst. **21**(4), 779–790 (2012)
7. E. Yeatman, Energy harvesting from motion using rotating and gyroscopic proof masses. Proc. Inst. Mech. Eng. J. Mech. Eng. Sci. **222**(1), 27–36 (2008)
8. S. Roundy, On the effectiveness of vibration-based energy harvesting. J. Intell. Mater. Syst. Struct. **16**(10), 809–823 (2005)
9. S. Beeby, M. Tudor, N. White, Energy harvesting vibration sources for microsystems applications. Meas. Sci. Technol. **17**(12), 175–195 (2006)
10. T. Sterken, K. Baert, C. Van Hoof, R. Puers, G. Borghs, P. Fiorini, I. MCP, B. Leuven, Comparative modelling for vibration scavengers [MEMS energy scavengers], in *Proceedings of IEEE Sensors* (2004), pp. 1249–1252
11. B. Andò, S. Baglio, C. Trigona, N. Dumas, L. Latorre, P. Nouet, Nonlinear mechanism in MEMS devices for energy harvesting applications. J. Microm Microeng **20**, 1–12 (2010)
12. F. Maiorca, F. Giusa, C. Trigona, B. Andò, A.R. Bulsara, S. Baglio, Diode-less mechanical H-bridge rectifier for "zero threshold" vibration energy harvesters, in *Sensors and Actuators A: Physical, Volume 201, 15 October 2013*

13. C. Trigona; F. Maiorca; B. Andò; S. Baglio, Tri-stable behavior in mechanical oscillators to improve the performance of vibration energy harvesters, in *2013 Transducers & Eurosensors XXVII: The 17th International Conference on Solid-State Sensors, Actuators and Microsystems (TRANSDUCERS & EUROSENSORS XXVII), June 2013*
14. B. Andò, S. Baglio, F. Maiorca, C. Trigona, Two dimensional bistable vibration energy harvester. Procedia Eng **47**, 1061–1064 (2012); *Proceedings of Euro Sensors XXVI, September 9–12, 2012, Kraków, Poland*
15. F. Giusa, A. Giuffrida, C. Trigona, B. Andò, A.R.Bulsara, S. Baglio, Random mechanical switching harvesting on inductor. A novel approach to collect and store energy from weak random vibrations with zero voltage threshold. *Sensors and Actuators A: Physical*, vol. 198 (2013)
16. S. Bradai, S. Naifar, C. Trigona, S. Baglio, O. Kanoun, Electromagnetic transducer with bistable-RMSHI for energy harvesting from very weak kinetic sources, in *2018 IEEE International Instrumentation and Measurement Technology Conference (I2MTC)* (2018)
17. F. Giusa, F. Maiorca, A. Noto, C. Trigona, B. Andò, S. Baglio, A diode-less mechanical voltage multiplier: a novel transducer for vibration energy harvesting. Sens. Actuators A **212**, 1 (2014)
18. A. Noto; C. Trigona; B. Andò; S. Baglio, Novel Switched Capacitor (SC) approach based on the bistable mechanical switches, in *SENSORS, 2013 IEEE* (2013)
19. M. Ferrari, V. Ferrari, M. Guizzetti, B. Andò, S. Baglio, C. Trigona, Improved energy harvesting from wideband vibrations by nonlinear piezoelectric converters. Sens. Actuators A **162**(2), 425–431 (2010)
20. F. Cottone, H. Vocca, L. Gammaitoni Nonlinear energy harvesting. Phys. Rev. Lett. **102**, 080601 (2009)
21. S. Roundy, P.K. Wright e, J. Rabaey, A study of low level vibrations as a power source for wireless sensor nodes. Comput. Commun. **26**(11), 1131–1144 (2003)
22. D. Guyomar, A. Badel, E. Lefeuvre, C. Richard, Toward energy harvesting using active materials and conversion improvement by nonlinear processing. IEEE Trans. Ultrason. Ferroelectr. Freq. Control **52**(4), 584–595 (2006)
23. C. Trigona; B Andò; S Baglio, Performance measurement methodologies and metrics for vibration energy scavengers. IEEE Trans. Instrum. Meas. **66**(12) (2017)

Chapter 9
The Effects of Amplification of Fluctuation Energy Scale by Quantum Measurement Choice on Quantum Chaotic Systems: Semiclassical Analysis

Y. Shi[1], S. Greenfield[1,2,3], J. K. Eastman[2,3,4], A. R. R. Carvalho[3], and A. K. Pattanayak[1(✉)]

[1] Department of Physics and Astronomy, Carleton College, 1 N. College Street, Northfield, MN 55057, USA
{shiy, greenfields, arjendu}@carleton.edu

[2] Centre for Quantum Computation and Communication Technology (Australian Research Council), Sydney, NSW, Australia
jessica.eastman@anu.edu.au

[3] Centre for Quantum Dynamics, Griffith University, Brisbane, QLD 4111, Australia
andre.carvalho@q-ctrl.com

[4] Department of Quantum Science, Research School of Physics and Engineering, Australian National University, Canberra, ACT, Australia

Abstract. Measurement choices in weakly-measured open quantum systems can affect quantum trajectory chaos. We consider this scenario semi-classically and show that measurement acts as nonlinear generalized fluctuation and dissipation forces. These can alter effective dissipation in the quantum spread variables and hence change the dynamics, such that measurement choices can enhance quantum effects and make the dynamics chaotic, for example. This analysis explains the measurement dependence of quantum chaos at a variety of parameter settings, and in particular we demonstrate that the choice of monitoring scheme can be more relevant than system scale β in determining the 'quantumness' of the system.

9.1 Introduction

Measuring a quantum system has an unavoidable effect on its state. This is a feature with no classical counterpart that introduces an entirely quantum pathway to manipulate quantum systems. In particular, the continuous monitoring of a quantum system provides the ability to implement real-time control, which can be used to enhance or suppress desirable effects in the system dynamics. Recent work has shown that continuously measured open quantum system trajectory

V. In et al. (Eds.): *Proceedings of the 5th International Conference on Applications in Nonlinear Dynamics*, Understanding Complex Systems, https://doi.org/10.1007/978-3-030-10892-2_9

dynamics can change between the qualitatively dramatic different regimes of chaos (with high dynamical algorithmic complexity) and regularity (with qualitatively different dynamical complexity) depending on parameter choices [3,12]. In particular, the phase ϕ setting on a laser used as the local oscillator for making a homodyne measurement of the signal from a driven dissipative nonlinear quantum oscillator was shown to considerably affect the system dynamics [3]. The back-action from this kind of measurement manifests as a generalized dissipation $\mathbf{F}(\phi)$ and 'noise' $\mathbf{N}(\phi)$ where changes in ϕ can strongly affect the quantum dynamics sometimes making them chaotic, depending in a puzzling way on a combination of system parameters, including size, and the behavior of the classical limit. Understanding this puzzle would help us use ϕ, an external experimentally accessible parameter, to control quantum trajectories in useful ways.

We consider this system in the semi-classical regime where the measurement localization allows us to accurately and efficiently simulate the quantum state as a wave packet described completely by the coupled dynamics of its expectation values (centroid) and variances (spread). We use a formalism [10] representing $|\psi(t)\rangle$ as the dynamics of two oscillators: the centroid (x, p) and the spread (χ, Π) of the wave packet (detailed definitions below). Without environmental coupling these evolve according to the Hamiltonian $H(x, p, \chi, \Pi) = p^2/2 + \Pi^2/2 + U(x, \chi) = H_1(x, p) + H_2(\chi, \Pi) + U_{12}(x, \chi)$ where the relative size of the 'quantum' Hamiltonian H_2 and the coupling U_{12} change with size, such that the influence of the quantum oscillator on the classical motion increases with β via $U_{12}(x, \chi)$. The environment acts with $\mathbf{N}$ coupling only to (x, p) and the $\phi-$dependent part of $\mathbf{F}$ coupling only to (χ, π). Energy analysis is useful to understand the non-trivial effect of changing $\mathbf{N}$ and $\mathbf{F}$ with ϕ. Small changes in the fluctuation and dissipation $\mathbf{N}, \mathbf{F}$ change how the nonlinear dynamics amplify the quantum fluctuations and significantly change the energy range for the dynamics for χ, Π. This change alters the $U_{12}(x, \chi)$ coupling and hence the influence of the quantum oscillator on the classical dynamics.

We consider several such (Γ, β, ϕ) combinations to consider the effects of changing these parameters on the various competing effects. Our simulations verify our energy-based explanation for ϕ-dependent quantum trajectory chaos. We also find that measurement angle ϕ can affect the relative quantum energy scale compared to classical one by orders of magnitude more than the system scale β.

Below, we review the coupled-oscillator formalism then focus on the ϕ dependence of $\mathbf{F}$ and $\mathbf{N}$ before presenting our results and analysis. We conclude with a discussion about adaptive control of quantum trajectories as well as prospects for experimental implementations of these ideas.

9.2 Semi-classical Coupled Oscillator Model

Our analysis starts with the quantum model of a damped driven Duffing oscillator [2,3,6,9,12]. The Hamiltonian $\hat{H}_D = \frac{1}{2}\hat{P}^2 + \frac{\beta^2}{4}\hat{Q}^4 - \frac{1}{2}\hat{Q}^2 - \frac{g}{\beta}\hat{Q}\cos(\omega t)$

describes the double-well oscillator driven sinusoidally with strength g in terms of dimensionless position ($\hat{Q}$) and momentum ($\hat{P}$) operators. β serves as a dimensionless effective Planck's constant [2,9]: larger β describe a smaller system and $\beta \to 0$ is the classical limit. Quantum mechanical damping is introduced via the interaction of the system with a zero-temperature Markovian bath, which corresponds to having $\hat{a} = (\hat{Q} + i\hat{P})/\sqrt{2}$ in the decoherence superoperator [4,8]. Furthermore, we consider that this dissipative quantum channel is being weakly and continuously monitored, such that the state of the system evolves conditioned on the measurement outcomes as given by the following Ito stochastic equation [13,15]

$$|d\psi\rangle = \left(-\frac{i}{\hbar}\hat{H} + \langle \hat{L}^\dagger \rangle \hat{L} - \frac{\hat{L}^\dagger \hat{L}}{2} - \frac{\langle \hat{L}^\dagger \rangle \langle \hat{L} \rangle}{2} \right) |\psi\rangle dt + (\hat{L} - \langle \hat{L} \rangle)|\psi\rangle d\xi. \quad (9.1)$$

Here, $\hat{L} = \sqrt{2\Gamma}\hat{a}$ represents the dissipative environmental interaction of strength Γ, and $\hat{H} = \hat{H}_D + \hat{H}_R$. Since the quantum dissipation is symmetric in $\hat{Q}$ and $\hat{P}$, the term $\hat{H}_R = \frac{\Gamma}{2}\left(\hat{Q}\hat{P} + \hat{P}\hat{Q}\right)$ is added to yield the correct classical limit where dissipation appears only in the momentum variable. The noisy dynamics is given in terms of a complex-valued Wiener process, $d\xi$, with $M(d\xi) = 0, M(d\xi d\xi^*) = dt$, and $M(d\xi d\xi) = u\,dt$, where $M(\cdot)$ denotes the mean over realizations and the complex parameter $u = |u|e^{-2i\phi}$ must satisfy the condition $|u| \leq 1$ [13, 15]. Here we will consider the situation where $|u| = 1$, which has been shown to correspond to monitoring the dissipative channel with a quantum optical homodyne measurement [3,15] with ϕ being the phase of the local oscillator. In this case, the noise can be written as $d\xi = e^{-i\phi}dW$, where dW is a real Wiener process. Recent analysis [7] shows that nano-electro-mechanical systems are well described by this model and current experiments are within range of the phenomena we report.

A semi-classical analysis starting with the dynamics of $\langle \hat{Q} \rangle = x, \langle \hat{P} \rangle = p$ proves very useful [5,9,12]; the centroid variables' dynamics depend on second moment terms V_{QQ}, V_{PP}, V_{PQ} where $V_{AB} = \langle (\hat{A}^\dagger - \langle \hat{A} \rangle^*)(\hat{B} - \langle \hat{B} \rangle) \rangle$. In this limit, $|\psi(t)\rangle$ is accurately and completely described by the $4D$ phase-space vector $\mathbf{X} = (x, p, \chi, \Pi)$ with dynamics given by

$$\dot{x} = p + \sqrt{\Gamma} N_x(\phi, \chi, \Pi)\, dW, \quad (9.2a)$$

$$\dot{p} = x - \beta^2 x^3 + \frac{g}{\beta}\cos\omega t + \Gamma F_p + 3x\beta^2\chi^2 + \sqrt{\Gamma} N_p(\phi, \chi, \Pi)\, dW, \quad (9.2b)$$

$$\dot{\chi} = \Pi + \Gamma F_\chi(\phi, \chi, \Pi), \quad (9.2c)$$

$$\dot{\Pi} = \chi(-3\beta^2(x^2 + \chi^2) + 1) + \frac{1}{4\chi^3} + \Gamma F_\Pi(\phi, \chi, \Pi), \quad (9.2d)$$

with the change of variables $V_x = \chi^2, V_{xp} = \chi\Pi, V_p = 1/4\chi^2 + \Pi^2$ for convenience below. The random effect of the continuous monitoring is given by the stochastic terms $\mathbf{N} = (N_x, N_p, N_\chi, N_\Pi)$ with

$$N_x = 2\left(\chi^2 - \frac{1}{2}\right)\cos(\phi) - 2\chi\Pi\sin(\phi), \tag{9.3a}$$

$$N_p = -2\left(\frac{1}{4\chi^2} + \Pi^2 - \frac{1}{2}\right)\sin(\phi) + 2\chi\Pi\cos(\phi), \tag{9.3b}$$

while $N_\chi = 0 = N_\Pi$. The dissipation $\mathbf{F} = (F_x, F_p, F_\chi, F_\Pi)$ has $F_x = 0, F_p = -2\Gamma$ and

$$F_\chi = \left[\chi - \chi^3 + \chi\Pi^2 - \frac{1}{4\chi}\right]\cos(2\phi) - \Pi\left[-1 + 2\chi^2\right]\sin(2\phi)$$
$$+\chi - \chi^3 - \chi\Pi^2 + \frac{1}{4\chi}, \tag{9.4a}$$

$$F_\Pi = \left[\Pi^3 - \Pi + \frac{3\Pi}{4\chi^2} - \Pi\chi^2\right]\cos(2\phi) + \left[-\frac{1}{4\chi^3} + \frac{1}{\chi} - \chi + 2\chi\Pi^2\right]\sin(2\phi)$$
$$+\left(-\Pi^3 - \Pi - \frac{3\Pi}{4\chi^2} - \Pi\chi^2\right). \tag{9.4b}$$

9.3 Coupling Between Centroid and Spread Oscillators

With the model from the previous section, we can now describe how the spread oscillator, given by the canonically conjugate pair (χ, Π), influences the dynamics of the classical oscillator, given by the centroid variables (x, p).

For $\Gamma \to 0$, Eqs. (9.2) have a Hamiltonian structure with

$$H(x, p, \chi, \Pi) = \frac{1}{2}p^2 + \frac{1}{2}\Pi^2 + U(x, \chi, t). \tag{9.5}$$

Thus, we can represent $\mathbf{X}(t)$ as a point trajectory traveling in a time-dependent $2D$ semi-classical potential, $U(x, \chi, t) = U_1(x, t) + U_2(\chi) + U_{12}(x, \chi)$, given in terms of

$$U_1(x, t) = -\frac{1}{2}x^2 + \frac{1}{4}\beta^2 x^4 + \frac{g}{\beta}x\cos\omega t, \tag{9.6}$$

$$U_2(\chi) = \frac{3}{4}\beta^2\chi^4 - \frac{1}{2}\chi^2 + \frac{1}{8\chi^2}, \tag{9.7}$$

$$U_{12}(x, \chi) = \frac{3}{2}\beta^2 x^2 \chi^2. \tag{9.8}$$

The $U(x, \chi, t)$ potential is shown in Fig. 9.1 for $g = 0$ and two different values of β. The driving sinusoidally tilts the potential along x, rocking the particle between the two classical wells depending on the amplitude.

The inter-oscillator coupling U_{12}, which allows the classical and quantum oscillators to influence each other, only exists for nonlinear systems. Different dynamical regimes can be quantified via the relative β dependence of $\overline{U}_1, \overline{U}_2, \overline{U}_{12}$ where the overbar represents a time average over the trajectory:

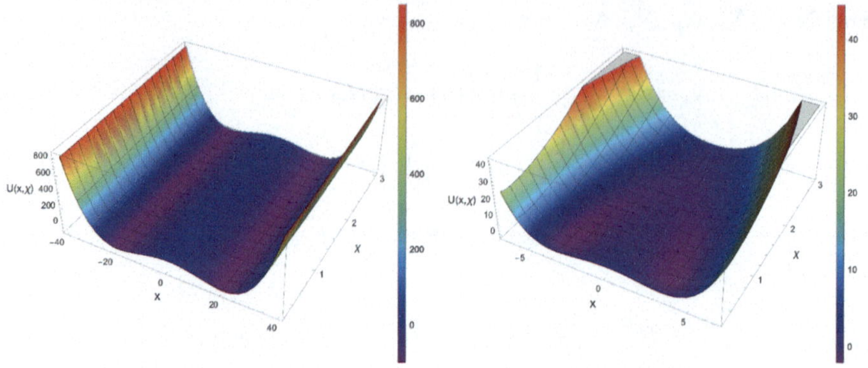

(a) $\beta = 0.05$ (b) $\beta = 0.3$

Fig. 9.1. Potential $U(x, \chi, t)$ for $g = 0$ and **a** $\beta = 0.05$, and **b** $\beta = 0.3$. For larger β, a path between the two wells is possible through higher values of χ

- For $\beta \to 0, U_{12} \to 0$ and the quantum (χ, Π) dynamics do not influence the classical (x, p) dynamics. These latter are invariant [2, 6, 9] under change of β. That is, the phase-space dynamics are identical except that the scale of x increases as β^{-1}, and $U_1 \sim \beta^{-2}$.
- The near-classical limit $\beta \ll 1$ has $\overline{U}_1 >> \overline{U}_{12} \simeq \overline{U}_2$. In Fig. 9.1 at $\beta = 0.05$ we can see that this results in a well where the classical double-well shape is seemingly barely altered by quantum effects in the typical dynamical range for χ, which is natural since $\overline{U}_1 >> \overline{U}_{12}$.
- As β increases, we get that $\overline{U}_1 \geq \overline{U}_{12} \simeq \overline{U}_2$. We see in Fig. 9.1 that for $\beta = 0.3$ this changes $U(x, \chi)$ in the χ direction, and creates a non-classical path from one x well minimum to the other that avoids the well maximum at increased χ, considerably altering the dynamics for (x, p) in the process.

This β regime where $\overline{U}_1 \geq \overline{U}_{12} \simeq \overline{U}_2$ is our focus. When $\overline{U}_1 \simeq \overline{U}_{12} \simeq \overline{U}_2$ we expect quantum effects to matter in a way that is not visible in semi-classical dynamics. It is important to realize that systems dynamics and dissipation can alter U_{12} dramatically. In particular, the time dependence of quantum spread variables depends on the components of the Jacobian of classical dynamics. That is, not only does the U_{12} coupling between the two oscillators only exist for non-linear systems, but as (χ, π) is being dragged around by (x, p) in this regime, the same dynamical properties that cause the chaotic separations of (x, p) trajectories in time causes the (χ, π) spread oscillators to grow and oscillate more rapidly; that is, chaotic dynamics can nonlinearly amplify U_{12} in principle. The constraining factor is the dissipation, as we see below.

9.4 Measurement-Dependent Dissipative Forces and Oscillator Energetics

To see how the measurement angle ϕ affects the dynamics, we rewrite the dissipative forces as $\mathbf{F} = (F_x, F_p, F_\chi, F_\Pi) = \mathbf{F_c}\cos 2\phi + \mathbf{F_s}\sin 2\phi + \mathbf{F_0}$, where the definitions of $\mathbf{F_c}, \mathbf{F_s}, \mathbf{F_0}$ are evident from the form of Eqs. (9.4). Defining these three components, which are shown in Fig. 9.2, is useful since all $\mathbf{F}$ are weighted superpositions of them. In particular, at $\phi = 0$, $\mathbf{F} = \mathbf{F_0} + \mathbf{F_c}$ and at $\phi = \pi/2, \mathbf{F} = \mathbf{F_0} - \mathbf{F_c}$. In the latter, the contributions of $\mathbf{F_0}$ and $\mathbf{F_c}$ along the $\Pi = 0$ axis are in opposite directions and tend to cancel out, while in the former, they add up, forcing the system towards small values of χ. Note that, in this case, by suppressing higher χ values, the dissipative force works against the non-classical mechanism for inter-well transitions explained in the previous section. In either case, while the size of the Γ governs how the driving energy absorbed is dissipated, it is the measurement angle ϕ that effectively alters the energy flow between the two oscillators.

To make the connection with energy flow more evident, we can look at how the input power, introduced by the external driving term, is distributed over the different available channels. From conservation of energy, we can write that

$$\frac{dE_g(\mathbf{X}(t))}{dt} + \frac{dE_\Gamma(\mathbf{X}(t))}{dt} + \frac{dE_{\sqrt{\Gamma}}(\mathbf{X}(t))}{dt} + \frac{dE_H(\mathbf{X}(t))}{dt} = 0, \tag{9.9}$$

where we used $g, \Gamma, \sqrt{\Gamma}, H$ to label the energy terms originated from driving, dissipation, noise, and the time-independent part of Eq. (9.5), respectively. For the time-independent Hamiltonian term, $\dot{E}_H = 0$. If we now take the time average, the contribution from the noise $\overline{\dot{E}_{\sqrt{\Gamma}}}$ also vanishes. This means that, focusing only on the average values, the input power from the drive $\overline{\dot{E}_g}$ balances the dissipated energy $\overline{\dot{E}_\Gamma}$. The dynamics, in particular the Lyapunov exponent λ for $\mathbf{X}(t)$, depends strongly on the Gaussian curvature of the $U(x, \chi)$ potential [1,11,14] along $\mathbf{X}(t)$, which can be sensitive to small changes in the steady-state mean ($\overline{H}$) and variance (ΔH) of the total oscillator energy given by Eq. (9.5).

9.5 Simulation Results

Finally, we put together all the understanding developed in the previous sections to explain the semiclassical mechanism responsible for the reported [3] effects of measurement angle on quantum trajectory chaos. While for some parameter values the underlying phenomenon was shown to be purely quantum, for others, semiclassical effects seemed to play a role, but remained unexplained [3].

We consider the same two dissipative couplings $\Gamma_1 = 0.05$, $\Gamma_2 = 0.10$ previously studied in [3]. It is important to understand the the difference in the classical limiting behavior at the two Γ values. Consider the Poincaré sections (shown on top of corresponding trajectories) in the (x, p) (classical) phase space in Fig. 9.3. We notice that at low dissipation Γ_1 yields a simple inter-well periodic

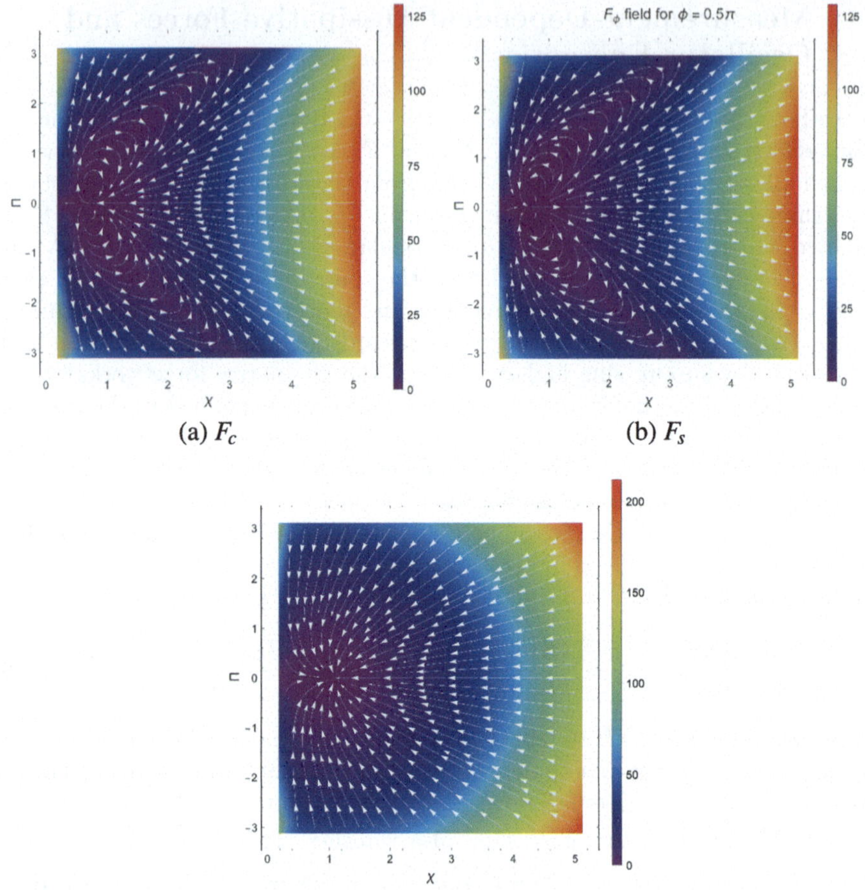

(a) F_c (b) F_s

(c) F_0

Fig. 9.2. Plots depicting the magnitude and direction of $\mathbf{F}_0, \mathbf{F}_c, \mathbf{F}_s$. Different measurement angles correspond to a weighted superposition. The differences in the $\mathbf{F}_c, \mathbf{F}_s$ components pushes the (χ, Π) orbit to different scales, changing the coupling to the classical (x, p) oscillator

orbit that never goes inside the classical separatrix defined by the $H_1(x, p) = 0$ curve and has $\lambda < 0$. Hence the energy absorbed is dissipated exactly over a single period (although $\Delta H \neq 0$). However, at higher Γ_2, even though the orbit must dissipate what it absorbs on average since it stays confined in energy, the time-dependence of the dissipation term $\dot{E}_{\mathbf{F}}$ does not synchronize with the driving $\dot{E}_g$, such that the orbit wanders chaotically in a bounded energy range spanning the separatrix with $\lambda > 0$.

To understand the semiclassical behavior, for each Γ we use both $\phi = 0, \pi/2$ settings, and examine all these cases at two different length scales β. For each of these parameter combinations, we show the Poincaré sections in (x, p) as well

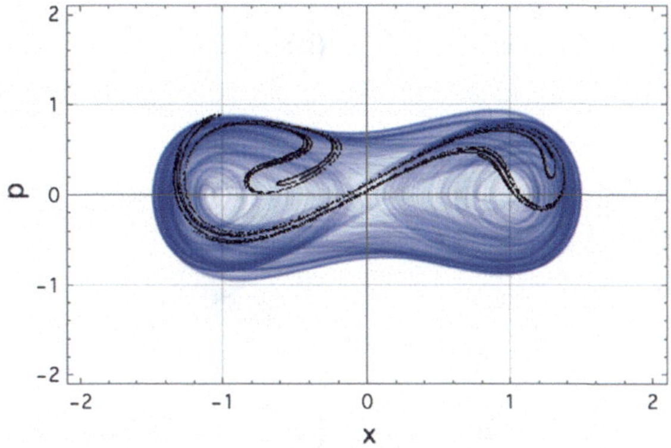

(a) $\Gamma = 0.10$

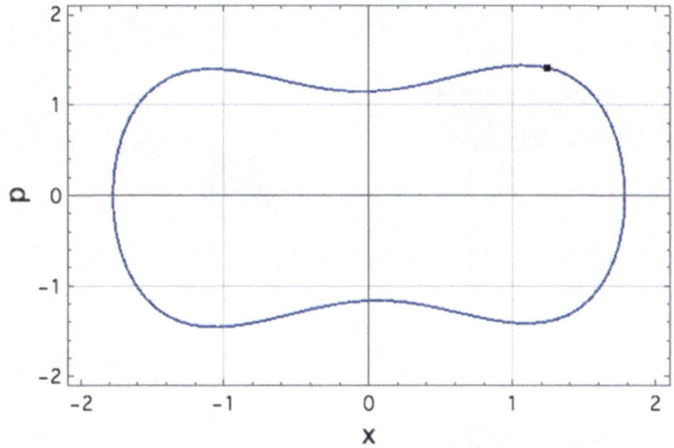

(b) $\Gamma = 0.05$

Fig. 9.3. Phase space trajectories (blue) superimposed with Poincaré sections (black) for the classical Duffing oscillator. Chaotic and regular behaviour are shown for $\Gamma = 0.1$ (top) and $\Gamma = 0.05$ (bottom), respectively

as the (x, χ) space, the latter demonstrating how the range of χ affects classical behavior.

The first case analysed was for $\Gamma = 0.1$. Here we see that for both $\beta = 0.01$ and $\beta = 0.05$, and irrespective of ϕ, the quantum perturbations do not seem to visibly change the chaotic (x, p) Poincaré sections. The (x, χ) Poincaré sections are very instructive, however. First note that the range of χ is essentially independent of β for both ϕ values. On the other hand, the $\beta-$independent χ range for $\phi = \pi/2$ is much greater than for $\phi = 0$, consistent with our analysis of the role of the dissipative force for different measurement angles. As already observed in [3], for this case, strong dependency of the Lyapunov exponent with the measurement angle is purely a quantum effect, with little contribution of semiclassical origin (Fig. 9.4).

On the other hand, the case shown in Fig. 9.5 for Γ_1 is emblematic of the interplay between the two competing factors analysed in this paper: the coupling

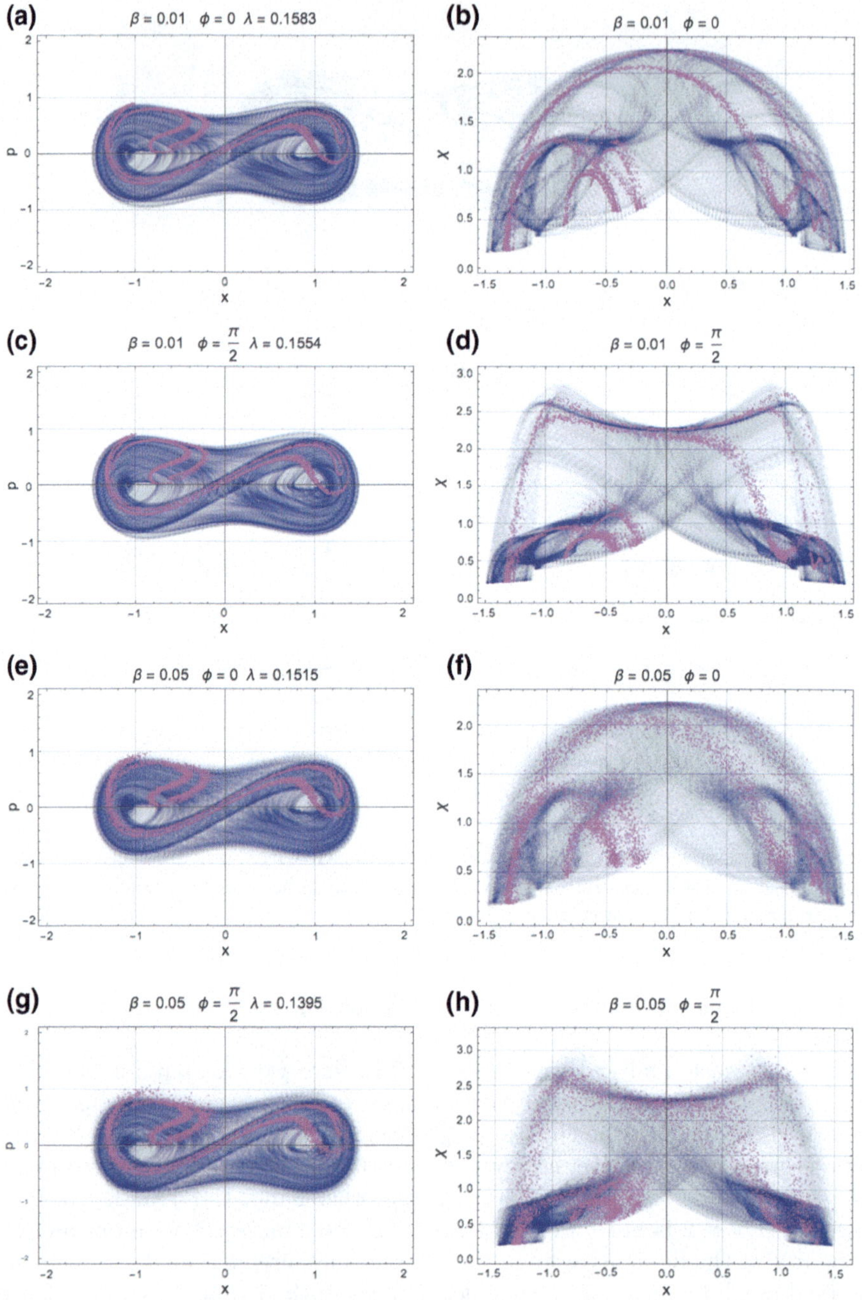

Fig. 9.4. x-p (left) and x-χ (right) trajectories for $\Gamma = 0.1$. The values of β were 0.01 (**a**–**d**) and 0.05 (**e**–**h**). For each case, the two measurement angles $\phi = 0$ (**a**, **b**, **e**, **f**) and $\phi = \pi/2$ (**c**, **d**, **g**, **h**) were considered

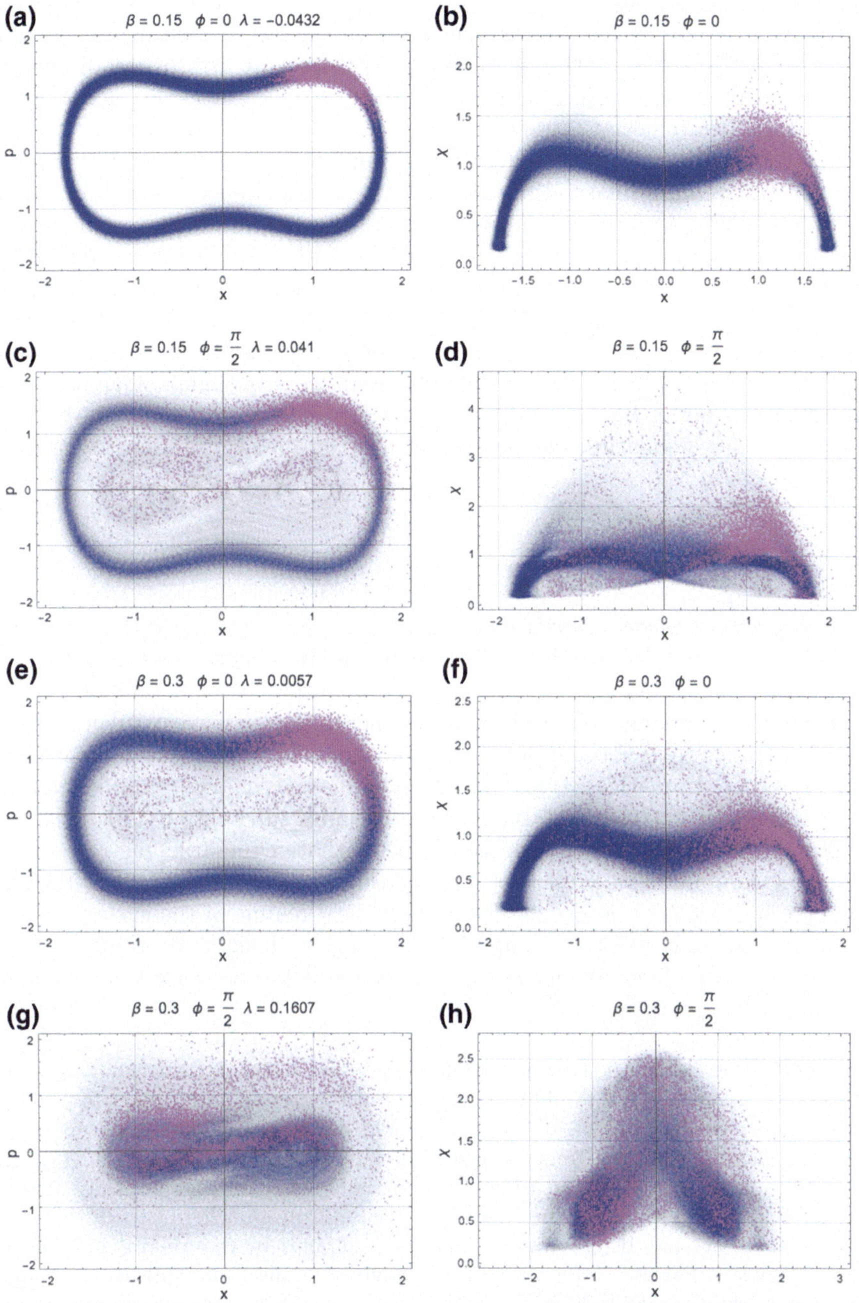

Fig. 9.5. x-p (left) and x-χ (right) trajectories for $\Gamma = 0.05$ and two values of β: 0.15 (**a**–**d**) and 0.3 (**e**–**h**). For each case, the two measurement angles $\phi = 0$ (**a**, **b**, **e**, **f**) and $\phi = \pi/2$ (**c**, **d**, **g**, **h**) were considered

between centroid and spread variables, and measurement-dependent dissipation. At $\beta = 0.15$, the $\phi = 0$ case has smaller $\overline{U}_2$, $\overline{U}_{12}$ (visible in the range in χ) than for $\phi = \pi/2$. Consistent with our previous discussion, for $\phi = 0$, the dissipative force pulls the system towards smaller values of χ, leading, therefore, to the observed smaller values of $\overline{U}_2$ and $\overline{U}_{12}$. For $\phi = \pi/2$, the dissipative force is not as effective in suppressing the effect of the nonlinear spread-centroid coupling, therefore the quantum corrections perturb the classical energy synchronization and induce chaos. At $\beta = 0.3$, the semiclassical approximation is in principle not valid, but we find the same qualitative behavior with a full quantum simulation. Semiclassically, $\overline{U}_2$, $\overline{U}_{12}$ for $\phi = 0$ is smaller than for $\phi = \pi/2$. But the larger value of β allows both angle settings to destroy the periodic motion although, again, chaos is stronger for $\phi = \pi/2$. It is worth noticing, from both the visual Poincaré sections as well as quantitatively from the λ obtained, that $\beta = 0.15$, $\phi = \pi/2$ shows larger $\overline{U}_2, \overline{U}_{12}$ values than for $\beta = 0.3$, $\phi = 0$ case such that it is effectively a more quantum system, and affects the classical motion to a greater extent.

9.6 Conclusion

In closing, we have shown that a semi-classical nonlinear oscillator that is weakly monitored and coupled to the environment can be accurately understood as a classical centroid oscillator coupled to a 'quantum' spread oscillator via a nonlinear U_{12} coupling. We find that the the choice of measurement angle ϕ should be understood through its change on the dissipative measurement back-action that can dramatically alter how the nonlinear dynamics amplifies the size of U_{12} to perturb the classical dynamics, sometimes substantially.

This leads to the remarkable observation that, comparing across all the parameter combinations presented, the measurement angle ϕ is more relevant than system scale β in determining the dynamical regime of the system.

We are currently working on applications of these insights deep in the quantum regime where different mechanisms apply, as well as to adaptive control and quantum thermodynamics.

Acknowledgements. All those at Carleton would like to thank Bruce Duffy for computational support, and AP would like to thank the Towsley and other Carleton College funds for support of students. AP and AC would like to thank the organizers of the Quantum Thermodynamics Conference 2018 in Santa Barbara for the excellent opportunity to learn and have conversations that partially led to this manuscript. AC also thanks AP's hospitality during his visits to Carleton College, where part of this work was developed. SG and JE gratefully acknowledge support by the Australian Research Council Centre of Excellence for Quantum Computation and Communication Technology (project number CE110001027).

References

1. P. Brumer, J.W. Duff, A variational equations approach to the onset of statistical intramolecular energy transfer. J. Chem. Phys. **65**(9), 3566–3574 (1976). https://doi.org/10.1063/1.433586
2. T.A. Brun, I.C. Percival, R. Schack, Quantum chaos in open systems: a quantum state diffusion analysis. J. Phys. A: Math. Gen. **29**(9), 2077–2090 (1996). http://stacks.iop.org/0305-4470/29/2077
3. J.K. Eastman, J.J. Hope, A.R. Carvalho, Tuning quantum measurements to control chaos. Sci. Rep. **7**, p. 44,684 (2017). https://doi.org/10.1038/srep44684
4. V. Gorini, A. Kossakowski, E.C.G. Sudarshan, Completely positive dynamical semigroups of n-level systems. J. Math. Phys. **17**, 821 (1976)
5. J. Halliwell, A. Zoupas, Quantum state diffusion, density matrix diagonalization, and decoherent histories: a model. Phys. Rev. D **52**, 7294–7307 (1995). https://doi.org/10.1103/PhysRevD.52.7294
6. A. Kapulkin, A.K. Pattanayak, Nonmonotonicity in the quantum-classical transition: chaos induced by quantum effects. Phys. Rev. Lett. **101**(7), 074101 (2008). https://doi.org/10.1103/PhysRevLett.101.074101
7. Q. Li, A. Kapulkin, D. Anderson, S.M. Tan, A.K. Pattanayak, Experimental signatures of the quantum-classical transition in a nanomechanical oscillator modeled as a damped-driven double-well problem. Physica Scripta **2012**(T151), 014055 (2012). http://stacks.iop.org/1402-4896/2012/i=T151/a=014055
8. G. Lindblad, On the generators of quantum dynamical semigroups. Math. Phys. **48**, 119 (1976)
9. Y. Ota, I. Ohba, Crossover from classical to quantum behavior of the duffing oscillator through a pseudo-lyapunov-exponent. Phys. Rev. E **71**, 015201 (2005). https://doi.org/10.1103/PhysRevE.71.015201.
10. A.K. Pattanayak, P. Brumer, Chaos and lyapunov exponents in classical and quantal distribution dynamics. Phys. Rev. E **56**, 5174–5177 (1997). https://doi.org/10.1103/PhysRevE.56.5174
11. A.K. Pattanayak, W.C. Schieve, Predicting two dimensional hamiltonian chaos. Z. Naturforsch. **52a**, 34 (1997)
12. B. Pokharel, M.Z.R. Misplon, W. Lynn, P. Duggins, K. Hallman, D. Anderson, A. Kapulkin, A.K. Pattanayak, Chaos and dynamical complexity in the quantum to classical transition. Sci. Rep. **8**(1), 2108 (2018). https://doi.org/10.1038/s41598-018-20507-w
13. M. Rigo, N. Gisin, Unravellings of the master equation and the emergence of a classical world. Quantum Semiclassical Opt. J. Eur. Opt. Soc. Part B **8**(1), 255 (1996). http://stacks.iop.org/1355-5111/8/i=1/a=018
14. M. Toda, Instability of trajectories of the lattice with cubic nonlinearity. Phys. Lett. A **48**(5), 335–336 (1974). https://doi.org/10.1016/0375-9601(74)90454-X, http://www.sciencedirect.com/science/article/pii/037596017490454X
15. H.M. Wiseman, L. Diósi, Complete parameterization, and invariance, of diffusive quantum trajectories for markovian open systems. Chem. Phys. **268**(1–3), 91–104 (2001). https://doi.org/10.1016/S0301-0104(01)00296-8

Chapter 10
Intentional Nonlinearity in Energy Harvesting Systems

Brian P. Mann[1(✉)], Samuel C. Stanton[2], and Brian P. Bernard[3]

[1] Duke University, Durham, NC, USA
brian.mann@duke.edu
[2] Army Research Office, Durham, NC, USA
samuel.c.stanton2.civ@mail.mil
[3] Shreiner University, Kerrville, TX, USA
bpbernard@schreiner.edu

Abstract. The success of portable electronics, remote sensing, and surveillance equipment is dependent upon the availability of remote power. While batteries can sometimes fulfill this role over short time intervals, batteries are often undesirable due to their finite life span, need for replacement and environmental impact. Instead, researchers have begun investigating methods of scavenging energy from the environment to eliminate the need for batteries or to simply prolong their life. While solar, chemical and thermal sources of energy transfer are sometimes viable, many have recognized the abundance of environmental disturbances that cause either rigid body motion or structural vibrations. This paper describes recent research efforts focused on the intentional use of nonlinearity to enhance the capabilities of energy harvesting systems. In addition, this paper identifies some of the primary challenges that arise in nonlinear harvesters and some new strategies to resolve these challenges. For example, nonlinearities can often result in multiple attractors with both desirable and undesirable responses that may co-exist. I will describe an approach that uses small perturbations to steer the dynamic response to the desirable attractor, thus leveraging the basins of attraction. Other examples will highlight the potential for nonlinear electromechanical transduction and comparisons for single frequency, multi-frequency, and stochastic environments.

10.1 Introduction

The success of portable electronics and remote sensing devices is dependent upon the availability of remote power. While batteries can sometimes fulfill this role over short time intervals, they are often undesirable due to their finite life span, need for replacement, and environmental impact. Instead, researchers are now investigating methods of scavenging energy from the environment to eliminate the need for batteries or to prolong their life [1]. While solar, chemical, and

V. In et al. (Eds.): *Proceedings of the 5th International Conference on Applications in Nonlinear Dynamics*, Understanding Complex Systems, https://doi.org/10.1007/978-3-030-10892-2_10

thermal sources of energy transfer are sometimes viable, many have recognized the abundance of environmental disturbances that cause either rigid body motion or structural vibrations. This has led to a dramatic increase in the number of studies for vibration-based energy harvesting [2–8].

Most prior works have focused on the power harvested when the response behavior is adequately characterized as a linear oscillator being driven by harmonic excitation. For this type of design, the optimal performance is realized when the natural frequency of the oscillator is nearly identical to a dominant frequency in the ambient environment. Thus, the prototypical approach is to frequency match or to design and fabricate energy harvesting devices to have a natural frequency that coincides with a dominant frequency in ambient environment [5,9–11]. This equates to building a vibrational harvesters with very specific mass-spring-damper properties that set the resonant frequency to a dominant frequency of their host environment. As such, they can be highly sensitive to uncertainties which may arise from the imprecise characterization of the host environment or, alternatively, from manufacturing defects and tolerances. This design-for-resonance approach places several performance limitations on the energy harvester. Specifically, a linear device will perform poorly when the system's resonance and excitation frequency do not coincide. Additionally, very little energy will be extracted from multi-frequency and/or random excitation sources. Problems also arise in applications where the excitation frequency drifts or changes over time [3,4].

The vast majority of past research has focused on inertial generators that operate in a linear regime [9,12–19]. However, it has recently been suggested that the intentional use of nonlinearity enable future harvesters to overcome the limitations of a linear device. More specifically, there is great interest in the concept of intentionally using nonlinearity to enhance performance. In fact, several recent works have suggested the intentional use of nonlinearity might be beneficial to energy harvesting systems [8,14,20,21]. More specifically, several studies have explored the use of nonlinearities broaden the frequency spectrum, to extend the bandwidth, engage nonlinear resonances, and/or to facilitate tuning [14,20–28]. These efforts take aim at overcoming the limitations of linear devices, which only perform well under very specific circumstances [8].

The content of this paper is organized as follows. The next section summarizes the limitations of a linear harvester by simply examining the response and uncertainty in the response of a linear oscillator. This is followed by a conceptual discussion prior attempts to use nonlinearity in energy harvesting devices. Section 10.3.2 describes several examples where researchers have explored bistability in both piezoelectric and electromagnetic harvesters. This is followed by a discussion of dynamic magnifiers and a summary of potential future research avenues.

10.2 Linear Energy Harvester Limitations

Oscillators are often designed to operate within a linear regime in vibratory energy harvesters. While restricting the oscillator to operate in a linear regime can greatly simplify the math analysis, it also limits the harvester's performance in several ways. To illustrate these points, we consider the following contrived example of a dimensionless linear oscillator

$$y'' + \mu y' + y = \Gamma \sin \eta \tau \,, \tag{10.1}$$

where y is the dimensionless displacement, a $()'$ denotes a derivative with respect to dimensionless time, μ is a damping coefficient, η is the ratio of the excitation frequency to the natural frequency, and Γ is the excitation level. For the typical case where $\mu > 0$, the steady-state response of Eq. (10.1) is given by

$$y = r \cos(\eta \tau - \phi) \,, \tag{10.2}$$

where the amplitude of the response, r, is given by

$$r = \frac{\Gamma}{\sqrt{\left(1 - \eta^2\right)^2 + \left(\mu \eta\right)^2}} \,. \tag{10.3}$$

Here, it is important to note that the power harvested will be proportional to the response amplitude. To both quantify and unveil the robustness of the linear oscillator's response to parameter variations, an expression for total uncertainty in the oscillator's response U_r is introduced

$$U_r^2 = \left(\frac{\partial r}{\partial \mu}\right)^2 U_\mu^2 + \left(\frac{\partial r}{\partial \eta}\right)^2 U_\eta^2 + \left(\frac{\partial r}{\partial \Gamma}\right)^2 U_\Gamma^2 \tag{10.4}$$

where U_{x_i} represents the uncertainty in the variable x_i at the same confidence level. It is common to express the uncertainty at the 95% confidence level (or 20:1 odds) and, consequently, 95% of the physical realizations can be expected to lie within the confidence intervals [29].

Figure 10.1 shows the nominal response amplitude and clearly affirms a large nominal response near resonance. A more in-depth study of Fig. 10.1 also reveals that the response away from this narrow-band peak is rather small. While these result highlight the importance of aligning the natural frequency with the excitation frequency, a more complete understanding of the robustness of the frequency matching strategy is obtained by also considering the uncertainty in the oscillator's response for uncertainties in the system's parameters. As noted previously, uncertainties in these parameters are quite common and arise from the imprecise characterization of the host environment or, alternatively, from imperfections in manufacturing and/or tolerances. The dashed lines of Fig. 10.1 show the confidence intervals or expected deviation in the oscillators response. Note that the dashed lines were obtained by first determining the uncertainty in the response U_r; next, the upper and lower confidence intervals were determined from $r_u = r + U_r$ and $r_l = r - U_r$ where r_u is the upper confidence interval and r_l is the lower. In essence, the confidence intervals provide a measure of the robustness in the response of the system when parameter uncertainty is considered.

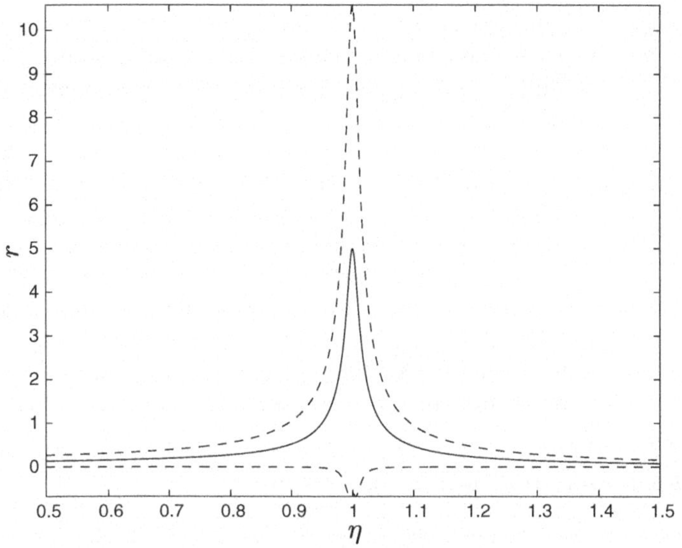

Fig. 10.1. Nominal response (solid line) and confidence intervals (dashed lines) of a linear oscillator for $\mu = 0.02$ and $\Gamma = 0.1$ and parameter uncertainties $U_\mu = \mu/5$, $U_\eta = 0.02$, and $U_\Gamma = \Gamma/10$. Confidence intervals show a lack of robustness in the nominal response in the vicinity of resonance

The confidence intervals highlight the lack of robustness in a frequency-matching strategy, since even small parameter variations, or uncertainty, can cause large differences in the expected response. More specifically, the upper and lower confidence intervals, dashed lines in Fig. 10.1, show the uncertainty in the oscillator response can sometimes be as large as the nominal value (solid line). Armed with this understanding, we now focus our attention on the intentional use of nonlinearity to address the limitations imposed by the linear oscillator.

10.3 Nonlinear Examples

Despite the fact that nonlinearities are inherent in many natural and engineered systems, it is common for engineers to remove, or attempt to remove, all nonlinearity from their designs. Although this simplifies the performance analyses, it also overlooks a wide array of phenomena, that could potentially enable the harvesting of more energy. Improving the performance of inertial harvesters requires that they become more robust to uncertainties and/or subtle changes in their environment. More specifically, the ideal harvester would perform well in a variety of settings and could scavenge energy from a broad range of frequencies. This means the harvester must be able to adjust, adapt, or tune into its current environment. Furthermore, it is essential that future harvesters have a broader frequency response - thus enabling energy to be scavenged over a wider range of frequencies.

This section will discuss select past works that sought to use nonlinear behavior to improve the performance of energy harvesting systems. The section starts with some examples of using some common structural nonlinearities and describes their potential benefits and pitfalls for different environments. This followed by a discussion of some past works that used nonlinearity in the electromechanical coupling of a harvester device. It is important to note that many of the provided examples will show a benefit to the intentional use of nonlinearity; however, as one might expect, nonlinearity must be intelligently designed into a device to reap these benefits. Furthermore, the mere introduction of nonlinearity into these systems also introduces new problems to consider, such as the presence multiple attractors, i.e. both a high and low energy response. Additional works, which have considered different types of random excitation, such as broadband white noise and colored noise, are also discussed in Sect. 10.3.2.

10.3.1 Hardening and Softening Systems

Several researchers have studied energy harvesting systems with either hardening or softening-spring-like behavior. For example, Ref. [8] considered a electromagnetic inductions system with nonlinear restoring forces that were created from a magnet levitation system. The restoring force in that system was a hardening type spring and it showed the ability to tune by peak in its frequency response by changing the relative magnet positions. However, a hardening system can only alter its peak response to one side of linear resonance. Systems displaying similar hardening type behavior have been investigated in many other references. Upon comparing the peak response of the linear oscillator to that of the hardening system, it may seem problematic that the linear oscillator has a larger response for single frequency excitation. However, an uncertainty analysis on the frequency response of the hardening system has shown its response is more robust [30].

To help cover a broader range of frequencies, some investigators have sought to combine hardening and softening type effects into a single device. For example, Fig. 10.2 shows a harvester that demonstrated the potential of adding nonlinearity from magnet-magnet interactions to create either a hardening and softening effect [21]. More specifically, positioning the adjustable magnets behind the tip mass creates a hardening frequency response - thus extending the region of a relatively large response to higher frequencies. If the adjustable magnets are pushed forward of the tip mass, a softening type behavior is created, thus the region of relatively large responses switches directions and extends to frequencies lower than the linear natural frequency.

10.3.2 Bistable Systems

The concept of a bistable system can be brought into focus by considering the motion of a small ball rolling on the surface under the influence of gravity, see Fig. 10.3, where the ball height is proportional to the potential energy. Consider first the potential energy of a linear oscillator, shown in Fig. 10.3a. This system has a linear relationship between the restoring force and deflection which

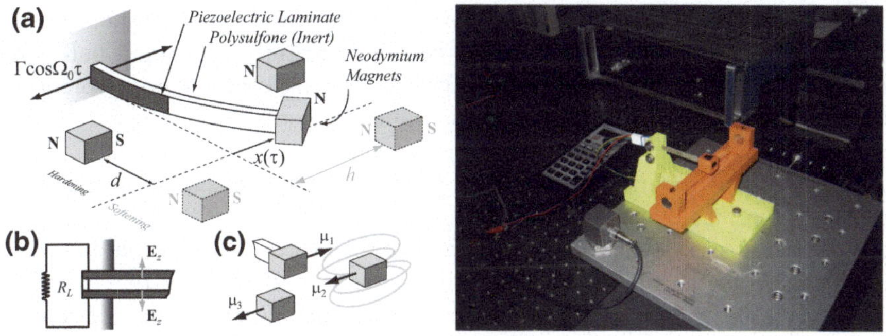

Fig. 10.2. Illustration of an experimental system from Ref. [21] that demonstrated that the nonlinear restoring forces enable tuning and a broader range of frequencies with a large amplitude response

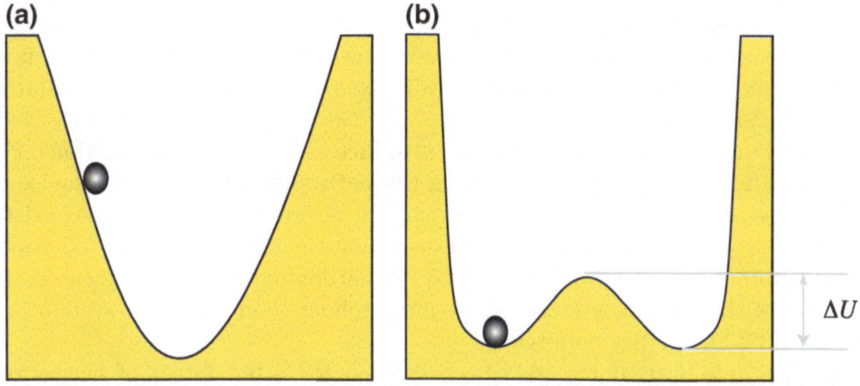

Fig. 10.3. Potential energy curves for: **a** the quadratic potential well of a linear oscillator and **b** a nonlinear oscillator with two stable equilibria separated by an unstable equilibrium position. The energy difference between the potential energy barrier and the stable equilibria, labeled ΔU, is an important factor for determining the threshold for an escape

results in a quadratic potential energy well with a single equilibrium. Regardless of where the ball placed, it will eventually come to rest at the bottom of the potential energy well. Shaking the parabola laterally yields the linear harmonic oscillator with the largest response occurring when it is shaken at its resonance frequency.

Consider next the same ball under the influence of a nonlinear restoring force where the potential energy description may be more complex - see Fig. 10.3b. Consider again the same ball under the influence of small lateral excitations. This results in a system that behaves linearly for small-amplitude motions with oscillations that remain confined to a single well. For increasingly large excitations, motion amplitudes grow until the threshold for a potential well escape occurs (i.e. where an escape is imminent for energy levels above the threshold criteria ΔU in Fig. 10.3b). Once exceeding the threshold criteria, the small ball would then escape from the potential well and traverse both potential wells, sometimes called well-mixing behavior, with large-amplitude displacements and velocities. Acknowledging the dramatic increase in the energetic response of the oscillator in the post-escape regime [31], several researchers have become interested in this type of system [32].

Figure 10.4 shows example responses from a prototypal bistable harvester. In contrast to the hardening and softening cases, the bistable system shows the emergence of additional solution branches. More specifically, these solutions are associated with the oscillations within a single potential well and those that cross the center potential well barrier and are the result of a potential well escape phenomenon. This system can exhibit similar P_a (dimensionless power) values to those of the linear system, but, as in the case of the softening and hardening system, displays more complex scaling in its response behavior as Γ, the dimensionless excitation, is increased. In addition, the plots of ρ vs. P_a, where ρ is the dimensionless electrical load, show the system can have even more local maxima. Further examples of bistable energy harvesters can be found in references [14,20,24,27,33,34].

As a summary, a bistable harvester introduces some new considerations. For example, while the strategy of matching the natural frequency of the device to a frequency in the environment still exists, an alternative strategy also exists. In particular, one can instead focus on designing the potential energy curves to ensure a potential well escape. Similar to the hardening and softening cases, the responses of the bistable system can be more robust than the linear system (see reference [30] for further details).

The bistable system has also been studied for other forms of excitation, such as random excitation [35–37]. One result worth mentioning is the finding of reference [35]. In this study, it was shown that a bistable harvester could outperform a linear harvester in an environment with colored noise.

10.3.3 Coupling Nonlinearity

The work of Ref. [22] was the first to consider the influence of nonlinear electromechanical coupling in PZT systems. Since then, the inherent nonlinearities in piezoelectric harvesters have been studied in greater detail [38]. Outside of piezoelectric systems, inherent nonlinearities have also been studied in electromagnetic induction systems [28]. One interesting finding worth mentioning is that nonlinear coupling appears to be particularly suited to multi-frequency excitation [28]. However, further research needs to be done to further explore the potential benefits and pitfalls of nonlinear coupling.

10.3.4 Dynamic Magnifier

The use of a dynamic magnifier has been another area of inquiry for linear and nonlinear systems. A dynamic magnifier is a dummy oscillator, essentially an

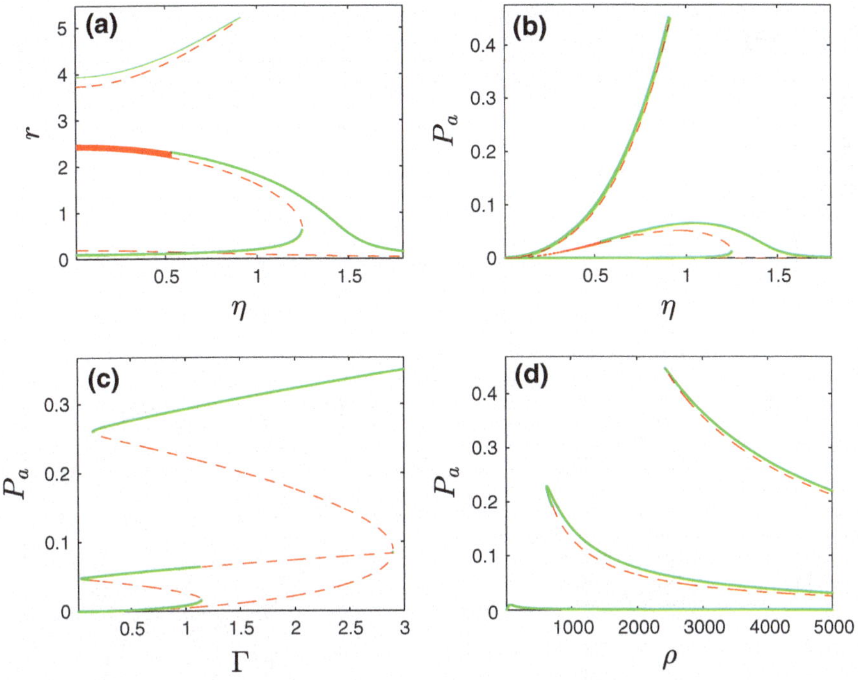

Fig. 10.4. Plots showing the stable (green dots) and unstable (red dots) response trends for a harvester with a bistable potential well. Graphs show frequency responses for **a** the oscillation amplitude of the mechanical system and **b** the dimensionless average power; graphs (**c**) and **d** plot the dimensionless average power for changes in Γ and ρ, respectively

oscillator without any electromechanical coupling, that is used to magnify the response of the primary oscillator, i.e. the one with electromechanical coupling. As a brief summary, several researchers have now shown that a dynamic magnifier can successfully increase the energy harvested from the primary oscillator and even be used to modify the corresponding basins of attraction [39].

10.4 Further Considerations

Many recent works have explored the use of nonlinearity in vibratory energy harvesters, e.g. see [8,14,18,22,30,34,35,38,40–42]. While these investigations, along with many other recent works, have advanced the current understanding on the beneficial use of nonlinearity, the introduction of nonlinearity can also cause many additional difficulties. Paramount amongst these challenges, and a common issue in nearly all nonlinear harvesting systems, is the presence of coexisting solutions. To illustrate the problem, Fig. 10.5a shows the frequency response for a Duffing Oscillator with coexisting solutions over the dimensionless frequency range of $\approx 1.25 < \eta < 2$. Assuming the environmental excitation

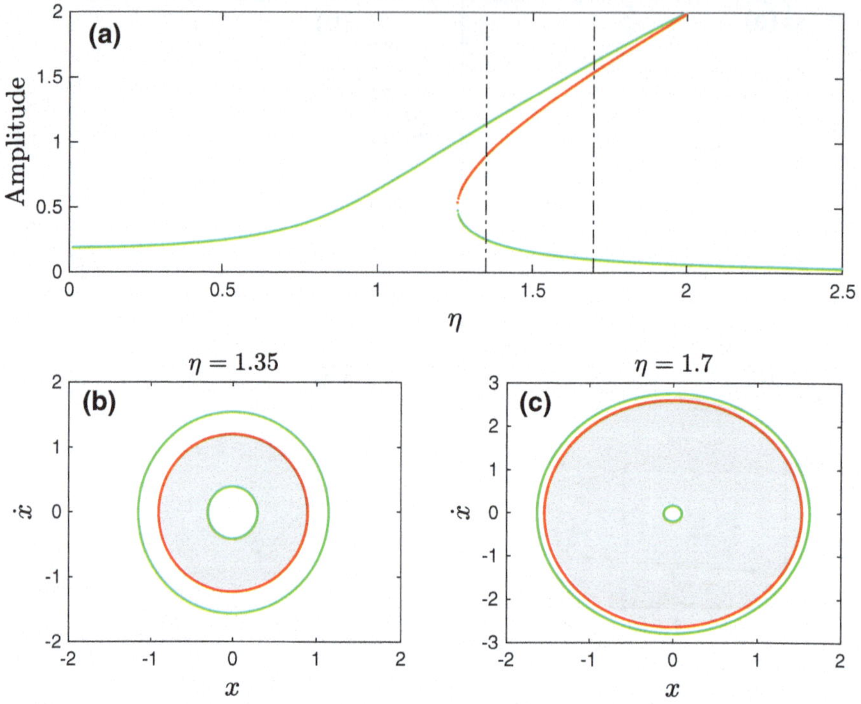

Fig. 10.5. Illustrative example of a system with coexisting periodic solutions, i.e. two or more stable periodic solutions for the same system parameters. Plots illustrate the challenge of attractor selection in energy harvesting systems. Plot **a** shows a bifurcation diagram illustrating a hardening spring nonlinearity in response to dimensionless frequency η, and plots (**b**) and **c** show the corresponding periodic attractors and repellers in phase space for two different values of η. Curves are labeled stable (green) and unstable (red)

remains constant, only the initial conditions determine whether a higher or lower energy solution is obtained. Furthermore, if the basins of attraction are studied for this range of η, one finds that the more desirable response (higher amplitude) is unlikely to be obtained when the excitation frequency is closer to the peak response. Thus a fundamental challenge prevalent in nearly all nonlinear energy harvesting approaches is a strategy to select a desired attractor.

Vibratory energy harvesters convert mechanical energy into electrical energy with electromechanical coupling, e.g. piezoelectric, electromagnetic, or capacitive. While these transduction schemes allow some form of control to be applied to alter the response of the mechanical system, a number of challenges prevent the use of continuous control. To elaborate, the power required to apply continuous control is typically larger than the power harvested. It is also common that the electromechanical coupling is not strong enough to drastically alter the response of the mechanical system in a single application of control, unless external energy is provided. Thus methods to choose the desired attractor present an on-going area of research.

10.5 Conclusions

This paper discusses select past works on the intentional use of nonlinear behavior in inertial energy harvesters. Many forms of nonlinearity have been investigated and many have shown some potential benefit. However, the fact remains that analyzing these nonlinear systems can be much more difficult than their linear counterpart. The introduction of nonlinearity adds an interesting feature that can allow effecting device tuning in a semi-active or passive way to overcome uncertainties in the environmental excitation or physical parameters of the system.

Nonlinear energy harvesting systems often have co-existing solutions. When one of the responses is desirable and the other undesirable, it becomes critically important to have methods to select the desired attractor with minimal energy expenditure. A great solution to this problem should be the target of future investigations.

Acknowledgements. Research support from the U.S. Army Research Office is gratefully acknowledged.

References

1. S. Roundy, P.K. Wright, J.M. Rabaey, *Energy Scavenging for Wireless Sensor Networks* (Springer, New York, 2003)
2. C.R. Saha, Optimization of and electromagnetic energy harvesting device. IEEE Trans. Mag. **42**(10), 3509–3511, 42
3. S.M. Shahruz, Limits of performance of mechanical band-pass filters used in energy scavenging. J. Sound Vib. **293**(1–2), 449–461 (2006)
4. S.M. Shahruz, Design of mechanical band-pass filters for energy scavenging. J. Sound Vib. **292**(3–5), 987–998 (2006)
5. N.G. Stephen, On energy harvesting from ambient vibration. J. Sound Vib. **293**, 409–425 (2006)
6. B. Yang, C. Lee, W. Xiang, J. Xie, J.H. He, R.K. Kotlanka, S.P. Low, H. Feng, Electromagnetic energy harvesting from vibrations of multiple frequencies. J. Micromech. Microeng. **19**(035001), 1–8 (2009)
7. B.C. Yen, J.H. Lang, A variable-capacitance vibration-to-electric energy harvester. IEEE Trans. Circuits Syst. 1 –Fundam. Theory Appl. **53**(2), 288–295 (2005)
8. B. Mann, N. Sims, Energy harvesting from the nonlinear oscillations of magnetic levitation. J. Sound Vib. **319**, 515–530 (2009)
9. G.A. Lesieutre, G.K. Ottman, H.F. Hofmann, Damping as a result of piezoelectric energy harvesting. J. Sound Vib. **269**(3–5), 991–1001 (2004)
10. H.A. Sodano, D.J. Inman, G. Park, Generation and storage of electricity from power harvesting devices. J. Intell. Mater. Syst. Struct. **16**, 67–75 (2005)
11. H.A. Sodano, D.J. Inman, G. Park, Comparison of piezoelectric energy harvesting devices for recharging batteries. J. Intell. Mater. Syst. Struct. **16**, 799–807 (2005)

12. S.P. Beeby, R.N. Torah, M.J. Tudor, P. Glynne-Jones, T. O'Donnell, C.R. Saha, S. Roy, A micro electromagnetic generator for vibration energy harvesting. J. Micromech. Microeng. **17**, 1257–1265 (2007)
13. S.B. Horowitz, M. Sheplak, L.N. Cattafesta, T. Nishida, A mems acoustic energy harvester. J. Micromech. Microeng. **16**, 174–181 (2006)
14. B. Mann, B. Owens, Investigations of a nonlinear energy harvester with a bistable potential well. J. Sound Vib. **329**, 1215–1226 (2010)
15. S.P. Beeby, M.J. Tudor, N.M. White, Energy harvesting vibration sources for microsystems applications. Meas. Sci. Technol. **17**, 175–195 (2006)
16. E.S. Leland, P.K. Wright, Resonance tuning of piezoelectric vibration energy scavenging generators using compressive axial load. Smart Mater. Struct. **15**, 1413–1420 (2006)
17. G. Poulin, E. Sarraute, F. Costa, Generation of electrical energy for portable devices comparative study of an electromagnetic and piezoelectric system. Sens. Actuators A **116**, 461–471 (2004)
18. J.M. Renno, M.F. Daqaq, D.J. Inman, On the optimal energy harvesting from a vibration source. J. Sound Vib. **320**, 386–405 (2009)
19. S. Roundy, On the effectiveness of vibration based energy harvesting. J. Intell. Syst. Struct. **16**, 809–823 (2005)
20. A. Erturk, J. Hoffmann, D.J. Inman, A piezomagnetoelastic structure for broadband vibration energy harvesting. Appl. Phys. Lett. **94**(254102), 1–4 (2009)
21. S.C. Stanton, C.C. McGehee, B.P. Mann, Reversible hysteresis for broadband magnetopiezoelastic energy harvesting. Appl. Phys. Lett. **95**, 174103–3 (2009)
22. A. Triplett, D.D. Quinn, The effect of nonlinear piezoelectric coupling on vibration-based energy harvesting. J. Intell. Mater. Syst. Struct. **20**(16), 1959–1967 (2009)
23. M.S. Soliman, E.M. Abdel-Rahman, E.F. El-Saadany, A wideband vibration-based energy harvester. J. Micromech. Microeng. **18**, 1–11 (2008)
24. S.C. Stanton, C.C. McGehee, B.P. Mann, Nonlinear dynamics for broadband energy harvesting: investigation of a bistable piezoelectric inertial generator. Phys. D: Nonlinear Phenom. **239**, 640–653 (2010)
25. V.R. Challa, M.G. Prasad, Y. Shi, F.T. Fisher, A vibration energy harvesting device with bidirectional resonance frequency tunability. Smart Mater. Struct. **17**(1), 015035 (2008)
26. D.A.W. Barton, S.G. Burrow, L.R. Clare, Energy harvesting from vibrations with a nonlinear oscillator. J. Vib. Acoust. **132**(2), 021009 (2010)
27. A. Cammarano, S.G. Burrow, D.A.W. Barton, Modelling and experimental characterization of an energy harvester with bi-stable compliance characteristic. J. Syst. Control Eng. **225**, 475–484 (2011)
28. B.A.M. Owens, B.P. Mann, Linear and nonlinear electromagnetic coupling models in vibration-based energy harvesting. J. Sound Vib. **331**, 922–937 (2012)
29. H.W. Coleman, W.G. Steele, *Experimentation and Uncertainty Analysis for Engineers*, 2nd edn. (Wiley, New York, 1999)
30. B.P. Mann, D.A.W. Barton, B.A.M. Owens, Uncertainty in performance for linear and nonlinear energy harvesting strategies. J. Intell. Mater. Syst. Struct. **23**, 1451–1460 (2012)
31. B.P. Mann, Energy criterion for potential well escapes in a bistable magnetic pendulum. J. Sound Vib. **323**, 864–867 (2009)
32. R.L. Harne, K.W. Wang, A review of the recent research on vibration energy harvesting via bistable systems. Smart Mater. Struct. **22**(023001), 1–12 (2013)
33. S.C. Stanton, B.P. Mann, B.A.M. Owens, Harmonic balance analysis of the bistable piezoelectric inertial generator. J. Sound Vib. (2012)

34. Z. Wu, R.L. Harne, K.W. Wang, Energy harvester synthesis via coupled linear-bistable system with multistable dynamics. J. Appl. Mech. **25**(8), 937–950 (2014)
35. S.C. Stanton, B.P. Mann, B.A.M. Owens, Melnikov theoretic methods for characterizing the dynamics of the bistable piezoelectric inertial generator in complex spectral environments. Phys. D **241**, 711–720 (2012)
36. L. Gammaitoni, I. Neri, H. Vocca, Nonlinear oscillators for vibration energy harvesting. Appl. Phys. Lett. **94**, pp. 164102 (2009)
37. M.F. Daqaq, R. Masana, A. Erturk, D.D. Quinn, On the role of nonlinearities in vibratory energy harvesting: a critical review and discussion. Appl. Mech. Rev. **66**(4), pp (2014)
38. S.C. Stanton, A. Erturk, B.P. Mann, D.J. Inman, Nonlinear piezoelectricity in electroelastic energy harvesters: modeling and experimental identification. J. Appl. Phys. **108**, 1–9 (2010)
39. B.P. Bernard, B.P. Mann, Increasing viability of nonlinear energy harvesters by adding an excited dynamic magnifier. J. Intell. Mater. Syst. Struct. **29**(6), 1196–1205 (2017)
40. T. Seuaciuc-Osório, M.F. Daqaq, Energy harvesting under excitations of time-varying frequency. J. Sound Vib. **329**, 2497–2515 (2010)
41. R. Ramlan, M. Brennan, B. Mace, I. Kovacic, Potential benefits of a non-linear stiffness in an energy harvesting device. Nonlinear Dyn. **59**, 545–558 (2010)
42. B.J. Bowers, D.P. Arnold, Spherical, rolling magnet generators for passive energy harvesting from human motion. J. Micromech. Microeng. **19**(094008), 1–7 (2009)

Chapter 11
Nonlinear Operation of Inertial Sensors

Andrew B. Sabater(✉), Kari M. Moran, Eric Bozeman, Andrew Wang, and Kevin Stanzione

SPAWAR Systems Center Pacific, 53560 Hull St, San Diego, CA 92152, USA
{andrew.b.sabater,kari.moran,eric.bozeman,andrew.wang, kevin.stanzione}@navy.mil

Abstract. It is often assumed, and has been shown experimentally, that nonlinear operation of inertial sensors–in particular gyroscopes–can degrade or trivially improve performance. As such, the standard practice is to operate below or near the threshold where nonlinear effects become significant. The limitation with this method is that the dynamic range, or the range of excitations where the sensor behaves linearly, shrinks as the dimensions of the sensor decrease. Thus, while relatively large mechanical gyroscopes, such as hemispherical resonator gyroscopes (HRGs), can achieve navigation-grade performance, microelectromechanical system (MEMS) gyroscopes, being orders of magnitude smaller, have orders of magnitude worse performance. A relatively new class mechanical gyroscope, the frequency modulated (FM) gyroscope, is able to address long-term noise performance issues. The trade-off with FM gyroscopes, compared to the standard amplitude modulated ones, is that short-term noise can be elevated. One means of improving short-term gyroscope performance is improving short-term frequency stability. It has been shown theoretically and experimentally that while most states within the nonlinear regime of an oscillator degrade frequency stability, a select few allow operation at a lower fundamental limit. This work describes and provides some preliminary experimental work on the constructive exploitation of nonlinear operation with FM gyroscopes.

11.1 Introduction

When cost, size, weight, and power (CSWaP) are not constrained, current technology utilizing inertial sensors allows for navigation in the absence of GPS. Classically inertial navigation relies on the fusion of measurements of acceleration and rotation rate to estimate position. Depending on the sensor technology, the position estimate exhibits a random drift that grows with time. While schemes that utilize velocity measurements instead of acceleration have been shown to reduce this drift [1], the drift is dominated by noise associated with the gyroscopes. Navigation-grade gyroscopes are often based on ones that exploit

V. In et al. (Eds.): *Proceedings of the 5th International Conference on Applications in Nonlinear Dynamics*, Understanding Complex Systems, https://doi.org/10.1007/978-3-030-10892-2_11

the Sagnac effect, but mechanical gyroscopes like hemispherical resonator gyroscopes (HRGs) are also in this class. Miniaturized mechanical gyroscopes based on silicon microelectromechanical systems (MEMS) technology perform much worse in part due to the scaling of noise processes with size and the differences between quartz and silicon.

Conventional mechanical gyroscopes employ an amplitude modulated (AM) scheme. Utilizing a structure with degenerate modes that can be coupled via the Coriolis effect during rotation, energy can be exchanged between these modes. A relatively new class of mechanical gyroscope, the frequency modulated gyroscope, uses the same structure as an AM gyroscope, however a frequency modulation (FM) effect associated with angular momentum conservation is employed. In a comparison study between operating the same structure in AM and FM modes, it was found that FM operation had superior long-term stability [2]. This may in part be due the ease of automatic mode matching with FM operation, as AM operation with similar matching capabilities is able to significantly reduce long-term drift [3]. Both results are significant as they afford a path towards reducing the elevated frequency random walk of silicon resonators as compared to quartz resonators.

One of the limitations with silicon MEMS technology, and in particular, FM gyroscopes, is the elevation of short-term noise processes. The associated gyroscope metric is angular random walk (ARW). With a well designed AM MEMS gyroscope, theromechanical noise is the limiting process for ARW that scales poorly with size [4]. For FM gyroscopes, oscillator instabilities as well as noise from frequency demodulation can also significantly contribute to ARW [5]. This work seeks to reduce ARW via nonlinear operation by enhancing frequency stability. Other recent works have shown significant ARW improvements with nonlinear AM gyroscopes [6]. This work is distinct in that instead of seeking to maximize displacement, frequency stability is optimized.

The following section describes nonlinear FM operation. It is a variant of Lissajous FM operation that accounts for nonlinear operation. Following this, regimes that optimize frequency stability are explored. It is shown experimentally that nonlinear operation provides significant improvements, and operation in nonlinear regimes with a linear design degrades stability. These regimes are then used to reduce gyroscope ARW. Concluding remarks are then made.

11.2 Nonlinear FM Gyroscope Operation

In order to implement an amplitude modulated gyroscope, a structure with degenerate modes that can be effectively coupled via the Coriolis effect is, typically, needed. Examples of such structures can be found in [7], but the one used in this study is discussed in the following section. A controller is implemented; it maintains the oscillations of one of these modes at a frequency close to its resonant frequency. Due to the Coriolis effect, in the presence of rotation, energy from the oscillation mode is transferred to the other. Thus, by measuring the amplitude of the other mode, rotation rate can be estimated following calibration. While conceptually simple, imperfections associated with the structure, temperature changes that effect the oscillating frequency, and variability of the gains of the needed amplifiers–to name just a few–can degrade performance. Adding complexity to the calibration process can help to mitigate some–but not all–of these issues.

In order to address some of the previously noted challenges, FM operation can be used. A variety of FM modes have been implemented, but the nonlinear FM mode described here is based upon a more generalized version of Lissajous FM operation [8]. Consider the equations of motion for the generic vibratory gyroscope with the modification of cubic stiffness terms [9]

$$\begin{aligned} z_1'' - \varepsilon A_g \Omega z_2' + \varepsilon c_1 z_1' + \varepsilon c_{12} z_2' + \left(\omega_0^2 - A_c \Omega^2\right) z_1 + \varepsilon \delta z_1 + \varepsilon q_c z_2 + \varepsilon \alpha_1 z_1^3 &= \varepsilon F_1(t), \\ z_2'' + \varepsilon A_g \Omega z_1' + \varepsilon c_2 z_2' + \varepsilon c_{21} z_1' + \left(\omega_0^2 - A_c \Omega^2\right) z_2 - \varepsilon \delta z_2 + \varepsilon q_c z_1 + \varepsilon \alpha_2 z_2^3 &= \varepsilon F_2(t), \end{aligned} \tag{11.1}$$

where z_1 and z_2 are the displacements of mode 1 and 2, respectively, $(\bullet)'$ denotes the time-derivative, and ε is a smallness parameter. Effects associated with rotation rate Ω are captured with A_g and A_c where A_g is the angular gain due to the Coriolis effect and A_c is the centripetal force coefficient. The modes are nominally assumed to be matched with a natural frequency of ω_0, but frequency mismatch is captured with δ. Damping is captured with the c terms that allow for a more general case of unequal cross-axis damping. Note that the damping and quality factor terms are inversely related and of the form $c = \omega_0/Q$. Quadrature error, or mechanical coupling between the two modes, is described by q_c. The forces that are used to excite and drive the modes, as well as noise, are captured by F_1 and F_2. Lastly, nonlinear effects (e.g. geometric and electrostatic nonlinearities [6]) are given by α_1 and α_2.

Unlike AM operation, in FM operation, both modes are operated with feedback to create self-sustaining oscillators at set amplitudes. Assuming that damping is weak, the forces needed to sustain the oscillations can be ignored in the present analysis. Noise is also ignored in the present analysis. However, as will be shown in Sect. 11.4, frequency stability can be dramatically improved or degraded based on the feedback structure. Moreover, while the optimal linear feedback structure seeks to maximize displacement, the optimal nonlinear feedback structure depends on the time scale under consideration. This distinction separates Lissajous FM operation from nonlinear FM operation. Using the method of averaging, one can analyze the oscillations of the coupled system in Eq. (11.1). Assuming the following coordinate transformation

$$\begin{aligned} z_1(t) &= a_1(t) \cos[\omega_0 t + \phi_1(t)], \\ z_1'(t) &= -a_1(t) \omega_0 \sin[\omega_0 t + \phi_1(t)], \\ z_2(t) &= a_2(t) \cos[\omega_0 t + \phi_2(t)], \\ z_2'(t) &= -a_2(t) \omega_0 \sin[\omega_0 t + \phi_2(t)], \end{aligned} \tag{11.2}$$

the slow-flow equations are given by

$$
\begin{aligned}
a_1' &= \varepsilon\left[-\frac{1}{2}c_1 a_1 + \frac{1}{2}a_2\left(A_g\Omega - c_{12}\right)\cos(\phi_1-\phi_2) + \frac{q_c}{2\omega_0}a_2\sin(\phi_1-\phi_2)\right] + O(\varepsilon^2),\\
\phi_1' &= \varepsilon\left[\frac{3}{8}\frac{\alpha_1}{\omega_0}a_1^2 + \frac{1}{2}\frac{\delta}{\omega_0} - \frac{1}{2}\frac{A_c\Omega^2}{\omega_0} + \frac{1}{2}\frac{q_c}{\omega_0}\frac{a_2}{a_1}\cos(\phi_1-\phi_2) - \frac{1}{2}(A_g\Omega - c_{12})\frac{a_2}{a_1}\sin(\phi_1-\phi_2)\right] + O(\varepsilon^2),\\
a_2' &= \varepsilon\left[-\frac{1}{2}c_2 a_2 + \frac{1}{2}a_1\left(A_g\Omega - c_{21}\right)\cos(\phi_1-\phi_2) + \frac{q_c}{2\omega_0}a_1\sin(\phi_1-\phi_2)\right] + O(\varepsilon^2),\\
\phi_2' &= \varepsilon\left[\frac{3}{8}\frac{\alpha_2}{\omega_0}a_2^2 - \frac{1}{2}\frac{\delta}{\omega_0} - \frac{1}{2}\frac{A_c\Omega^2}{\omega_0} + \frac{1}{2}\frac{q_c}{\omega_0}\frac{a_1}{a_2}\cos(\phi_1-\phi_2) - \frac{1}{2}(A_g\Omega + c_{21})\frac{a_1}{a_2}\sin(\phi_1-\phi_2)\right] + O(\varepsilon^2).
\end{aligned}
\tag{11.3}
$$

The equations that describe the amplitude dynamics are given for completeness, but effectively can be ignored for low rotation rates as amplitude controllers are used. If the oscillation frequencies of z_1 and z_2 are given by $\omega_0 + \phi_1'$ and $\omega_0 + \phi_2'$, respectively, then the sum of the oscillation frequencies Σ_{12} to first-order is

$$
\begin{aligned}
\Sigma_{12} = 2\omega_0 + \varepsilon&\left[\frac{3}{8}\frac{\alpha_1}{\omega_0}a_1^2 + \frac{3}{8}\frac{\alpha_2}{\omega_0}a_2^2 - \frac{A_c\Omega^2}{\omega_0} + \frac{1}{2}\frac{q_c}{\omega_0}\left(\frac{a_1}{a_2} + \frac{a_2}{a_1}\right)\cos(\phi_1-\phi_2)\right.\\
&\left.-\frac{1}{2}\left[A_g\Omega\left(\frac{a_1}{a_2} + \frac{a_2}{a_1}\right) + c_{12}\frac{a_2}{a_1} - c_{21}\frac{a_1}{a_2}\right]\sin(\phi_1-\phi_2)\right].
\end{aligned}
\tag{11.4}
$$

A rate estimate can be produced by demodulating the $\sin(\phi_1 - \phi_2)$ term. To be more explicit in regards to the needed signal processing to estimate rate, first the signals produced by modes 1 and 2 are frequency demodulated. Next these demodulated signals are summed together and are multiplied by $\sin(\phi_1 - \phi_2)$. The product is then low-pass filtered to reject the high-frequency component produced by multiplication.

Equation (11.4) implies several features of FM operation. If the cross-axis damping terms are equal, as is classically assumed [9], then the estimate is unbiased. The rate estimate requires tracking the relative phase, as quadrature error ($\cos(\phi_1 - \phi_2)$ term) is often greater than the rate signal. However, under certain conditions, tracking the relative phase can be simplified as it can be a linear function of time. Consider the difference of the oscillation frequencies Δ_{12}

$$
\begin{aligned}
\Delta_{12} = \varepsilon&\left[\frac{\delta}{\omega_0} + \frac{3}{8}\frac{\alpha_1}{\omega_0}a_1^2 - \frac{3}{8}\frac{\alpha_2}{\omega_0}a_2^2 + \frac{1}{2}\frac{q_c}{\omega_0}\left(\frac{a_2}{a_1} - \frac{a_1}{a_2}\right)\cos(\phi_1-\phi_2)\right.\\
&\left.-\frac{1}{2}\left[A_g\Omega\left(\frac{a_2}{a_1} - \frac{a_1}{a_2}\right) + c_{12}\frac{a_2}{a_1} + c_{21}\frac{a_1}{a_2}\right]\sin(\phi_1-\phi_2)\right].
\end{aligned}
\tag{11.5}
$$

If a_1 and a_2 are assumed to be equal and cross-axis damping is ignored, then the frequency difference is independent of relative phase and is set by the uncoupled dynamics. More simply,

$$
\phi_1 - \phi_2 \approx \varepsilon\left[\frac{\delta}{\omega_0} + \frac{3}{8}\frac{\alpha_1}{\omega_0}a_1^2 - \frac{3}{8}\frac{\alpha_2}{\omega_0}a_2^2\right]t. \tag{11.6}
$$

Thus, for the purposes of providing a simplified understand of FM operation, rate information is encoded as a frequency modulation at approximately the uncoupled frequency difference.

11.3 Resonator Design

The resonator designed for this effort is a variant of the quadruple mass gyroscope (QMGs) with internal levers [10] that has been miniatured to fit the 2×2 mm form factor of the Episeal process [6,11]. The use of internal levers alters the ordering of the modes such that lowest modes can be utilized for gyroscope operation. This is an advantage with AM gyroscopes as sensitivity increases with decreasing operating frequency. Careful design of the levers can also increase the relative separation between the modes utilized for gyroscope operation and parasitic modes. This, in turn, aids in improving the quality factor. In practical applications, the sensitivities to linear acceleration are important metrics. It has been shown that internal levers aid in rejecting in-plane acceleration (Fig. 11.1).

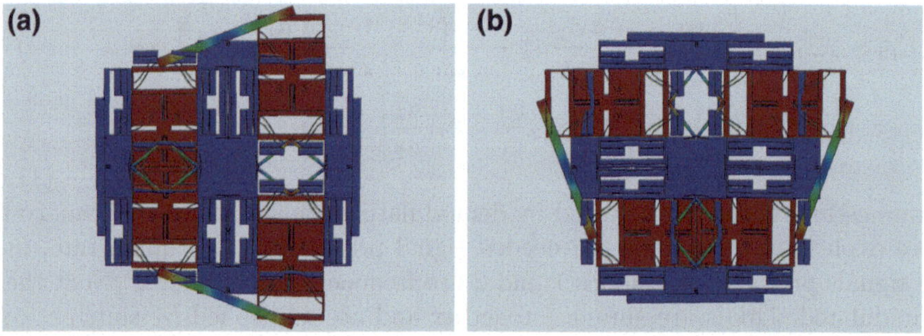

Fig. 11.1. Degenerate modes utilized for gyroscope operation have simulated natural frequencies close to 24 kHz

Compared to similarly sized QMGs with spring coupling, the dynamic range, or the range of excitations where the response of the system is linear, is dramatically decreased with the use of lever coupling [6]. This is experimentally shown in the following section. Thus, optimum operation, from a frequency stability perspective, requires careful consideration of nonlinear effects [12–14]. A challenge with designing for the Episeal process, in particular a QMG with internal levers, is careful attention to the fabrication process. The minimum feature size allowed is 3 μm, but some blowout, or over-etching, is expected. The QMG described in this effort was designed to the minimum feature size, and as a result, there is significant variability between the designed and measured operating frequencies. The device selected for study in the following sections has natural frequencies in the 16 kHz range. The sub-micron blowout is believed to be the suspect for the significant difference between design and measurement. Even in the absence of blowout, the anisotropic nature of silicon contributes to breaking the degeneracy of the modes. The native split between the modes utilized for gyroscope operation is close to 300 Hz, but electrostatic actuation can be utilized to decrease it. The minimum frequency split is limited by synchronization effects during FM gyroscope operation.

11.4 Frequency Stability

Closed-loop frequency response measurements [6] were combined with frequency stability measurements to find regimes that optimize frequency stability [12–14]. In open-loop testing, the frequency of the oscillator used to excite a resonator is swept in proximity of a resonant frequency. In the case that nonlinear effects are significant, hysteresis can be observed (i.e. the steady-state response switches between two different branches). Repeating the experiment allows one to provide different initial conditions to the system such that a more complete picture of the bifurcation structure can be recorded. However, operation near a saddle-node bifurcation point can slow the dynamics of the system and delay the switch between branches [15]. Closed-loop frequency response measurements overcome this limitation as they allows for the complete stabilization of the steady-state response.

The closed-loop frequency response method is very similar to experimental continuation [16] in that feedback control is used is used to stabilize states that are unstable in the open-loop configuration. Using a Zurich Instruments HF2LI, a phase locked-loop (PLL) was used to control the relative phase between the input and output of the resonator. This converts the dynamics of the system to an autonomous one such that period of oscillation is a measured instead of specified. Thus, while in open-loop testing one sweeps frequency and measures amplitude and phase, in closed-loop testing one sweeps phase and measures amplitude and frequency. By measuring frequency for a long enough period of time, one can quantify frequency stability. For a confident measure of fractional frequency stability using an Allan deviation method at a given integration time, the measurement period is typically greater than the given integration time by an order of magnitude. The challenge with combining closed-loop frequency response measurements with frequency stability measurements is that parameters of the resonator, such as the natural frequency, may drift. Frequency random walk was shown to be significant on time scales greater than 10 s, so the measurement duration was set at 10 s to allow for accurate frequency stability measurements on time scales less than 1 s. A delay of 2 s was used to allow for the system to settle between set phase values.

Experimental results showing the normalized steady-state amplitude and fractional frequency stability for relatively low excitation cases are shown in Fig. 11.2. The device was biased at 10 V and in the rest of the experimental result shown here. The fractional frequency stability is shown at 0.2 s. The integration time that would optimize gyroscope operation would be the one that corresponds to the frequency split. The selected integration time was the one that balanced white noise and frequency random walk. As mentioned in the previous section, even for the lowest excitation case, nonlinear effects are significant. Dots have been added to the figures to show common amplitude and phase states. The primary results that these figures display are that frequency stability can be further improved as nonlinear effects become significant and that optimum frequency stability does not always correspond with the peak amplitude. The improvement is subtle at the selected integration time, but the contour diagrams of the Allan deviation in Fig. 11.3 shows that the improvement is particularly significant on shorter time scales associated with gyroscope operation.

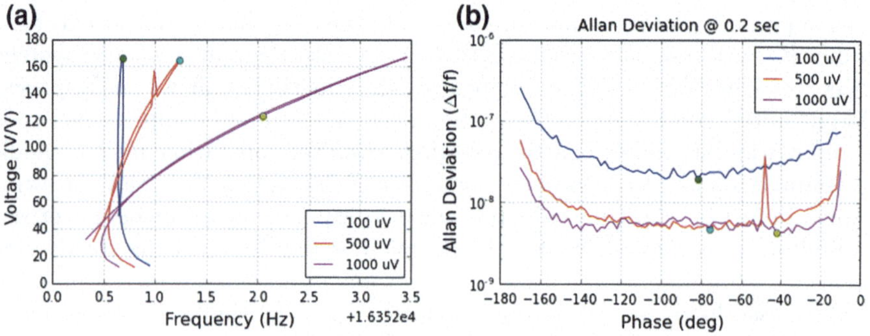

Fig. 11.2. Normalized steady-state amplitude response (**a**) and Allan deviation at an averaging time of 0.2 s (**b**). For even very low excitations, the device exhibits nonlinear behavior

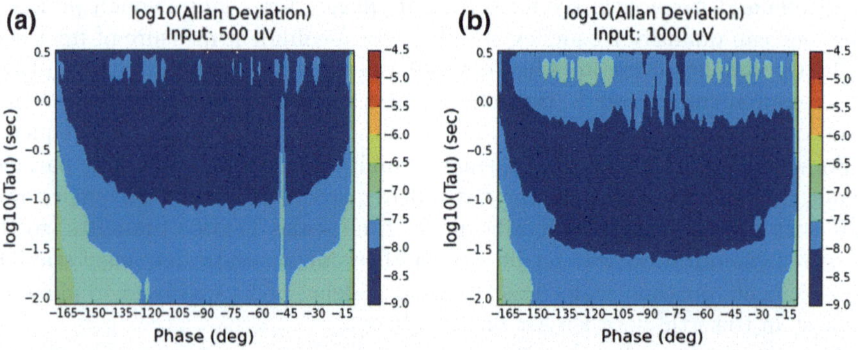

Fig. 11.3. Contour diagrams of the Allan deviation measurements for 500 μV (**a**) and 1000 μV (**b**) excitations. The larger excitation case provides a significant improvement on the short-term time scales associated with gyroscope operation

There are however, limits to the improvements that nonlinear operation can provide [13]. Figure 11.4 shows the steady-state amplitude and fractional frequency stability for larger excitation cases. At an integration time of 0.2 s, the lower excitation case is more stable. While at the shorter time-scales associated with gyroscope operation, the larger excitation case is more stable, there are other limits to nonlinear operation such as resonant pull-in effects. Ignoring phase delays associated with the electronics, linear design principals dictate that frequency stability is optimized at the peak amplitude; based on the employed configuration, that corresponds to a feedback phase of $-90°$. Operation at this point in the nonlinear regime can dramatically degrade stability. Lastly, it is important to note that between trials, the natural frequency of the device shifted, but the phase values that minimized the associated Allan deviation values stayed relatively constant.

Utilizing the test results discussed in this section, and similarly for the other mode of the gyroscope, phase and excitation values for gyroscope operation were selected based on those that minimized frequency fluctuations at a time scale of 0.2 s.

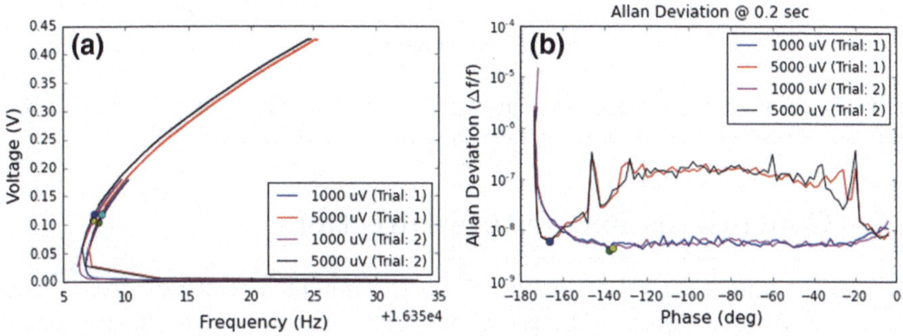

Fig. 11.4. Steady-state amplitude response (**a**) and Allan deviation at an averaging time of 0.2 s (**b**). While there is significant natural frequency drift between the trials, the Allan deviation measurements are much more stable

11.5 Gyroscope Operation

Using the results from the previous section, preliminary test results (see Fig. 11.5) of the nonlinear FM gyroscope are discussed. Over the selected time scale, white noise is the dominant noise process. This is characterized by the approximately $1/\sqrt{\tau}$ slopes of the Allan deviation measurements. As mentioned in the introduction, the gyroscope metric associated with white noise processes is angular random walk (ARW). Both trials were conducted in regimes where nonlinear effects are observed, but these result show that operation with significant nonlinear effects can be used reduce ARW. These findings are in accord with other recent works [6].

The frequency split used in these experiments was selected to be the minimum before synchronization between the two modes was observed. For the 100 μV trial, the frequency split was approximately 12 Hz. For the larger excitation case, the frequency split was 33 Hz. It has been shown that decreasing the frequency split, until effects associated with close to carrier noise become significant, decreases ARW [5]. Thus, while it is possible that nonlinear operation may increase the effective coupling between the two modes, implementation of techniques to decrease this coupling while operating with significant nonlinear effects may further reduce ARW.

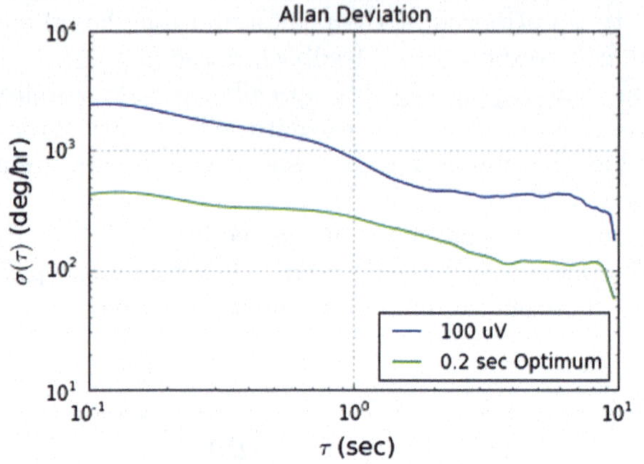

Fig. 11.5. Allan deviation of rate output for two different excitation cases. These results show that nonlinear operation can be utilized to decrease ARW

11.6 Conclusions and Future Directions

This work documents preliminary work on the nonlinear FM gyroscope. Compared to AM gyroscopes, FM gyroscopes have been shown to have excellent long-term stability, but degraded short-term stability. By operating in regimes that improve frequency stability, it was demonstrated that ARW, the gyroscope metric associated with short-term noise, can be reduced. While theoretically there are limits to the improvements that nonlinear operation can bring, the limit for the associated time scale for gyroscope operation has yet to be reached with the tested device. Future work will focus on reaching that limit.

References

1. I.P. Prikhodko, B. Bearss, C. Merritt, J. Bergeron, C. Blackmer, Towards self-navigating cars using MEMS IMU: challenges and opportunities, in *Proceedings of 2018 IEEE INERTIAL* (IEEE, Moltrasio, 2018), pp. 1–4
2. B. Eminoglu, Y.-C. Yeh, I.I. Izyumin, I. Nacita, M. Wireman, A. Reinelt, B.E. Boser, Comparison of long-term stability of AM versus FM gyroscopes. In: Proceedings of 2016 IEEE MEMS (IEEE, Shanghai, 2016), pp. 954–957
3. I.P. Prikhodko, S. Nadig, J.A. Gregory, W.A. Clark, M.W. Judy, Half-a-month stable 0.2 degree-per-hour mode-matched MEMS gyroscope, in *Proceedings of IEEE MEMS* (IEEE, Kauai, 2017), pp. 1–4
4. R.P. Leland, Mechanical-thermal noise in MEMS gyroscopes. IEEE Sens. J. **5**(3), 493–500 (2005)
5. B. Eminoglu, B.E. Boser, (2018) Chopped rate-to-digital FM gyroscope with 40 ppm scale factor accuracy and 1.2 dph bias. In: Proceedings of 2018 IEEE ISSCC (IEEE, San Francisco, 2018), pp. 178–180
6. P. Taheri-Tehrani, M. Defoort, D.A. Horsley, Operation of a high quality-factor gyroscope in electromechanical nonlinearities regime. J. Micromech. Microeng. **27**, 075015 (2017)

7. IEEE Standard Specification Format Guide and Test Procedure for Coriolis Vibratory Gyros. IEEE Standard 1431-2004 (2004)
8. I.I. Izyumin, M.H. Kline, Y.-C. Yeh, B. Eminoglu, C.H. Ahn, V.A. Hong, Y. Yang, E.J. Ng, T.W. Kenny, B.E. Boser, A 7ppm, 6 deg/hr frequency-output MEMS gyroscope, in *Proceedings of 2015 IEEE MEMS* (IEEE, Estoril, 2015), pp. 33–36
9. D.D. Lynch, *Vibratory Gyro Analysis by the Method of Averaging* (1995), pp. 26–34
10. B.R. Simon, S. Khan, A.A. Trusov, A.M. Shkel, mode ordering in tuning fork structures with negative structural coupling for mitigation of common-mode g-sensitivity, in *Proceedings of 2015 IEEE SENSORS* (IEEE, Busan, 2015), pp. 1–4
11. R.N. Candler, M.A. Hopcroft, B. Kim, W.-T. Park, R. Melamud, M. Agarwal, G. Yama, A. Partridge, M. Lutz, T.W. Kenny, Long-Term and accelerated life testing of a novel single-wafer vacuum encapsulation for MEMS resonators. J. Microelectromech. Syst. **15**(6), 1446–1456 (2006)
12. D.S. Greywall, B. Yurke, P.A. Busch, A.N. Pargellis, R.L. Willett, Evading amplifier noise in nonlinear oscillators. Phys. Rev. Lett. **72**(19), 2992–2995 (1994)
13. J. Juillard, A. Brenes, Impact of excitation waveform on the frequency stability of electrostatically-actuated micro-electromechanical oscillators. J. Sound Vib. **422**, 79–91 (2018)
14. L.G. Villanueva, E. Kenig, R.B. Karabalin, M.H. Matheny, R. Lifshitz, M.C. Cross, M.L. Roukes, Surpassing fundamental limits of oscillators using nonlinear resonators. Phys. Rev. Lett. **110**, 177208 (2013)
15. N. Miller, Noise in nonlinear micro-resonators. Ph.D. Thesis, Michigan State University, Lansing, Michigan, (2012)
16. J. Sieber, A. Gonzalez-Buelga, S.A. Neild, D.J. Wagg, B. Krauskopf, Experimental continuation of periodic orbits through a fold. Phys. Rev. Lett. **100**, 244101 (2008)

Chapter 12
Microtransitions in a $2-d$ Load Bearing Hierarchical Network

Anupama Roy and Neelima Gupte(✉)

Department of Physics, Indian Institute of Technology Madras, Chennai 600036, India
anupama@physics.iitm.ac.in, gupte@physics.iitm.ac.in

Abstract. The prediction of the critical point of a phase transition is useful in many practical contexts. Therefore, the identification of precursors, or early warning signals of the critical point, has become the focus of current interest. Recent model studies have shown that a series of small transitions, which have been called microtransitions, act as precursors to the percolation transition. Here, we identify the existence of microtransitions in the process of avalanche transmission on a specific realisation of branching hierarchical networks. We note that microtransitions are seen clearly in this realization, which we call the $V-$ lattice. Additionally, the positions of the microtransitions show scaling behaviour here. This can be used to calculate the position of the critical point, which is seen to be in agreement with the observed result. The correlation function of the time series of the weight transmission also shows interesting behaviour, which can be used to draw inferences about the structure and behaviour of the system. Additionally we utilise the structure factor, and the ratio of the heights of the peaks of the Fourier transform of the correlation function to infer information about the structure of the lattices. We discuss the utility of our results and generalisability to other contexts.

12.1 Introduction

The identification of precursors of phase transitions has been a topic of current research interest. The existence of phase transitions in real life situations such as congestion in road and internet traffic [1], blackouts in power grids [2], and monsoon dynamics [3] has lead to the realisation that the the prediction of phase transitions is a problem of great practical utility. It is in these contexts that the identification of precursors to the transitions assumes great importance. Early warning signals of transitions have been found in diverse phenomena ranging from the medical sciences, to ecosystems and climate phenomena [4]. In the context of theoretical models, microtransitions, where functions of the order parameter show small, but abrupt changes, have been used as precursors and predictors of the phase transition in the case of the percolation transition [5].

V. In et al. (Eds.): *Proceedings of the 5th International Conference on Applications in Nonlinear Dynamics*, Understanding Complex Systems, https://doi.org/10.1007/978-3-030-10892-2_12

The present paper discusses a set of microtransitions which serve as precursors of a phase transition, in the context of the transmission of avalanches on a $2-d$ branching hierarchical lattice.

The specific model considered here, is a $2-d$ load bearing hierarchical network which can serve as a model of diverse systems ranging from natural systems such as river networks [6] and granular media [7], as well as for social systems [8] and is also similar to models that arise in biological contexts as models of lung inflation [9]. Studies of packet transmission and avalanche transmission on such networks have been carried out to understand phenomena like internet traffic congestion, and jamming.

Here, we investigate the microtransitions in avalanche transmission on a special realisation of $2-d$ load bearing hierarchical network, where the network shows a transition from a state where most of the transmissions are successful, i.e. all test weights get absorbed, to one where most of the transmissions fail. This transition has been seen to be a discontinuous transition for this special realization (which we call the $V-$ lattice). We see the presence of microtransitions, signalled by oscillations in the relative variance of the order parameter in avalanche transmission for the $V-$ lattice. We have seen that the positions of microtransitions act as precursors to the transition point for this case. These follow a scaling law, and the critical point can be predicted with good accuracy, using the scaling behavior.

The microtransitions of the system can also be analysed using the correlation function. We analyse the peaks in the Fourier spectrum of the correlation function of the absorbed weight. The positions of the peaks follow a scaling relation with a power similar to the scaling relation for the microtransitions in the order parameter. We also analyse the structure factor of the $V-$ lattice network. A comparison of the ratio of the peaks in the structure factor, and that of the peaks in the Fourier transform of the correlation function can be used to infer information about the actual cluster geometry and capacity distribution. We discuss the implications of our results.

12.2 The Network

The 2-dimensional load bearing hierarchical network considered here, is based on a regular triangular lattice [10]. Every node can connect with its nearest neighbours in the layer below with probability $\frac{1}{2}$. Thus a site i in the layer L can connect to either of its nearest neighbours in the layer $L+1$. Each node is assigned a number, which represents its capacity. Every node in the topmost layer has unit capacity. The capacity w_i^L of a site i in the layer L is sum of the capacities of sites to which it is connected in the layer above and its own capacity one. The capacities obey the following equation;

$$w_i^L = l(i_l^{L-1}, i^L)w(i_l^{L-1}) + l(i_r^{L-1}, i^L)w(i_r^{L-1}) + 1 \tag{12.1}$$

L $= 1, \ldots, N$, where N is the total number of layers in the network. The link $l(i_l^{L-1}, i^L)$ takes value 1 if a connection exists between i_l^{L-1} and i^L, otherwise

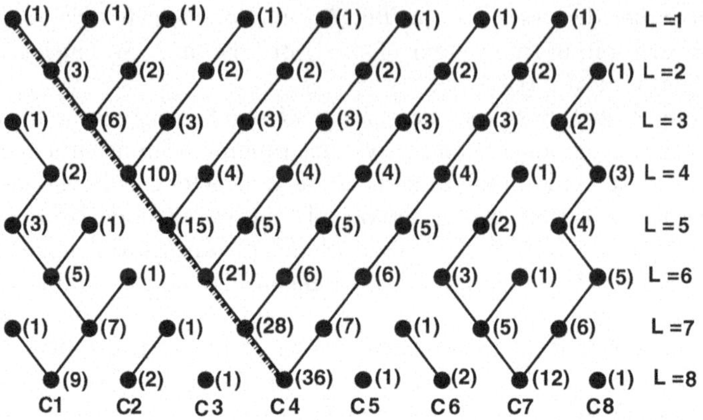

Fig. 12.1. The critical realisation of the $2-d$ load bearing hierarchical network, the $V-$ lattice of size $M = 8 \times 8$. The solid circles are nodes and solid lines are links of the network. The beaded line is the trunk of the network. C1, C2 are the clusters. The capacity of each node is indicated next to it in the bracket

it takes the value zero. The link $l(i_r^{L-1}, i^L)$ for the right connection has similar behaviour. Here, i_r^{L-1} and i_l^{L-1} are the sites which lie in the $L-1$th layer, and lie to the right and the left of the site i in the Lth layer.

The network consists of many clusters, where a cluster is the collection of connected sites of the network. The size of a cluster is defined as the total number of connected sites in that cluster. The cluster having the largest number of connected sites is the maximal cluster. The strongest path from the topmost layer to the bottommost layer in the maximal cluster is called the trunk of the network.

Figure 12.1 shows the critical realisation of the $2-d$ load bearing hierarchical network. All the sites in the topmost row and the $(N-L+1)$ sites of the Lth row constitute a $V-$ shaped maximal cluster. One of its arms constitutes the trunk and other arms run parallel to each other opposite to the trunk. This structure is called the $V-$ lattice because of its "V" shaped maximal cluster. Similar structures have been seen in a model of river deltas [11], as well as in Martian gullies [12,13].

This realization is called the critical realisation because the distribution of avalanche times shows power law behaviour for this realisation, as do other quantities [14]. On the other hand, the original lattice, i.e. typical realizations show Gaussian behavior for the avalanche distribution, and non power law behavior for other quantities [15]. We discuss here the microtransitions seen for the critical realisation of the $2-d$ load bearing hierarchical network, the $V-$ lattice network, and the behavior of the structure factors and correlation functions. A similar analysis can be carried out for typical realizations and will be discussed elsewhere.

12.3 The Avalanche Transmission

The avalanche transmission process, or the process of weight or packet transmission along the connected paths of the network [10,15] is defined as follows: when a weight W is deposited on a site in the first layer it retains a weight equal to its capacity W_c and transmits the rest $W - W_c$ to the site it is connected to in the layer below. Thus the weight is transmitted in the downward direction and the sites involved in this process constitute the path of connection. If there is still excess weight left at the bottommost layer of the network it is then transmitted to a randomly chosen unoccupied site of the first layer. Let P_L be the site on such path P. We can write, $W^{ex}(P_L) = W - \sum_{K=1}^{L} W_c(P_K)$. If a test weight transmitted in this way encounters a fully saturated site, and also has no alternate path to take, then the transmission is considered to have failed. If the transmitted weight is absorbed at some site in the network then the transmission corresponds to a successful transmission. The order parameter in an avalanche process on the $2-d$ load bearing hierarchical networks is defined as the fraction of transmissions that are successful. The order parameter for the typical realisations, i.e. the original lattice varies continuously with the test weight, whereas it shows a discontinuous variation with the test weight for the critical realisation i.e. the $V-$ lattice case (Fig. 12.2) [16]. Thus the critical geometry of the $V-$ lattice, leads to a situation where there is a discontinuous phase transition. We note that the transmission of messages on these base substrates shows a percolation transition on the original lattices, and an explosive percolation transition on the $V-$ lattices [16].

We have seen that microtransitions appear very clearly in the case of the $V-$ lattice network and the positions of microtransitions follow a power law

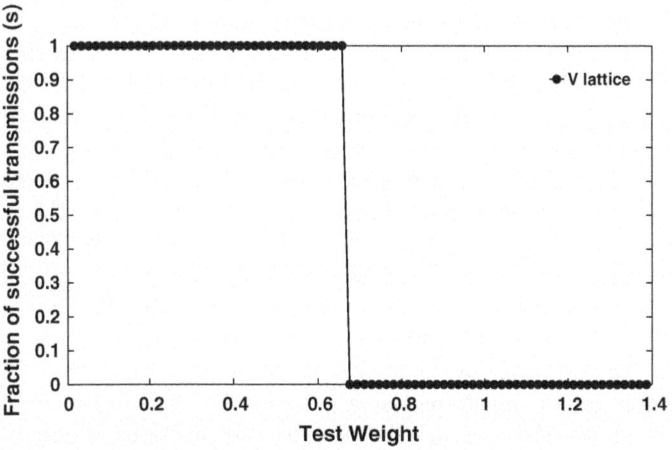

Fig. 12.2. The order parameter as a function of test weight for the $V-$ lattice of size $M = 50 \times 50$. Here the order parameter shows a discontinuous transition

behaviour, which helps to calculate the critical point of the phase transition with good accuracy. All these results are discussed in detail in the next section.

12.4 Microtransition in Avalanche Transmission for the $V-$ Lattice

We now study microtransitions for this case. The microtransitions are signalled by microscopic changes in the order parameter. The major transition in the system, is the transition of the order parameter at the critical point, from values of order zero to values of order one, whereas the microtransitions are small changes in the order parameter well before the transition point. The study of the avalanche transmission on the network shows a transition of the network from the state where all avalanche transmissions are successful, i.e. all the test weights get absorbed, to a state where almost all transmissions fail, as the test weight, which is placed on the top layer, increases. In order to study the microtransitions which occur before the transition, we look at the variance and relative variance of the absorbed weight, which is the weight absorbed by the occupied nodes of the network for a given test weight. The absorbed weight has nonzero value in the free flow state and it is zero in the state where all avalanche transmissions fail. The variance V and relative variance RV of the absorbed weight is defined as, $V = (\langle O^2\rangle - \langle O\rangle^2)$ and $RV = \frac{(\langle O^2\rangle - \langle O\rangle^2)}{\langle O\rangle^2}$ where, O is the weight absorbed by the occupied nodes of the network for a given test weight. The average is taken over the total number of nodes which are occupied. If a node is partially occupied we consider that as a occupied node.

We see that the variance of the absorbed weight shows a set of sharp peaks before the transition point (Fig. 12.3a). These peaks arise when a node with very high capacity becomes occupied in the course of the transmission. For the $V-$ lattice, these peaks occur when the nodes which belong to the trunk of the maximal cluster of the network become occupied.

The details of the simulation are as follows. We start our simulation by putting the test weight at the right most channel of the $V-$ lattice. When the nodes of a given channel becomes occupied (i.e. have absorbed all the weight that their capacity permits), we deposit the excess test weight (if any) on the nearest unoccupied node in the topmost row next to the channel. This process continues until the entire test weight gets absorbed by the network or the unabsorbed or residual weight does not have any alternate path to take. Clearly, as the nodes of the trunk become occupied, the available capacity of unoccupied nodes decreases, and the fluctuation in absorbed weight decreases, hence the amplitude of the peaks in the variance of the absorbed weight decreases. The positions of the peaks are the positions of the microtransitions (See Fig. 12.3a). The relative positions of the microtransitions follow a power law with a power close to -1. Figure 12.3b shows the scaling relations for the positions of microtransitions for different lattice sizes. The exponent does not depend on the lattice size for larger lattice sizes. Figure 12.4 shows the log-log plot of the scaling relation. We calculate the critical point from the scaling relation for the $V-$ lattice. From

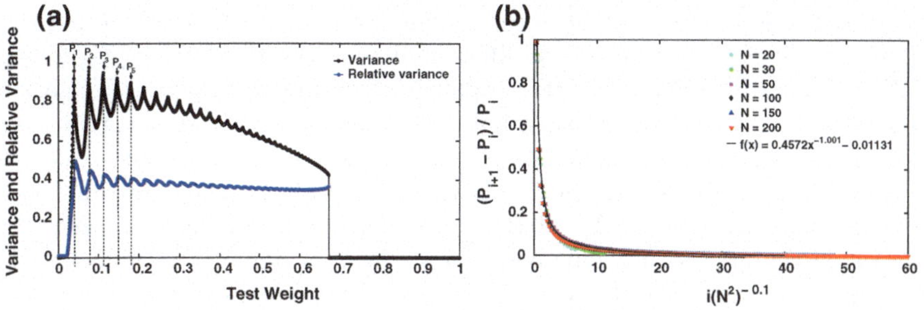

Fig. 12.3. **a** The variance (black line) and the relative variance (blue line) of total absorbed weight as function of test weight for the $V-$ lattice of size $M = 50 \times 50$. The variance shows sharp peaks before the actual transition whereas the relative variance shows jumps at the same positions of peaks. **b** The scaling law for the relative positions of the peaks for different lattice sizes is a power law $\sim ax^b + c$ with $a = 0.4572 \pm 0.0004$, $b = -1.001 \pm 0.001$, $c = -0.01131 \pm 0.00015$. Data for different lattice sizes collapse nicely

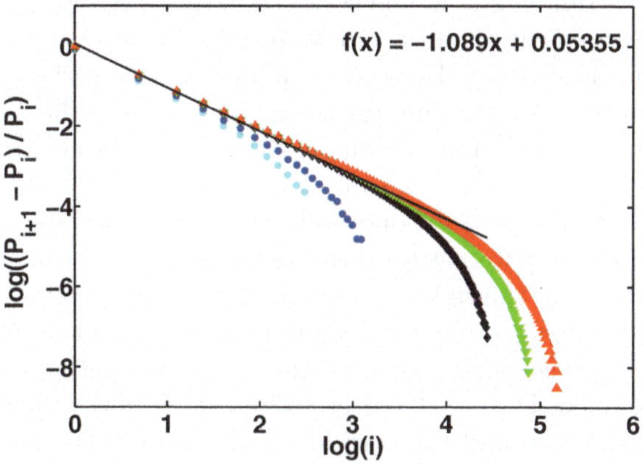

Fig. 12.4. The log log plot (e base) of the scaling relations for the $V-$ lattice of different sizes shows a linear behaviour $ax+b$ with $a = -1.089 \pm 0.017$ and $b = 0.05355 \pm 0.04120$ for $N = 20, 30, 50, 100, 150, 200$

the scaling law we can write, $P_{i+1} = \prod_{j=1}^{i} (aj^b + c + 1)P_1$. Using this equation, we can calculate the position of any peak if we know the value of the position of the first peak. As the peaks in the variance plot arise because of the nodes that belong to the trunk, the number of peaks are the same as the total number of sites of the trunk. The number of such sites in a $N \times N$ site $V-$lattice is N, so there will be at most N peaks in the variance plot. Hence the parameter value which corresponds to the P_{50} peak is the critical point, for a 50×50 lattice. For this case we calculate $P_{50} = P_c = 0.669$ using $P_1 = 0.039$. The value of P_c from the order parameter plot is 0.675. Both the results are in good agreement. We also study the finite size scaling of the scaling relation, which shows a nice data collapse for the scaling relations of different sizes of the $V-$ lattice (Fig. 12.3b).

We see microtransitions in the avalanche transmission on the $V-$ lattice, where we see a discontinuous transition from a free flow state to a jamming like state. We note that the microtransitions act as precursors of the phase transition in avalanche transmissions on the $V-$ lattice. The peaks i.e. the microtransitions are sharp here because the $V-$ lattice has more nodes of high capacity. The positions of the microtransitions helps to predict the parameter value which is very close to actual critical point. Therefore, the microtransitions are a good predictor of the critical point of a phase transition.

12.5 Microtransitions in Correlation Function for the $V-$ Lattice

Microtransitions can also be identified in other quantities. In this section, we analyse microtransitions in the correlation function which is a function of the test weight as well as a time lag (τ). The correlation function in any variable relates the value of the variable at any instant t with the value after certain time interval $t+\tau$ of the time series data. Here, we generate a time series of the absorbed weight for a given test weight. The test weight we choose is the maximum weight that the network can bear. The weight on the topmost layer now propagates in the network using the weight transmission process defined in Sect. 12.3. The hopping of the test weight from one node to another node is considered as a unit time step. If the test weight reaches the lowest layer of the network starting from the topmost row of the rightmost channel, we deposit the weight remaining unabsorbed on the nearest unoccupied node in the topmost row next to the channel. This process continues until all the connected paths from the topmost layer become occupied. During this process we calculate the weight absorbed by a node at each time step and the correlation function of the time series, which is defined as, $\rho(\tau, W_{test}) =< W(t, W_{test})W(t+\tau, W_{test}) > - < W(t, W_{test}) >^2$.

Here, $< W(t, W_{test})W(t+\tau, W_{test}) >= \frac{1}{t_{max}} \sum_{t=1}^{t_{max}-\tau} \Big(W(t, W_{test}) \times W(t+\tau, W_{test})\Big)$ and $< W(t, W_{test}) >^2 = \Big(\frac{1}{t_{max}} \sum_{t=1}^{t_{max}-\tau} W(t, W_{test})\Big)^2$.

Here the avalanche transmission, takes place as described earlier. The correlation function for the $V-$lattice shows oscillatory behavior (See Fig. 12.5a) and the amplitude of oscillation varies with the increase of the time lag. We also evaluate the discrete Fourier transform of the correlation function. The Fourier spectrum shows several peaks for the $V-$ lattice (Fig. 12.5b). We analyse the frequencies corresponding to the peaks in the Fourier spectrum. For the $V-$lattice, the relative frequencies $\frac{f_{i+1}-f_i}{f_i}$ show power law behavior, with a power close to -1 (Fig. 12.6). This exponent value is close to the exponent of the scaling relation for the microtransitions in the avalanche transmission. Therefore the peaks in the Fourier spectrum scale in a way which is similar to the microtransition peaks.

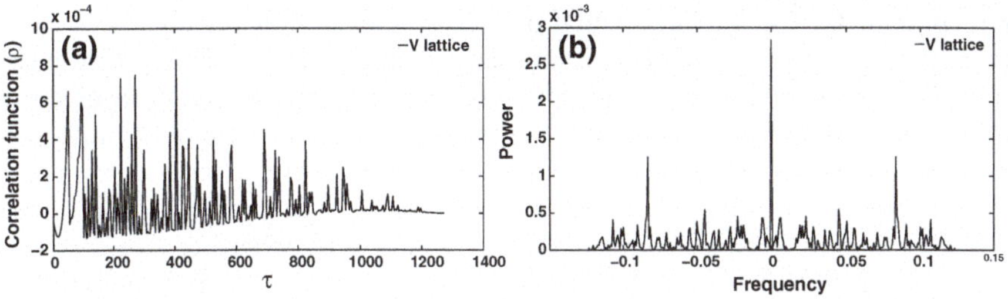

Fig. 12.5. **a** The correlation function of the absorbed weight as a function of the time lag(τ). This shows oscillatory behaviour. **b** The Fourier spectrum of the correlation function for the $V-$ lattice of size $M = 50 \times 50$. It shows peaks at different frequencies. We choose a test weight which is the same as the total capacity of the network and change the τ in steps $\Delta\tau = 4$

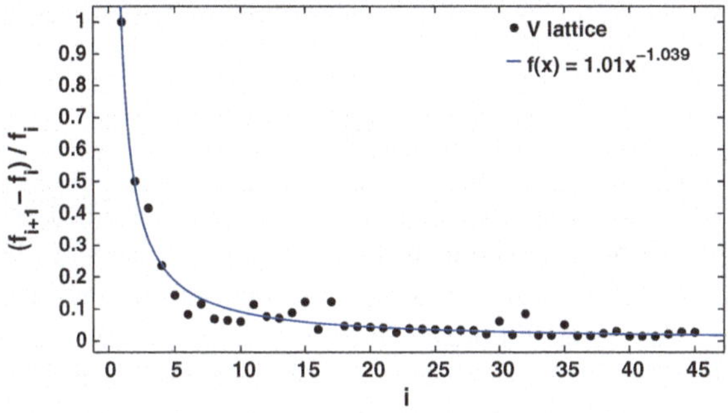

Fig. 12.6. The scaling relation for the peak positions of the Fourier spectrum for the $V-$ lattice of size $M = 50 \times 50$. The scaling relation behaves as a power law $\frac{(f_{i+1}-f_i)}{f_i} \sim ai^b$ with constant $a = 1.01 \pm 0.0567$ and exponent $b = -1.039 \pm 0.0626$

12.6 Structure Factors

In the previous section we have analysed the positions of the peaks in the Fourier spectrum of the correlation function and seen that they follow the same scaling relation as the positions of microtransitions seen in the variance. We have discussed in Sect. 12.4 that the microtransitions appear when the channels of the network become occupied via the avalanche process. The nodes of the trunk become occupied when the parallel arms of the $V-$ lattice are occupied. These results indicate that the channels of the network can be identified from the peaks of the Fourier spectrum. In a more general sense, information about the structure of the network can be extracted from the knowledge of the processes that occur on the networks. We test this notion in the context of the structure factor of the network, and the peaks of the Fourier spectrum of the correlation function.

We calculate the structure factor for the parallel arms of the $V-$ shaped unit cell. The basis of the network is the $2-d$ triangular lattice. The translation lattice vectors of the $V-$ shaped unit cell are chosen as the basis vectors of the triangular lattice. In the calculation of the structure factor for the V lattice the reciprocal lattice vectors corresponding to a node i are the reciprocal lattice vectors of the links through which it is connected to the nodes in layer above. We have defined a reciprocal lattice vector G_i which is the sum of the reciprocal lattice vectors of the links through which ith node in the layer L is connected to the nodes in the layer $L-1$. Each node can have at most two connections in the layer above. So we can write $G_i = G_{i1} + G_{i2}$, where G_{i1} and G_{i2} are the reciprocal lattice vectors for the left and right connections respectively. If there is a node for which either the left or right connection exists then G_i is equal to G_{i1} or G_{i2} depending on the existence of the left or right connection. G_{i1} and G_{i2} can be written as the linear combination of the basis vectors of the reciprocal lattice:$G_{i1} = \nu_{11}b_1 + \nu_{12}b_2$ and $G_{i2} = \nu_{21}b_1 + \nu_{22}b_2$, where, ν_{11}, ν_{12} are the intercepts of the G_{i1} on the basis vectors and similarly ν_{21}, ν_{22} are intercepts of G_{i2}. If G_{i1} and G_{i2} are parallel to b_2 and b_1 respectively, then we choose the corresponding intercept to be zero. The structure factor of the $V-$ shaped unit cell is given by,

$$S = \sum_{i=1}^{n} f_i \exp\left(ir_i \cdot G_i\right) \tag{12.2}$$

where i is the node index and runs from 1 to n, the total number of nodes in the V-shaped unit cell. Here, f_i is the form factor of the ith node, r_i is the position of the ith node and G_i is the corresponding reciprocal lattice vector. The reciprocal lattice vectors of the topmost nodes are found by using the connection pattern; e.g. for our $V-$ lattice the leftmost node in the first layer will have both left and right connections and the rest of the nodes in the first layer will have only right connections.

We need to identify the form factor f_i in the usual definition of the structure factor, in the context of the network. This quantity can be compared with the connectivity of the nodes. A node with high connectivity is connected to many paths in the network and is also accessible with greater probability. Therefore, it will have a larger contribution in the structure factor. In the case of our network the capacity of a node is dependent on the connectivity of the nodes to which it is connected in the layer above it. Thus, if the form factor is identified with the capacity, it can encode information about the structure and connectivity of the network. To test this idea, we calculate the structure factor considering equal as well as unequal form factors for each node and compare the result with the Fourier transform of the structure function.

We first discuss the structure factor of the V lattice with equal form factor f for each node. This situation represents the regular V shaped unit cell. We see that for an even lattice the structure factor of each odd arm is $n_i f$, where n_i is the number of nodes in the ith arm and for each even arm, it turns out to be f. For the odd lattice, the structure factor for each odd arm is $n_i f$ but for each even arm it is zero. We arrange the structure factors for the each arm in decreasing order and calculate the ratios of the consecutive values. We also use the Fourier spectrum of the correlation function defined earlier, and compare the ratios of the peaks seen here, with the structure factor ratios of the arms of

Table 12.1. The ratios of the peak values of the Fourier spectrum, the ratios of the calculated structure factors considering equal form factors for each node of the $V-$ lattice and the ratios of the peaks of the correlation function for the $M = 100 \times 100$ $V-$ lattice

FFT-peak ratios	SF ratios	Correlation function peak ratios
0.965080	0.9655	0.965517
0.964187	0.9643	0.964283
0.803984	0.8000	0.8000
0.855609	0.8571	0.8571
0.985813	–	–
0.956847	0.9565	0.956522
0.906445	0.9091	0.909091
0.982580	–	0.962456
0.853807	0853	–
0.946950	–	0.947368
0.834599	0.833	0.8333
0.981879	–	0.981818
0.976152	0.9762	0.976190
0.972843	0.9730	0.972973
0.937364	0.9373	0.9375
0.985540	–	0.985507

the V lattice network, after arranging both sets of peaks in decreasing order. A comparison of these ratios shows that some of the peak ratios of the correlation function and its Fourier spectrum match with the structure factor ratios for the $V-$ lattice (Table 12.1). We have seen that microtransitions in the avalanche transmission on the $V-$ lattice correspond to transmission along each parallel arm of the lattice. These parallel arms can be identified from the ratios of the structure factor for the arms.

However, it is clear that the form factor of each node is not equal. For the $V-$ lattice, the nodes have different capacities at different layers, and so have different form factors. Our calculation shows that for the $V-$ lattice of both even and odd size, the structure factors for the odd arm is proportional to the total capacity of the nodes which constitute the arm; whereas for the even arm it is $\frac{(N-i+1)(N-i+2)}{2} - \frac{N-i}{2}$ for the ith arm, where $i = 1$ for the right most arm, and N is the total number of layers of the network. We again compare the ratios of the structure factor of the parallel arms of the $V-$ lattice with the ratios of the peaks of the Fourier spectrum after arranging both in decreasing order. We have shown some of the ratios of the structure factors in the Table 12.2 which match with the peak ratios of the Fourier spectrum. It is clear from the table that the two sets of ratios match well (Fig. 12.7).

Table 12.2. Comparison of the FFT peak ratios with the ratios of the structure factors of the parallel arms of the $M = 100 \times 100$ $V-$ lattice, considering the unequal form factor for each node, where the form factor is equal to node capacity. These ratios are chosen from the ratios of the structure factors which match with the peak ratios of the Fourier spectrum (which are about 48% of the total no, of calculated ratios)

FFT-peak ratios	SF ratios	FFT- peak ratios	SF ratios
0.9568	0.957	0.9652	0.9651
0.9559	0.956	0.9929	0.9929
0.9469	0.9467	0.9891	0.9891
0.945	0.9454	0.9669	0.967
0.9998	0.9998	0.9706	0.9705
0.9827	0.9828	0.9847	0.9847
0.976	0.976	0.9964	0.9964
0.9562	0.9565	0.9752	0.9753

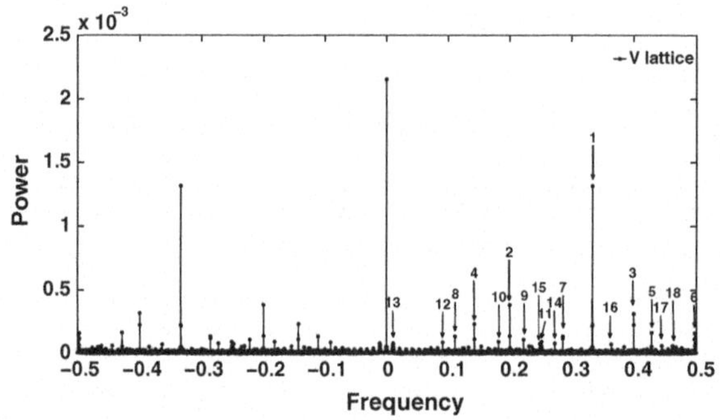

Fig. 12.7. Fourier spectrum for the 100×100 $V-$ lattice. Peak indices are according to their values in decreasing order

The ratios of the structure factor that correspond to the arms with low capacity are not seen in the ratios of the amplitude of the peaks in Fourier spectrum. The set of ratios that match for the unequal case is much larger than that seen for the equal form factor case, reflecting the true structure of the capacity distribution. Here, 26% of the ratios of the structure factor match with the Fourier spectrum peak ratios for the equal form factor case whereas this percentage increases to 48% when we consider unequal form factors for each node, reflecting the fact that the actual capacity distribution is now taken into account. This kind of comparison can thus be used to infer the capacity distribution for actual situations, e.g. in the case of force chains in granular media.

12.7 Conclusions

We have seen that microtransitions appear in the process of avalanche transmission on the $V-$ lattice network. The relative positions of these microtransitions obey a power law with power nearly equal to -1. The calculated value of the critical point from the scaling relation is close to the value of critical point from the order parameter for the $V-$lattice. Thus, the microtransitions behave as precursors of the phase transition and can be used to predict the point of transition.

We also analyze the microtransitions in the correlation functions of the time series arising from weight transmission. The correlation function here is an oscillatory function of τ. We observe that the peaks in the Fourier spectrum of the correlation function scale in a manner similar to the peaks in the variance of order parameter. Thus we see microtransitions in another quantity also.

We also calculate the structure factor to identify the channels corresponding to the microtransitions of the $V-$ lattice, considering both equal and unequal form factors of the nodes. We compare the ratios of the structure factors of the parallel arms for both the cases with the ratios of the peaks of the Fourier spectrum. It is seen that for the case where the form factors reflect the capacities, the structure factor can pick up many more ratios seen in the Fourier spectrum in the correlation function, compared to the equal form factor case. Thus, the structure factor can be used to estimate the way in which capacities are distributed in the lattice. This can be useful in many practical cases, e.g. in the identification of force chains in granular media, the jamming nodes of communication networks and the vulnerable nodes of power grids. We hope to pursue some of the applications in future work.

References

1. B. Tadić, G.J. Rodgers, Adv. Complex Syst. **05**, 445 (2002)
2. I. Dobson, B.A. Carreras, V.E. Lynch, D.E. Newman, Chaos **17**, 026103 (2007)
3. T.M. Lenton, H. Held, E. Kriegler, J.W. Hall, W. Lucht, S. Rahmstorf, Hans J. Schellnhuber, Proc. Natl. Acad. Sci. U.S.A. **105**, 1786–1793 (2008)
4. M. Scheffer, J. Bascompte, W.A. Brock, V. Brovkin, S.R. Carpenter, V. Dakos, H. Held, E.H. van Nes, M. Rietkerk, G. Sugihara, Nature **461**, 53–59 (2009)
5. W. Chen, M. Schröder, R.M. D'Souza, D. Sornette, J. Nagler, Phys. Rev. Lett. **112**, 1–5 (2014)
6. A.E. Scheidegger, International association of scientific hydrology. Bulletin **12**, 15–20 (1967)
7. S. Coppersmith, C. Liu, S. Majumdar, O. Narayan, T. Witten, Phys. Rev. E **53**, 4673–4685 (1996)
8. D. Griffeath, 1st edn. (Springer, Berlin, 1979)
9. B. Suki, A.L. Barabási, Z. Hantos, F. Peták, H.E. Stanley, Nature **368**, 615–618 (1994)
10. T.M. Janaki, N. Gupte, Phys. Rev. E **67**, 021503 (2003)
11. H. Seybold, J.S. Andrade, H.J. Herrmann, Proc. Natl. Acad. Sci. U.S.A. **104**, 16804–16809 (2007)

12. D. Reiss, G. Erkeling, K.E. Bauch, H. Hiesinger, Geophys. Res. Lett. **37**, 1–7 (2010)
13. T. Shinbrot, N.-H. Duong, L. Kwan, M.M. Alvarez, Proc. Natl. Acad. Sci. U.S.A. **101**, 8542–8546 (2004)
14. N. Gupte, A.D. Kachhvah, in *International Conference on Theory and Application in Nonlinear Dynamics (ICAND 2012)*, pp. 193–202
15. A.D. Kachhvah, N. Gupte, Pramana **77**, 873–879 (2011)
16. A.D. Kachhvah, N. Gupte, Phys. Rev. E **86**, 026104 (2012)

Chapter 13
Pseudospin-1 Systems as a New Frontier for Research on Relativistic Quantum Chaos

Ying-Cheng Lai(✉)

School of Electrical, Computer and Energy Engineering, Arizona State University, Tempe, AZ 85287, USA
Ying-Cheng.Lai@asu.edu

Abstract. Pseudospin-1 systems are characterized by the feature that their band structure consists of a pair of Dirac cones and a topologically flat band. Such systems can be realized in a variety of physical systems ranging from dielectric photonic crystals to electronic materials. Theoretically, massless pseudospin-1 systems are described by the generalized Dirac-Weyl equation governing the evolution of a three-component spinor. Recent works have demonstrated that such systems can exhibit unconventional physical phenomena such as revival resonant scattering, superpersistent scattering, super-Klein tunneling, perfect caustics, vanishing Berry phase, and isotropic low energy scattering. We argue that investigating the interplay between pseudospin-1 physics and classical chaos may constitute a new frontier area of research in relativistic quantum chaos with significant applications.

13.1 Introduction: What Are Pseudospin-1 Systems and Where Do They Arise?

Solid state materials whose energy bands contain a Dirac cone structure have been an active area of research since the experimental realization of graphene [1,2]. From the standpoint of quantum transport, the Dirac cone structure and the resulting pseudospin characteristic of the underlying quasiparticles can lead to unconventional physical properties/phenomena such as high carrier mobility, anti-localization, chiral tunneling, and negative refractive index, which are not usually seen in traditional semiconductor materials. Moreover, due to the underlying physics being effectively governed by the Dirac equation, relativistic quantum phenomena such as Klein tunneling, Zitterbewegung, and pair creations can potentially occur in solid state devices and be exploited for significantly improving or even revolutionizing conventional electronics. Uncovering/developing alternative materials with a Dirac cone structure has also been

V. In et al. (Eds.): *Proceedings of the 5th International Conference on Applications in Nonlinear Dynamics*, Understanding Complex Systems, https://doi.org/10.1007/978-3-030-10892-2_13

extremely active [3,4]. In this regard, the discovery of topological insulators [5,6] indicates that Dirac cones with a topological origin can be created, leading to the possibility of engineering materials to generate remarkable physical phenomena such as zero-field half-integer quantum Hall effect [7], topological magnetoelectric effect [8], and topologically protected wave transport [9,10].

A parallel line of research has concentrated on developing photonic materials with a Dirac cone structure, due to the natural analogy between electromagnetic and matter waves. For example, photonic graphene [11,12] and photonic topological insulators [13–18] have been realized, where novel phenomena of controlled light propagation have been demonstrated. Due to the much larger wavelength in optical materials as compared with the electronic wavelength, synthetic photonic devices with a Dirac cone structure can be fabricated at larger scales with a greater tunability through modulations. The efforts have led to systems with additional features in the energy band together with the Dirac cones, opening possibilities for uncovering new and "exotic" physics with potential applications that cannot even be conceived at the present.

The materials to be discussed in this article are those whose energy bands consist of a pair of Dirac cones and a topologically flat band, electronic or optical. For example, in a dielectric photonic crystal, Dirac cones can be induced through accidental degeneracy that occurs at the center of the Brillouin zone. This effectively makes the crystal a zero-refractive-index metamaterial at the Dirac point where the Dirac cones intersect with another flat band [19–23]. Alternatively, configuring an array of evanescently coupled optical waveguides into a Lieb lattice [24–27] can lead to a gapless spectrum consisting of a pair of common Dirac cones and a perfectly flat middle band at the corner of the Brillouin zone. As demonstrated more recently, loading cold atoms into an optical Lieb lattice provides another experimental realization of the gapless three-band spectrum at a smaller scale with greater dynamical controllability of the system parameters [28]. With respect to creating materials whose energy bands consist of a pair of Dirac cones and a topologically flat band, there have also been theoretical proposals on Dice or $\mathcal{T}_3$ optical lattices [29–34] and electronic materials such as transition-metal oxide $SrTiO_3/SrIrO_3/SrTiO_3$ trilayer heterostructures [35], 2D carbon or MoS_2 allotropes with a square symmetry [36,37], $SrCu_2(BO_3)_2$ [38] and graphene-In_2Te_2 bilayer [39]. Dirac cones with a flat band can also arise in a class of mechanical lattices [40].

In spite of the diversity and the broad scales to realize the band structure that consists of two conical bands and a characteristic flat band intersecting at a single point in different physical systems, there is a unified underlying theoretical framework: generalized Dirac-Weyl equation for massless spin-1 particles [31]. Comparing with the conventional Dirac cone systems with massless pseudospin/spin-1/2 quasiparticles (i.e., systems without a flat band), pseudospin-1 systems can exhibit quite unusual physics such as super-Klein tunneling for the two conical (linear dispersive) bands [23,32,41,42], diffraction-free wave propagation and novel conical diffraction [24–27], flat band rendering divergent dc conductivity with a tunable short-range disorder [43], unconventional

Anderson localization [44,45], flat band ferromagnetism [28,46,47], and peculiar topological phases under external gauge fields or spin-orbit coupling [35,48–50]. Especially, the topological phases arise due to the flat band that permits a number of degenerate localized states with a topological origin (i.e., "caging" of carriers) [51]. Most existing works, however, focused on the physics induced by the additional flat band, and the scattering/transport dynamics in pseudospin-1 systems have begun to be studied [52–54].

13.2 Generalized Dirac-Weyl Equation

The effective low-energy Hamiltonian associated with pseudospin-1 Dirac cones can be written, in the unit $\hbar = 1$, as [23,24,41]

$$H_0 = v_g \boldsymbol{S} \cdot \boldsymbol{k}, \tag{13.1}$$

where v_g is the magnitude of the group velocity associated with the Dirac cone, $\boldsymbol{k} = (k_x, k_y)$ denotes the wavevector, and $\boldsymbol{S} = (S_x, S_y)$ is a vector of matrices with components

$$S_x = \frac{1}{\sqrt{2}} \begin{pmatrix} 0 & 1 & 0 \\ 1 & 0 & 1 \\ 0 & 1 & 0 \end{pmatrix} \quad \text{and} \quad S_y = \frac{1}{\sqrt{2}} \begin{pmatrix} 0 & -i & 0 \\ i & 0 & -i \\ 0 & i & 0 \end{pmatrix}. \tag{13.2}$$

Along with another matrix

$$S_z = \begin{pmatrix} 1 & 0 & 0 \\ 0 & 0 & 0 \\ 0 & 0 & -1 \end{pmatrix},$$

the three matrices form a complete representation of spin-1, which satisfies the angular momentum commutation relations $[S_l, S_m] = i\epsilon_{lmn} S_n$ with three eigenvalues: $s = \pm 1, 0$, where ϵ_{lmn} is the Levi-Civita symbol. It follows from Eq. (13.1) that the energy spectrum consists of three bands that intersect at the Dirac point: a dispersionless flat band $E_0(\boldsymbol{k}) = 0$ and two linearly dispersive bands $E_\tau(\boldsymbol{k}) = \tau v_g |\boldsymbol{k}|$ with $\tau = \pm 1$ being the band index. The corresponding eigenfunctions in the position representation $\boldsymbol{r} = (x, y)$ are

$$\psi_{\boldsymbol{k},\tau}(\boldsymbol{r}) = \langle \boldsymbol{r} | \boldsymbol{k}, \tau \rangle = \frac{1}{2} \left[e^{-i\theta}, \sqrt{2}\tau, e^{i\theta} \right]^T e^{i\boldsymbol{k}\cdot\boldsymbol{r}}, \tag{13.3}$$

for the dispersive bands and

$$\psi_{\boldsymbol{k},0}(\boldsymbol{r}) = \langle \boldsymbol{r} | \boldsymbol{k}, 0 \rangle = \frac{1}{\sqrt{2}} \left[-e^{-i\theta}, 0, e^{i\theta} \right]^T e^{i\boldsymbol{k}\cdot\boldsymbol{r}}, \tag{13.4}$$

for the flat band, where $\theta = \tan^{-1}(k_y/k_x)$. The current operator is defined from Eq. (13.1) as

$$\hat{\jmath} = \nabla_{\boldsymbol{k}} H_0 = v_g \boldsymbol{S}. \tag{13.5}$$

The local current in a given state $\psi(\boldsymbol{r}) = [\psi_1, \psi_2, \psi_3]^T$ can thus be expressed as

$$\begin{aligned} \boldsymbol{j}(\boldsymbol{r}) &= v_g \psi^\dagger \boldsymbol{S} \psi \equiv (j_x, j_y) \\ &= \sqrt{2} v_g \left(\Re[\psi_2^*(\psi_1 + \psi_3)], -\Im[\psi_2^*(\psi_1 - \psi_3)] \right), \end{aligned} \tag{13.6}$$

which satisfies the common continuity equation

$$\frac{\partial}{\partial t}\rho + \nabla \cdot \boldsymbol{j} = 0, \tag{13.7}$$

where $\rho = \psi^\dagger \psi$ is the probability density associated with state ψ. From Eqs. (13.3) and (13.4), it can be seen that the associated local current density satisfies $\boldsymbol{j}_0 = \boldsymbol{0}$ for the flat band plane-wave, and

$$\boldsymbol{j}_\tau = v_g(\cos\theta, \sin\theta) = \tau v_g \frac{\boldsymbol{k}}{|\boldsymbol{k}|}, \tag{13.8}$$

for the dispersive band plane-wave. In terms of the Berry phase associated with the band structure, one obtains from Eqs. (13.3) and (13.4) the corresponding Berry connections

$$\begin{aligned} \mathscr{A}_{\boldsymbol{k}}^{\tau} &= \langle \boldsymbol{k}, \tau | i \nabla_{\boldsymbol{k}} | \boldsymbol{k}, \tau \rangle = 0, \\ \mathscr{A}_{\boldsymbol{k}}^{0} &= \langle \boldsymbol{k}, 0 | i \nabla_{\boldsymbol{k}} | \boldsymbol{k}, 0 \rangle = -2 \mathscr{A}_{\boldsymbol{k}}^{\tau} = 0 \end{aligned}$$

for all three bands. The Berry phase is thus given by

$$\Phi_B^{\tau,0} = \oint_{\mathscr{C}_{\boldsymbol{k}_d}^{\tau,0}} d\boldsymbol{k} \cdot \mathscr{A}_{\boldsymbol{k}}^{\tau,0} = 0, \tag{13.9}$$

for any closed path $\mathscr{C}_{\boldsymbol{k}_d}^{\tau,0}$ encircling the degeneracy point $\boldsymbol{k}_d$ of the momentum space defined in each band. It should be noted that the vanishing or 2π quantized Berry phase is consistent with the fundamental properties of spin-1 particles.

A remarkable phenomenon for pseudospin-1 Dirac cone systems, which is not usually seen in conventional Dirac cone systems such as graphene and topological insulators, is super-Klein tunneling [23]. Specifically, following the standard treatment of Klein tunning for graphene systems [55], one can consider the basic problem of wave scattering from a rectangular scalar (electrostatic) potential barrier defined as $V(x, y) = V_0 \Theta(x) \Theta(D - x)$ with barrier width D and height V_0. The transmission probability based on the effective Hamiltonian Eq. (13.1) for incident energy $E \neq 0, V_0$ is given by

$$T = \frac{(1-\gamma^2)(1-\gamma'^2)}{(1-\gamma^2)(1-\gamma'^2) + \frac{1}{4}(\gamma+\gamma')^2 \sin^2(q_x D)}, \tag{13.10}$$

where $\gamma = \tau \sin\theta$, $\gamma' = \tau' \sin\theta'$ with $\tau = \mathrm{sgn}(E), \tau' = \mathrm{sgn}(E - V_0)$, $\theta = \tan^{-1}(k_y/k_x)$ is the incident angle, and $\theta' = \arctan(k_y/q_x)$ with $q_x = \sqrt{(E - V_0)^2 - k_y^2}$. A striking feature of Eq. (13.10) is that, when the incident wave energy is one half of the potential barrier height, i.e., $E = V_0/2$, one has $\tau = -\tau', \theta = \theta'$ and, consequently, perfect transmission with $T \equiv 1$ for *any* incident angle θ - hence the term "super-Klein tunneling."

13.3 Transport Properties of Pseudospin-1 Systems

A recent work [52] addressed the following question: what types of transport properties can arise form pseudospin-1 systems whose band structure is characterized by coexistence of a pair of Dirac cones and a flat band? To address this question in the simplest possible setting while retaining the essential physics, ballistic wave scattering from a circularly symmetric potential barrier was studied. For conventional Dirac cone systems with pseudospin or spin-1/2 quasiparticles, there has been extensive work on scattering [56–58] with phenomena such as caustics [59], Mie scattering resonance [60], birefringent lens [61], cloaking [62], spin-orbit interaction induced isotropic transport and skew scattering [63,64], and electron whispering gallery modes [65]. However, there had been no corresponding studies for pseudospin-1 Dirac cone systems prior to the work in Ref. [52].

More specifically, scattering was studied [52] of pseudospin-1 particle from a circularly symmetric scalar potential barrier of height V_0 defined by $V(r) = V_0\Theta(R-r)$, where R is the scatterer radius and Θ denotes the Heaviside function. To characterize the scattering dynamics quantitatively, the scattering efficiency can be used, which is defined as the ratio of the scattering to the geometric cross sections [60]:

$$Q = \sigma/(2R), \tag{13.11}$$

where the scattering cross section σ can be calculated through the far field radial reflected current [52].

There were three main results [52]: revival resonant scattering, super-Klein tunneling induced perfect caustics, and universal low-energy isotropic transport without broken symmetries for massless quasiparticles. First, for small scatterer size, the effective three-component spinor wave exhibits revival resonant scattering as the incident wave energy is varied continuously - a phenomenon that has not been reported in any known wave systems. Strikingly, the underlying revival resonant modes show a peculiar type of boundary trapping profile in their intensity distribution. While the profile resembles that of a whispering gallery mode, the underlying mechanism is quite different: these modes occur in the wave dominant regime through the formation of fusiform vortices around the boundary in the corresponding local current patterns, rather than being supported by the gallery type of orbits through total internal reflections. Second, for larger scatterer size where the scattering dynamics are semiclassical, a perfect caustic phenomenon arises when the incident wave energy is about half of the barrier height, as a result of the super-Klein tunneling effect. A consequence is that the scatterer behaves as a lossless Veselago lens with effective negative refractive index resulting from the Dirac cone band structure. Compared with conventional Dirac cone systems for pseudospin-1/2 particles, the new caustics possess remarkable features such as significantly enhanced focusing, vanishing of the second and higher order caustics, and a well-defined static cusp. Third, in the far scattering field, an isotropic behavior arises at low energies. Considering that there is no broken symmetry so the quasiparticles remain massless, the phenomenon is quite surprising as conventional wisdom would suggest that the scattering be anisotropic. An analysis of the characteristic ratio of the transport to the elastic time as a function of the scatterer size revealed that the phenomenon of scattering isotropy can be attributed to vanishing of the Berry phase for massless pseudospin-1 particles that results in constructive interference between the time-reversed backscattering paths. Because of the isotropic structure, the emergence of a Fano-type resonance structure in the function of the ratio versus the scatterer size can be exploited to realize effective switch of wave propagation from a forward dominant state to a backward dominant one, and vice versa. In Ref. [52], an analytic theory with physical reasoning was developed to understand the three novel phenomena.

It is possible to conduct experimental test of the phenomena. For example, in a recent work [23], it was demonstrated for a class of two-dimensional dielectric photonic crystals with Dirac cones induced accidentally [19–22] that the Maxwell's equations can lead to an effective Hamiltonian description sharing the same mathematical structure as that of massless pseudospin-1 particles. Especially, the photonic analogy of the gate potential in the corresponding electronic system can be realized by manipulating the scaling properties of Maxwell's equations. Recent experimental realizations of photonic Lieb lattices consisting of evanescently coupled optical waveguides implemented through the femtosecond laser-writing technique [24–27] make them prototypical for studying the physics of pseudospin-1 Dirac systems. With a particular design of the refractive index profile across the lattice to realize the scattering configuration, the phenomena can be experimentally tested. Loading ultracold atoms into an optical Lieb lattice fabricated by interfering counter-propagating laser beams [28] provides another versatile platform to test the phenomena, where appropriate holographic masks can be used to implement the desired scattering potential barrier [32,66]. In electronic systems, the historically studied but only recently realized 2D magnetoplasmon system [67] is described by three-component linear equations with the same mathematical structure of massless pseudospin-1 particles, which can serve as a 2D electron gas system to test the phenomena.

From an applied perspective, the phenomenon of revival resonant scattering can be a base for articulating a new class of microcavity lasers based on the principles of relativistic quantum mechanics. It may also lead to new discoveries in condensed matter physics through exploiting the phenomenon in electronic systems. The phenomenon of perfect caustics can have potential applications in optical imaging defying the diffraction limit as well as in optical cloaking.

13.4 Superscattering of Pseudospin-1 Wave in Photonic Lattice

Another phenomenon is superscattering of pseudospin-1 wave from weak scatterers in the subwavelength regime where the scatterer size is much smaller than wavelength [53]. The phenomenon manifests itself as unusually strong scattering characterized by extraordinarily large values of the cross section even for arbitrarily weak scatterer strength. The physical origin of superscattering is revival resonances [53], for which the conventional Born theory breaks down. The phenomenon can be experimentally tested using synthetic photonic systems.

In wave scattering, a conventional and well accepted notion is that weak scatterers lead to weak scattering. This can be understood by resorting to the Born approximation. In particular, consider a simple 2D setting where particles are scattered from a circular potential of height V_0 and radius R. In the low energy (long wavelength) regime $kR < 1$ (with k being the wavevector), the Born approximation holds for weak potential: $(m/\hbar^2)|V_0|R^2 \ll 1$. Likewise, in the high energy (short wavelength) regime characterized by $kR > 1$, the Born approximation still holds in the weak scattering regime: $(m/\hbar^2)|V_0|R^2 \ll (kR)^2$. In general, whether scattering is weak or strong can be quantified by the scattering cross section. For scalar waves governed by the Schrödinger equation, in the Born regime the scattering cross section can be expressed as polynomial functions of the effective potential strength and size [68]. For spinor waves described by the Dirac equation (e.g., graphene systems), the 2D transport cross section is given by [58] $\Sigma_{tr}/R \simeq (\pi^2/4)(V_0R)^2(kR)$ (under $\hbar v_F = 1$). In light scattering from spherically dielectric, "optically soft" scatterers with relative refractive index n near unity, i.e., $kR|n-1| \ll 1$, the Born approximation manifests itself as an exact analog of the Rayleigh-Gans approximation [69], which predicts that the scattering cross section behaves as $\Sigma/(\pi R^2) \sim |n-1|^2(kR)^4$ in the small scatterer size limit $kR \ll 1$. In wave scattering, the conventional wisdom is then that a weak scatterer leads to a small cross section and, consequently, to weak scattering, and this holds regardless of nature of the scattering particle/wave, i.e., vector, scalar or spinor.

Superscattering of pseudospin-1 wave defies exactly the conventional wisdom [53]. The striking and counterintuitive phenomenon is that extraordinarily strong scattering can emerge from arbitrarily weak scatterers at sufficiently low energies (i.e., in the deep subwavelength regime). Accompanying this phenomenon is a novel type of resonances that can persist at low energies for weak scatterers. An analytic understanding of the resonance was obtained [53] and the resulting cross section was derived, with excellent agreement with results from direct numerical simulations.

13.5 Non-equilibrium Transport in the Pseudospin-1 Dirac-Weyl System

Quantum transport beyond the linear response and equilibrium regime is of great practical importance, especially in device research and development. There have been studies of nonlinear and non-equilibrium transport of relativistic pseudospin-1/2 particles in Dirac and Weyl materials. For example, when graphene is subject to a constant electric field, the dynamical evolution of the current after the field is turned on exhibits a remarkable minimal conductivity behavior [70]. The scaling behavior of nonlinear electric transport in graphene due to the dynamical Landau–Zener tunneling or the Schwinger pair creation mechanism has also been investigated [71, 72]. Under a strong electrical field, due to the Landau–Zener transition, a topological insulator or graphene can exhibit a quantization breakdown phenomenon in the spin Hall conductivity [73]. In addition, non-equilibrium electric transport beyond the linear response regime in 3D Weyl semimetals has been studied [74]. In these works, the quasiparticles are relativistic pseudospin-1/2 fermions arising from the Dirac or Weyl system with a conical type of dispersion in their energy momentum spectrum.

Recently, the transport dynamics of pseudospin-1 quasiparticles were studied [75]. Under the equilibrium condition and in the absence of disorders, the flat band acts as a perfect "caging" of carriers with zero group velocity and hence it contributes little to the conductivity [43, 76, 77]. However, the flat band can have a significant effect on the non-equilibrium transport dynamics. Through numerical and analytic calculation of the current evolution for both weak and strong electric fields, it was found [75] that the general phenomenon can arise of current enhancement as compared with that associated with non-equilibrium transport of pseudospin-1/2 particles. In particular, for a weak field, the interband current is twice as large as that for pseudospin-1/2 system due to the interference between particles from the flat band and from the negative band, the scaling behavior of which agrees with that determined by the Kubo formula. For a strong field, the intraband current is $\sqrt{2}$ times larger than that in the pseudospin-1/2 system, as a result of the additional contribution from the particles residing in the flat band. In this case, the physical origin of the scaling behavior of the current-field relation can be attributed to Landau–Zener tunneling. These findings suggested that, in general, the conductivity of pseudospin-1 materials can be higher than that of pseudospin-1/2 materials in the nonequilibrium transport regime. Indeed, the interplay between the flat band and the Dirac cones can lead to interesting physics that has just begun to be understood and exploited.

13.6 Discussion: Relativistic Quantum Chaos in Pseudospin-1 Systems

The field of quantum chaos aims to uncover the quantum manifestations or fingerprints of classical chaotic behaviors in the semiclassical limit [78, 79]. A vast majority of the works were for nonrelativistic quantum systems described by the Schrödinger equation. Recent years have witnessed a rapid development of Dirac materials [80, 81] such as graphene and topological insulators, which are described by the Dirac equation in relativistic quantum mechanics. A new field has thus emerged: relativistic quantum chaos [82, 83]. To study the unique physics of classical chaos in relativistic quantum systems is fundamental with potentially significant applications.

Existing works on relativistic quantum chaos [82, 83] focused on pseudospin-1/2 systems such as graphene, which are described by the conventional Dirac equation for two-component spinors. Pseudospin-1 systems, due to their unusual physics, can present a new platform to study relativistic quantum chaos. A technical difficulty that must be overcome is to solve the generalized Dirac-Weyl equation for three-component spinors in arbitrary geometrical domains that generate classical chaos. For example, while scattering of pseudospin-1 particles from a circular potential can be analytically solved [52], at the present there exists no method to solve the scattering problem for a chaotic geometry, e.g., a stadium shaped potential. At the time of writing, author's group is developing a multiple multipole technique to solve the generalized Dirac-Weyl equation for pseudospin-1 system with any given piecewise homogeneous potential, where the multipoles (or "fictitious" sources) are defined in terms of the analytic three-component spinor cylindrical wave basis of eigen-solutions in each sub-region separated by the potential boundaries. In addition, a wave-function matching based scattering matrix approach is being developed to deal with potential of the eccentric annular shape. Both methods are semi-analytic, while the former is more powerful for near-field calculations and is in principle applicable to arbitrary shape of the scattering potential. Preliminary studies have revealed that the methods are highly efficient and accurate, enabling unexpected phenomena to be uncovered such as the existence of an energy range in which pseudospin-1 chaotic cavities defy well known phenomena in quantum chaos such as Q-spoiling [84–86]. It is likely that uncovering, understanding, and exploiting the interplay between pseudospin-1 physics and classical chaos can represent a new frontier in relativistic quantum chaos.

Acknowledgments. This Review is based on Refs. [52–54]. I thank my former student and current post-doctoral fellow Dr. H.-Y. Xu - the main contributor of these works. I would like to acknowledge support from the Pentagon Vannevar Bush Faculty Fellowship program sponsored by the Basic Research Office of the Assistant Secretary of Defense for Research and Engineering and funded by the Office of Naval Research through Grant No. N00014-16-1-2828.

References

1. K.S. Novoselov et al., Electric field effect in atomically thin carbon films. Science **306**, 666–669 (2004)
2. C. Berger et al., Ultrathin epitaxial graphite: 2D electron gas properties and a route toward graphene-based nanoelectronics. J. Phys. Chem. B **108**, 19912–19916 (2004)
3. T. Wehling, A. Black-Schaffer, A. Balatsky, Dirac materials. Adv. Phys. **63**, 1–76 (2014)
4. J. Wang, S. Deng, Z. Liu, Z. Liu, The rare two-dimensional materials with Dirac cones. Natl. Sci. Rev. **2**(1), 22–39 (2015)
5. M.Z. Hasan, C.L. Kane, Colloquium: topological insulators. Rev. Mod. Phys. **82**, 3045–3067 (2010)
6. X.-L. Qi, S.-C. Zhang, Topological insulators and superconductors. Rev. Mod. Phys. **83**, 1057–1110 (2011)
7. X.-L. Qi, T.L. Hughes, S.-C. Zhang, Topological field theory of time-reversal invariant insulators. Phys. Rev. B **78**, 195424 (2008)

8. A.M. Essin, J.E. Moore, D. Vanderbilt, Magnetoelectric polarizability and axion electrodynamics in crystalline insulators. Phys. Rev. Lett. **102**, 146805 (2009)
9. C.-Z. Chang et al., Zero-field dissipationless chiral edge transport and the nature of dissipation in the quantum anomalous hall state. Phys. Rev. Lett. **115**, 057206 (2015)
10. Y.H. Wang et al., Observation of chiral currents at the magnetic domain boundary of a topological insulator. Science **349**, 948–952 (2015)
11. M.C. Rechtsman et al., Topological creation and destruction of edge states in photonic graphene. Phys. Rev. Lett. **111**, 103901 (2013)
12. Y. Plotnik et al., Observation of unconventional edge states in photonic graphene. Nat. Mater. **13, 57–62**, (2014) (Article)
13. Z. Wang, Y.D. Chong, J.D. Joannopoulos, M. Soljačić, Reflection-free one-way edge modes in a gyromagnetic photonic crystal. Phys. Rev. Lett. **100**, 013905 (2008)
14. Z. Wang, Y. Chong, J.D. Joannopoulos, M. Soljacic, Observation of unidirectional backscattering-immune topological electromagnetic states. Nature (London) **461**, 772–775 (2009)
15. M. Hafezi, E.A. Demler, M.D. Lukin, J.M. Taylor, Robust optical delay lines with topological protection. Nat. Phys. **7**, 907–912 (2011)
16. K. Fang, Z. Yu, S. Fan, Realizing effective magnetic field for photons by controlling the phase of dynamic modulation. Nat. Photonics **6**, 782–787 (2012)
17. A.B. Khanikaev et al., Photonic topological insulators. Nat. Mater. **12**, 233–239 (2013)
18. L. Lu, J.D. Joannopoulos, M. Soljaclc, Topological photonics. Nat. Photonics **8**, 821–829 (2014)
19. X. Huang, Y. Lai, Z.H. Hang, H. Zheng, C.T. Chan, Dirac cones induced by accidental degeneracy in photonic crystals and zero-refractive-index materials. Nat. Mater. **10**, 582–586 (2011)
20. J. Mei, Y. Wu, C.T. Chan, Z.-Q. Zhang, First-principles study of Dirac and Dirac-like cones in phononic and photonic crystals. Phys. Rev. B **86**, 035141 (2012)
21. P. Moitra et al., Realization of an all-dielectric zero-index optical metamaterial. Nat. Photonics **7**, 791–795 (2013)
22. Y. Li et al., On-chip zero-index metamaterials. Nat. Photonics **9**, 738–742 (2015)
23. A. Fang, Z.Q. Zhang, S.G. Louie, C.T. Chan, Klein tunneling and supercollimation of pseudospin-1 electromagnetic waves. Phys. Rev. B **93**, 035422 (2016)
24. D. Guzmán-Silva et al., Experimental observation of bulk and edge transport in photonic Lieb lattices. New J. Phys. **16**, 063061 (2014)
25. S. Mukherjee et al., Observation of a localized flat-band state in a photonic Lieb lattice. Phys. Rev. Lett. **114**, 245504 (2015)
26. R.A. Vicencio et al., Observation of localized states in Lieb photonic lattices. Phys. Rev. Lett. **114**, 245503 (2015)
27. F. Diebel, D. Leykam, S. Kroesen, C. Denz, A.S. Desyatnikov, Conical diffraction and composite Lieb bosons in photonic lattices. Phys. Rev. Lett. **116**, 183902 (2016)
28. S. Taie et al., Coherent driving and freezing of bosonic matter wave in an optical Lieb lattice. Sci. Adv. **1**, e1500854 (2015)
29. M. Rizzi, V. Cataudella, R. Fazio, Phase diagram of the Bose-Hubbard model with T_3 symmetry. Phys. Rev. B **73**, 144511 (2006)
30. A.A. Burkov, E. Demler, Vortex-peierls states in optical lattices. Phys. Rev. Lett. **96**, 180406 (2006)
31. D. Bercioux, D.F. Urban, H. Grabert, W. Häusler, Massless Dirac-Weyl fermions in a T_3 optical lattice. Phys. Rev. A **80**, 063603 (2009)

32. B. Dóra, J. Kailasvuori, R. Moessner, Lattice generalization of the Dirac equation to general spin and the role of the flat band. Phys. Rev. B **84**, 195422 (2011)
33. A. Raoux, M. Morigi, J.-N. Fuchs, F. Piéchon, G. Montambaux, From dia- to paramagnetic orbital susceptibility of massless fermions. Phys. Rev. Lett. **112**, 026402 (2014)
34. T. Andrijauskas et al., Three-level Haldane-like model on a dice optical lattice. Phys. Rev. A **92**, 033617 (2015)
35. F. Wang, Y. Ran, Nearly flat band with Chern number $c = 2$ on the dice lattice. Phys. Rev. B **84**, 241103 (2011)
36. J. Wang, H. Huang, W. Duan, Z. Liu, Identifying Dirac cones in carbon allotropes with square symmetry. J. Chem. Phys. **139**, 184701 (2013)
37. W. Li, M. Guo, G. Zhang, Y.-W. Zhang, Gapless MoS_2 allotrope possessing both massless Dirac and heavy fermions. Phys. Rev. B **89**, 205402 (2014)
38. J. Romhanyi, K. Penc, R. Ganesh, Hall effect of triplons in a dimerized quantum magnet. Nat. Commun. **6**, 6805 (2015)
39. G. Giovannetti, M. Capone, J. van den Brink, C. Ortix, Kekulé textures, pseudospin-one Dirac cones, and quadratic band crossings in a graphene-hexagonal indium chalcogenide bilayer. Phys. Rev. B **91**, 121417 (2015)
40. G.-L. Wang, H.-Y. Xu, Y.-C. Lai, Mechanical topological semimetals with massless quasiparticles and a finite berry curvature. Phys. Rev. B **95**, 235159 (2017)
41. R. Shen, L.B. Shao, B. Wang, D.Y. Xing, Single Dirac cone with a flat band touching on line-centered-square optical lattices. Phys. Rev. B **81**, 041410 (2010)
42. D.F. Urban, D. Bercioux, M. Wimmer, W. Häusler, Barrier transmission of Dirac-like pseudospin-one particles. Phys. Rev. B **84**, 115136 (2011)
43. M. Vigh et al., Diverging dc conductivity due to a flat band in a disordered system of pseudospin-1 Dirac-Weyl fermions. Phys. Rev. B **88**, 161413 (2013)
44. J.T. Chalker, T.S. Pickles, P. Shukla, Anderson localization in tight-binding models with flat bands. Phys. Rev. B **82**, 104209 (2010)
45. J.D. Bodyfelt, D. Leykam, C. Danieli, X. Yu, S. Flach, Flatbands under correlated perturbations. Phys. Rev. Lett. **113**, 236403 (2014)
46. E.H. Lieb, Two theorems on the Hubbard model. Phys. Rev. Lett. **62**, 1201–1204 (1989)
47. H. Tasaki, Ferromagnetism in the Hubbard models with degenerate single-electron ground states. Phys. Rev. Lett. **69**, 1608–1611 (1992)
48. H. Aoki, M. Ando, H. Matsumura, Hofstadter butterflies for flat bands. Phys. Rev. B **54**, R17296–R17299 (1996)
49. C. Weeks, M. Franz, Topological insulators on the Lieb and perovskite lattices. Phys. Rev. B **82**, 085310 (2010)
50. N. Goldman, D.F. Urban, D. Bercioux, Topological phases for fermionic cold atoms on the Lieb lattice. Phys. Rev. A **83**, 063601 (2011)
51. J. Vidal, R. Mosseri, B. Douçot, Aharonov-Bohm cages in two-dimensional structures. Phys. Rev. Lett. **81**, 5888–5891 (1998)
52. H.-Y. Xu, Y.-C. Lai, Revival resonant scattering, perfect caustics, and isotropic transport of pseudospin-1 particles. Phys. Rev. B **94**, 165405 (2016)
53. H.-Y. Xu, Y.-C. Lai, Superscattering of a pseudospin-1 wave in a photonic lattice. Phys. Rev. A **95**, 012119 (2017)
54. H.-Y. Xu, L. Huang, D. Huang, Y.-C. Lai, Geometric valley Hall effect and valley filtering through a singular Berry flux. Phys. Rev. B **96**, 045412 (2017)
55. M.I. Katsnelson, K.S. Novoselov, A.K. Geim, Chiral tunnelling and the Klein paradox in graphene. Nat. Phys. **2**, 620–625 (2006)

56. D.S. Novikov, Elastic scattering theory and transport in graphene. Phys. Rev. B **76**, 245435 (2007)
57. M.I. Katsnelson, F. Guinea, A.K. Geim, Scattering of electrons in graphene by clusters of impurities. Phys. Rev. B **79**, 195426 (2009)
58. J.-S. Wu, M.M. Fogler, Scattering of two-dimensional massless Dirac electrons by a circular potential barrier. Phys. Rev. B **90**, 235402 (2014)
59. J. Cserti, A. Pályi, C. Péterfalvi, Caustics due to a negative refractive index in circular graphene *p*-n junctions. Phys. Rev. Lett. **99**, 246801 (2007)
60. R.L. Heinisch, F.X. Bronold, H. Fehske, Mie scattering analog in graphene: Lensing, particle confinement, and depletion of Klein tunneling. Phys. Rev. B **87**, 155409 (2013)
61. M.M. Asmar, S.E. Ulloa, Rashba spin-orbit interaction and birefringent electron optics in graphene. Phys. Rev. B **87**, 075420 (2013)
62. B. Liao, M. Zebarjadi, K. Esfarjani, G. Chen, Isotropic and energy-selective electron cloaks on graphene. Phys. Rev. B **88**, 155432 (2013)
63. M.M. Asmar, S.E. Ulloa, Spin-orbit interaction and isotropic electronic transport in graphene. Phys. Rev. Lett. **112**, 136602 (2014)
64. A. Ferreira, T.G. Rappoport, M.A. Cazalilla, A.H. Castro Neto, Extrinsic spin Hall effect induced by resonant skew scattering in graphene. Phys. Rev. Lett. **112**, 066601 (2014)
65. Y. Zhao et al., Creating and probing electron whispering-gallery modes in graphene. Science **348**, 672–675 (2015)
66. W.S. Bakr, J.I. Gillen, A. Peng, S. Folling, M. Greiner, A quantum gas microscope for detecting single atoms in a Hubbard-regime optical lattice. Nature **462**, 74–77 (2009)
67. Jin, D., et al., Topological magnetoplasmon (2016). arXiv:1602.00553
68. L.I. Schiff, *Quantum Mechanics*, 3rd edn. (McGraw-Hill, New York, 1968)
69. R. Newton, *Scattering Theory of Waves and Particles.* Dover Books on Physics (Dover Publications, New York, 1982)
70. M. Lewkowicz, B. Rosenstein, Dynamics of particle-hole pair creation in graphene. Phys. Rev. Lett. **102**, 106802 (2009)
71. B. Rosenstein, M. Lewkowicz, H.-C. Kao, Y. Korniyenko, Ballistic transport in graphene beyond linear response. Phys. Rev. B **81**, 041416 (2010)
72. B. Dóra, R. Moessner, Nonlinear electric transport in graphene: quantum quench dynamics and the Schwinger mechanism. Phys. Rev. B **81**, 165431 (2010)
73. B. Dóra, R. Moessner, Dynamics of the spin Hall effect in topological insulators and graphene. Phys. Rev. B **83**, 073403 (2011)
74. S. Vajna, B. Dóra, R. Moessner, Nonequilibrium transport and statistics of Schwinger pair production in Weyl semimetals. Phys. Rev. B **92**, 085122 (2015)
75. C.-Z. Wang, H.-Y. Xu, L. Huang, Y.-C. Lai, Nonequilibrium transport in the pseudospin-1 Dirac-Weyl system. Phys. Rev. B **96**, 115440 (2017)
76. W. Häusler, Flat-band conductivity properties at long-range Coulomb interactions. Phys. Rev. B **91**, 041102 (2015)
77. T. Louvet, P. Delplace, A.A. Fedorenko, D. Carpentier, On the origin of minimal conductivity at a band crossing. Phys. Rev. B **92**, 155116 (2015)
78. H.-J. Stöckmann, *Quantum Chaos: An Introduction* (Cambridge University Press, New York, 1999)
79. Haake, F. *Quantum Signatures of Chaos*, 3rd edn.. Springer Series in Synergetics (Springer, Berlin, 2010)
80. A.H.C. Neto, K. Novoselov, Two-dimensional crystals: beyond graphene. Mater. Exp. **1**, 10–17 (2011)

81. P. Ajayan, P. Kim, K. Banerjee, Two-dimensional van der Waals materials. Phys. Today **69**, 38–44 (2016)
82. Y.-C. Lai, L. Huang, H.-Y. Xu, C. Grebogi, Relativistic quantum chaos - an emergent interdisciplinary field. Chaos **28**, 052101 (2018)
83. L. Huang, H.-Y. Xu, C. Grebogi, Y.-C. Lai, Relativistic quantum chaos. Phys. Rep. **753**, 1–128 (2018)
84. A. Mekis, J.U. Nöckel, G. Chen, A.D. Stone, R.K. Chang, Ray chaos and Q spoiling in lasing droplets. Phys. Rev. Lett. **75**, 2682–2685 (1995)
85. J.U. Nöckel, A.D. Stone, Ray and wave chaos in asymmetric resonant optical cavities. Nature **385**, 45–47 (1997)
86. C. Gmachl et al., High-power directional emission from microlasers with chaotic resonators. Science **280**, 1556–1564 (1998)

Chapter 14
Revealing Network Symmetries Using Time-Series Data

Ethan T.H.A. van Woerkom[1], Joseph D. Hart[2,3](✉), Thomas E. Murphy[2,4], and Rajarshi Roy[2,3,5]

[1] School of Physics and Astronomy, University of Edinburgh, James Clerk Maxwell Building, Peter Guthrie Tait Road, Edinburgh EH9 3FD, UK
ethanvanwoerkom@yahoo.com
[2] Institute for Research in Electronics and Applied Physics, University of Maryland, College Park, MD 20742, USA
{jhart12,tem,rroy}@umd.edu
[3] Department of Physics, University of Maryland, College Park, MD 20742, USA
[4] Department of Electrical and Computer Engineering, University of Maryland, College Park, MD 20742, USA
[5] Institute for Physical Science and Technology, University of Maryland, College Park, MD 20742, USA

Abstract. Complex dynamical networks may exhibit graph symmetries. These symmetries leave an imprint on network behaviour and statistics. This effect is first demonstrated in a small opto-electronic network. We then present the general conditions under which network statistics become invariant under the action of network symmetries. Statistical analyses can help reveal the symmetry group of a network graph without knowledge of the underlying network model. Finally, results from numerical experiments additionally demonstrate this.

14.1 Introduction

Network symmetries exist in many networks and give essential information about their structure. The difficult problem of reconstructing the graph of a complex network by studying its dynamics has been studied before [1,2]. In this article we demonstrate a tool that instead of resolving individual network connections, reconstructs the symmetry group of a network using time-series statistics. This article details how network symmetries manifest themselves in network behaviour and time-series statistics and gives methods to infer network symmetries from statistics in the general case. In this article, unless otherwise stated, we study a dynamical system as in Definition 1. In this article we study networks which contain symmetries. Definition 2 details the conditions for such a symmetry.

V. In et al. (Eds.): *Proceedings of the 5th International Conference on Applications in Nonlinear Dynamics*, Understanding Complex Systems, https://doi.org/10.1007/978-3-030-10892-2_14

Definition 1 A dynamical network is defined by the dynamics of N nodes, each represented by state vectors, adhering to the following equation:

$$\dot{\boldsymbol{x}}_i = F_i(\boldsymbol{x}_1(t), \ldots, \boldsymbol{x}_N(t)) \quad \text{for} \quad i \in \{1, \ldots, N\}. \tag{14.1}$$

Here, $\boldsymbol{x}_i$ are the state vectors of the nodes in the dynamical system and $F_i(\boldsymbol{x}_1, \ldots, \boldsymbol{x}_N)$ represents the functions that give the derivative of each node as a function of all other nodes.

Definition 2 A dynamical system is defined to have symmetry g if $F_{g(i)}(\boldsymbol{x}_1, \ldots, \boldsymbol{x}_N) = F_i(\boldsymbol{x}_{g(1)}, \ldots, \boldsymbol{x}_{g(N)})$, where g is a permutation $g : \{1, \ldots, N\} \to \{1, \ldots, N\}$.

A consequence of the above definition of a symmetry of a dynamical system is that if there is a solution $\boldsymbol{s}_i(t)$, then $\boldsymbol{s}_{g(i)}(t)$ is also a solution.

This article is organised as follows. In Sect. 14.2 we discuss an opto-electronic network experiment which demonstrates symmetries in time-averaged behaviour. In Sect. 14.3 we present a theorem stating the conditions required for symmetries in network dynamics to appear in network statistics. We give a more elaborate example of the consequences of this theorem in Sect. 14.4. In Sect. 14.5 we conclude and discuss the possibility of using the presented methods to retrieve general network symmetries.

14.2 Statistical Symmetries in a Small Opto-Electronic Network

In this section we detail an experiment with a small opto-electronic network that demonstrates how network graph symmetries affect network behaviour and present themselves in network statistics.

14.2.1 Experimental Setup

A network of four coupled opto-electronic time-delayed feedback systems is used [3]. Each of these systems, 'nodes', depicted in Fig. 14.1, consists of a laser diode which passes a light signal through an integrated Mach–Zehnder modulator, altering the intensity of the signal with a $\cos^2(x + \phi_0)$ nonlinearity, where x is the normalised input voltage to the modulator. This signal is then passed on to other nodes through optical fibres and returned for self-feedback.

The two input signals, being the self-feedback signal and the inputs from the other nodes respectively, are measured in separate photoreceivers. A Digital Signal Processing (DSP) board is then used to apply a feedback and coupling delay, apply a digital filter, and amplify the signal. The signal is then fed back into the Mach–Zehnder modulator. In this way, a coupled nonlinear chaotic oscillator with time delays is produced incorporating both coupling and self-feedback time delays. A two-pole digital Butterworth filter is used to filter the signal, with a high-pass frequency of $\omega_H/2\pi = 100$ Hz and a low-pass frequency

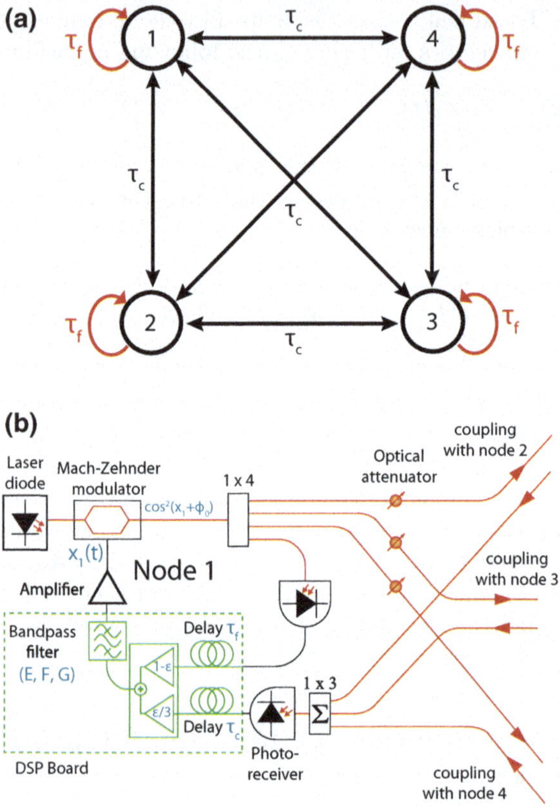

Fig. 14.1. a Possible connections in network, with self-feedback time delays τ_f and coupling time delays τ_c included. **b** Schematic of an opto-electronic node with connections to neighbours. Red connections are through optical fibres, whereas black connections are electronic. Figure from [3]

of $\omega_L/2\pi = 2.5$ kHz, operating at a sampling rate of 24 kSamples/s. Connections between nodes can be controlled by variable fibre-based attenuators.

Equations 14.2–14.5 well describe the dynamics of the network. In reality, due to the digital sampling of the DSP board, the system dynamics is partitioned into discrete time steps. In these equations, $\boldsymbol{u}_i$ represents the state of the digital filter corresponding to each node, ε the coupling strength (ranging from 0-1), and β the round-trip gain. Furthermore, τ_f is the self-feedback time delay of each node, whereas τ_c is the coupling time delay, which controls the input delay from other nodes. The DC offset of the Mach-Zehnder modulator is given by ϕ_0, and ω_H and ω_L are the high-pass and low-pass filter constants respectively. Finally, A_{ij} is the network connectivity matrix and n^i_{in} is the number of input nodes per node.

$$\dot{\boldsymbol{u}}_i(t) = \boldsymbol{E}\boldsymbol{u}_i(t) - \boldsymbol{F}\beta\cos(x_i(t) + \phi_0), \tag{14.2}$$

$$x_i(t) = \boldsymbol{G}\Big(\boldsymbol{u}_i(t-\tau_f) + \frac{\varepsilon}{n_{in}^i}\sum_j A_{ij}(\boldsymbol{u}_j(t-\tau_c) - \boldsymbol{u}_i(t-\tau_f))\Big), \tag{14.3}$$

where,

$$\boldsymbol{E} = \begin{bmatrix} -(\omega_L+\omega_H) & -\omega_L \\ \omega_H & 0 \end{bmatrix}, \boldsymbol{F} = \begin{bmatrix} \omega_L \\ 0 \end{bmatrix}, \boldsymbol{G} = \begin{bmatrix} 1 \; 0 \end{bmatrix},$$

$$\tau_f = \tau_c = 1.9 \text{ ms}, \quad \phi_0 = \pi/4, \quad \omega_H/2\pi = 100 \text{ Hz}, \quad \text{and} \quad \omega_L/2\pi = 2.5 \text{ kHz}.$$

14.2.2 Method

Trials on two different network configurations were conducted: the bidirectionally coupled star and chain networks in Fig. 14.2a, c. Consecutive runs were done on each network configuration, where ε, the coupling strength was increased in steps of 0.025, from 0 to 1. Measurement runs of length 2 s were recorded on an oscilloscope. During the first 0.5 s, the nodes were allowed to oscillate with only self-feedback, and no coupling, in order to set them in a random, independent state. Coupling was then enabled and the next 0.1 s of data discarded. The remaining 1.4 s were used for data analysis.

14.2.3 Results

Root mean square differences between nodes were calculated as a function of ε for each possible combination of nodes in both networks ($\sqrt{\langle||\boldsymbol{x}_i - \boldsymbol{x}_j||^2\rangle}$ for $i < j$) and then plotted in Fig. 14.2b, d. As is visible, the star network achieved synchronisation between outer nodes for high values of coupling, whereas the chain network did not synchronise in any way, since the difference did not approach zero.

It can be seen in Fig. 14.2b that the RMS differences for node combinations 2-1, 2-3 and 2-4, and 1-3, 3-4 and 4-1 line up for all values of ε. This is due to the fact that there exist symmetries in the graph permuting these nodes to each other, since interchanging the 'arms' of the star graph does not alter the topology of the graph. Therefore one would expect that their general behaviour, and so any generic statistics, such as the one used here, would line up and give equal results. If these results were not equal, then that would suggest that symmetry had been broken in the experiment.

Similarly, it can be seen that in the chain network diagram 1-4 and 3-4 line up, as well as 1-3 and 2-4. This is due to the fact that the reflection symmetry in the chain graph permutes nodes 1 and 2 to 4 and 3, meaning that they have similar behaviour, and so one expects $RMS_{12} = RMS_{43} = RMS_{34}$. For the same reason, $RMS_{13} = RMS_{24}$. Combinations 1-4 and 2-3 can not be permuted to any other combination of nodes without changing the graph topology, and so they stand alone and do not cluster. We have now effectively identified the orbits of all two-node combinations under the action of the symmetry group in this graph.

These results may seem trivial. However, they are not. Many factors exist in this real-world setup that break the symmetry of the network. The round-trip

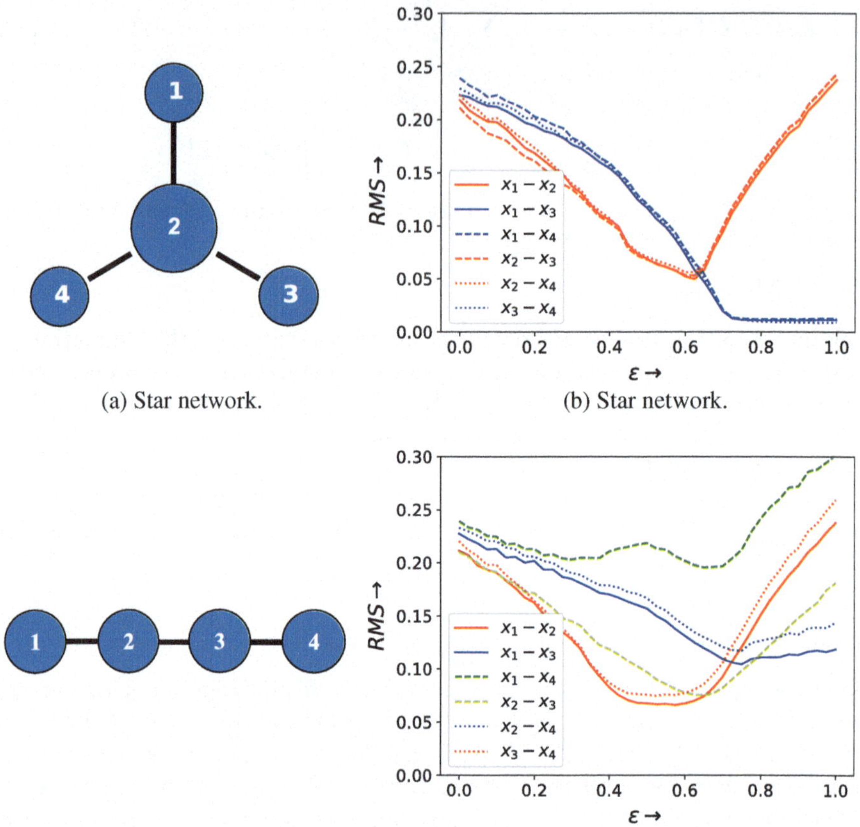

(a) Star network. (b) Star network.

(c) Chain network. (d) Chain network.

Fig. 14.2. **a**, **c** Networks used in experimental setup. **b**, **d** RMS difference calculated for each possible node combination as a function of coupling strength ε

gain β varies slightly in the different systems and can not be fixed exactly. The phase of the Mach–Zehnder modulators may shift slightly from the rest position and often needs to be recalibrated. Different lasers operate at slightly different intensities. These are all reasons for why the systems used in this setup are nominally homogeneous, but in reality only approximately the same. It can be concluded from this that network graph symmetries can robustly present themselves as symmetries in the statistics of real-world experimental setups where different factors break exact symmetries.

14.3 Statistical Symmetries in the General Case

The previous section shows that network symmetries can cause symmetries in statistical data to arise. In this section we will detail which conditions are necessary for this to happen in the general case. This effort culminates in the 'Main

Theorem' presented at the end of this section. We make use in the following of a so-called statistic of shape $S(\boldsymbol{a}_1, \ldots, \boldsymbol{a}_N) = H(\boldsymbol{s}_1(t), \ldots, \boldsymbol{s}_N(t))$. H can be imagined to be any calculation of the properties of a solution $s_i(t)$, such as the root mean square of the first node: $H(\boldsymbol{s}_1(t), \ldots, \boldsymbol{s}_N(t)) = \sqrt{\langle ||\boldsymbol{s}_1||^2 \rangle}$, or a cross-correlation between two nodes. First we define when a symmetry is present in a statistic using Definition 3.

Definition 3 Define as a statistic a function $S(\boldsymbol{a}_1, \ldots, \boldsymbol{a}_N) = H(\boldsymbol{s}_1(t), \ldots, \boldsymbol{s}_N(t))$ that when applied to a particular initial condition $(\boldsymbol{a}_1, \ldots, \boldsymbol{a}_N)$, applies a function H to the corresponding solution of the initial conditions $(\boldsymbol{s}_1(t), \ldots, \boldsymbol{s}_N(t))$. Call a statistic S invariant under the action of symmetry g when it is averaged over some distribution of initial conditions p, if:

$$\int S(\boldsymbol{x}_1, \ldots, \boldsymbol{x}_N) p(\boldsymbol{x}_1, \ldots, \boldsymbol{x}_N)(dx)^{n \times N} = \int S(\boldsymbol{x}_{g(1)}, \ldots, \boldsymbol{x}_{g(N)}) p(\boldsymbol{x}_1, \ldots, \boldsymbol{x}_N)(dx)^{n \times N}. \tag{14.4}$$

In order to prove the main theorem we must first prove the following Lemmas 1 and 2.

Lemma 1 (Symmetries in initial conditions continue down the line)
Let $\phi(\boldsymbol{x}_1, \ldots, \boldsymbol{x}_N)$ be a function of the variables of a dynamical system with symmetry g. Examine the initial conditions $\mathbf{c}_i$ and $\mathbf{c}_{g(i)}$, with $\boldsymbol{s}_i(t)$ and $\boldsymbol{z}_i(t)$ as respective solutions. Then:

- $\boldsymbol{z}_i(t) = \boldsymbol{s}_{g(i)}(t)$
- $\phi(\boldsymbol{z}_{g^{-1}(1)}(t), \ldots, \boldsymbol{z}_{g^{-1}(N)}(t)) = \phi(\boldsymbol{s}_1(t), \ldots, \boldsymbol{s}_N(t))$

Proof Due to the symmetry of the system, $\boldsymbol{s}_{g(i)}(t)$ is also a solution. Since $\boldsymbol{s}_{g(i)}(0) = \mathbf{c}_{g(i)} = \boldsymbol{z}_i(0)$, it follows from the uniqueness theorem that $\boldsymbol{s}_{g(i)}(t) = \boldsymbol{z}_i(t)$ and so, inserting $g^{-1}(i)$, we get $\phi(\boldsymbol{z}_{g^{-1}(1)}(t), \ldots, \boldsymbol{z}_{g^{-1}(N)}(t)) = \phi(\boldsymbol{s}_1(t), \ldots, \boldsymbol{s}_N(t))$.

□

Lemma 2 *Let $\boldsymbol{a}_i$ be the initial conditions for a dynamical system with symmetry g and corresponding solution $\boldsymbol{s}_i(t)$. Let $\mathbf{b}_i = \boldsymbol{a}_{g(i)}$ and $\boldsymbol{z}_i(t)$ be the corresponding solution to the initial conditions $\mathbf{b}_i$. Define the statistics $S(\boldsymbol{a}_1, \ldots, \boldsymbol{a}_N) = H(\boldsymbol{s}_1(t), \ldots, \boldsymbol{s}_N(t))$ and $T(\boldsymbol{b}_1, \ldots, \boldsymbol{b}_N) = H(\boldsymbol{z}_{g^{-1}(1)}(t), \ldots, \boldsymbol{z}_{g^{-1}(N)}(t))$.
Then $S(\boldsymbol{a}_1, \ldots, \boldsymbol{a}_N) = T(\boldsymbol{b}_1, \ldots, \boldsymbol{b}_N)$.*

Proof

$$\begin{aligned} T(\boldsymbol{b}_1, \ldots, \boldsymbol{b}_N) &= H(\boldsymbol{z}_{g^{-1}(1)}(t), \ldots, \boldsymbol{z}_{g^{-1}(N)}(t)) \\ &= H(\boldsymbol{s}_1(t), \ldots, \boldsymbol{s}_N(t)) = S(\boldsymbol{a}_1, \ldots, \boldsymbol{a}_N). \end{aligned} \tag{14.5}$$

□

Main Theorem
Let $p(\boldsymbol{x}_1,\ldots,\boldsymbol{x}_N)$ be a distribution over the possible initial conditions. Let statistics $S(\boldsymbol{a}_1,\ldots,\boldsymbol{a}_N) = H(\boldsymbol{s}_1(t),\ldots,\boldsymbol{s}_N(t))$ and $T(\boldsymbol{b}_1,\ldots,\boldsymbol{b}_N) = H(\boldsymbol{z}_{g^{-1}(1)}(t),\ldots,\boldsymbol{z}_{g^{-1}(N)}(t))$. If p is invariant under the symmetry g, that is, $p(\boldsymbol{x}_1,\ldots,\boldsymbol{x}_N) = p(\boldsymbol{x}_{g(1)},\ldots,\boldsymbol{x}_{g(N)})$, then:

$$\int_{\mathbb{R}^{N\times n}} p(\boldsymbol{x}_1,\ldots,\boldsymbol{x}_N)S(\boldsymbol{x}_1,\ldots,\boldsymbol{x}_N)(dx)^{N\times n} = \int_{\mathbb{R}^{N\times n}} p(\boldsymbol{x}_1,\ldots,\boldsymbol{x}_N)T(\boldsymbol{x}_1,\ldots,\boldsymbol{x}_N)(dx)^{N\times n}. \tag{14.6}$$

In other words, if a network has symmetry g and its initial conditions are invariant under g, then the statistic over the initial conditions will also be invariant under the action of g when averaged using distribution p.

Proof

$$\begin{aligned} &\int_{\mathbb{R}^{N\times n}} p(\boldsymbol{x}_1,\ldots,\boldsymbol{x}_N)T(\boldsymbol{x}_1,\ldots,\boldsymbol{x}_N)(dx)^{N\times n} \qquad (14.7)\\ &= \int_{\mathbb{R}^{N\times n}} p(\boldsymbol{y}_{g(1)},\ldots,\boldsymbol{y}_{g(N)})T(\boldsymbol{y}_{g(1)},\ldots,\boldsymbol{y}_{g(N)})(dy)^{N\times n} \; * \\ &= \int_{\mathbb{R}^{N\times n}} p(\boldsymbol{y}_1,\ldots,\boldsymbol{y}_N)S(\boldsymbol{y}_1,\ldots,\boldsymbol{y}_N)(dy)^{N\times n} \\ &= \int_{\mathbb{R}^{N\times n}} p(\boldsymbol{x}_1,\ldots,\boldsymbol{x}_N)S(\boldsymbol{x}_1,\ldots,\boldsymbol{x}_N)(dx)^{N\times n}. \end{aligned}$$

* Here the change of variables $\boldsymbol{x}_i = \boldsymbol{y}_{g(i)}$ has been used, which has $|J| = 1$. □

This theorem is the main result of this article. It states that if an experiment on a network with a symmetry g is done in such a way that the choice of initial conditions for all trials p does not break the symmetry of the experiment, then all statistics which are averaged over the initial conditions will also be invariant under g. If the system is ergodic then the conditions can be much more lax: any time-averaged statistic will be invariant under g when averaged over sufficiently long time-series. In an ergodic system all symmetries will therefore be present in the data of a single run, instead of having to require that the statistics are averaged over the intial conditions.

14.4 Numerical Results

The results from the previous theoretical section have been demonstrated in two different, small opto-electronic networks. The experimental setup is limited to experiments with at most 4 nodes. Simulations allow the results to be demonstrated in larger networks, with more control over the circumstances. The 8-node opto-electronic network from Fig. 14.3a was simulated using Equations 14.2–14.5. Simulation data from 100 trials with randomised initial conditions were combined. The root mean square amplitude ($\sqrt{\langle||\boldsymbol{x}_i||^2\rangle}$) of each node was calculated for every trial simulation of the opto-electronic network, as compared

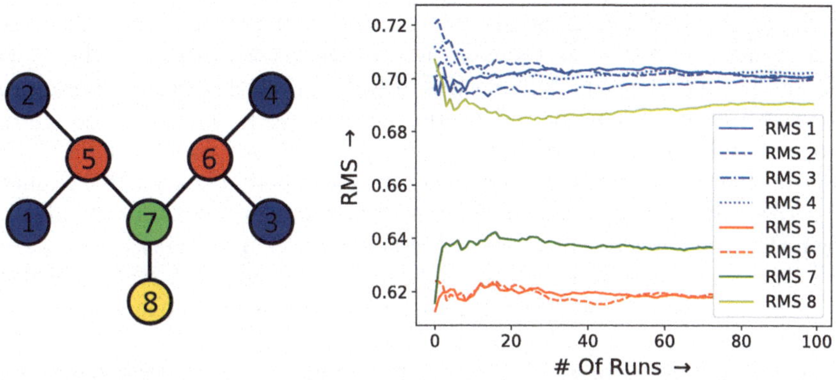

(a) 8-node opto-electronic network

(b) Average RMS after N runs

Fig. 14.3. **a** Network used in simulation trials with nodes grouped into orbits by colour. **b** Combined root mean square amplitudes from 100 trials of 2000 timesteps each

to the root mean square difference in Sect. 14.2. This was done to identify the orbits of the nodes of the graph.

The nodes of any graph can be partitioned into distinct orbits O_i. An orbit O_i is a subset of the nodes of a graph with symmetry group G, where for each node $a, b \in O_i$, $\exists g \in G$, such that $a = g(b)$, and each orbit is closed under the action of the G. As is visible in Fig. 14.3a, there are 4 orbits $\{1, 2, 3, 4\}$, $\{5, 6\}$, $\{7\}$, and $\{8\}$. Based on the main theorem, if network statistics are averaged over initial conditions in such a way that the conditions are invariant under the action of any symmetry, then one would expect these statistics to be invariant under the action of the network symmetries. Polling a network randomly over many different initial conditions is a close approximation to a continuous integral as stated in the main theorem, which is impossible to do in reality. Define a statistic RMS_i, which gives the root mean square of the signal from node i. If this network has symmetry g, then one expects that $RMS_i = RMS_{g(i)}$. This implies that any two nodes in the same orbit will have the same root mean square. We therefore expect in this numerical experiment to see the root mean squares of the separate nodes to cluster along the lines of their respective orbital partitions as a consequence of the main theorem.

The data were found to converge into distinct groups. As is visible in Fig. 14.3b, the four separate orbits of the graph in Fig. 14.3a can clearly be identified. We therefore confirm that the prediction of the main theorem holds in this case.

14.5 Conclusion

The main theorem in this work states that, when sampled under a distribution of initial conditions which is invariant under a symmetry g, any statistics calculated on these data must also be invariant under the action of the symmetry g. Symmetric networks therefore imply symmetric statistics. The converse was not shown to be true. Symmetries in time-series statistics are however still strong indicators of symmetries existing in a network. The work presented in both real and numeric experiments has shown that this result is indeed robust under ordinary real-world circumstances in real experiments, where symmetries are necessarily broken by small differences in the experiment. It has also been shown that the converse also reasonably holds for small networks. The question is whether the converse generally holds given that sufficient statistical testing and comparisons are done, and whether this can easily be extended to larger networks.

Assuming that the presence of network symmetries can be verified with a simple test, then in theory, the symmetry group of a network can be retrieved from experimental data. Since the symmetry group of a network with N nodes has up to $N!$ symmetries, it is unreasonable to check every symmetry. The question therefore arises how many tests need to be done to retrieve the full symmetry group, and what the algorithmic complexity is of this calculation. Regardless of the complexity of identifying the full group, this method can readily identify the orbits under the action of the entire network symmetry group, thereby already giving vital information about the network, and which clusters of synchrony may form [4,5].

Acknowledgments. This paper is a result of work performed as part of the TREND NSF REU program and the University of Maryland, College Park. JDH and RR are thankful for support from ONR Grant No. N000141612481. EvW is thankful for the aid of Olwen Enright and Timea Vitos for constructive input on the manuscript and Keshav Rakesh for his patient aid writing the proofs.

References

1. S. Pajevic, D. Plenz, Efficient network reconstruction from dynamical cascades identifies small-world topology of neuronal avalanches. PLoS Comput. Biol. **5**(1), e1000271 (2009)
2. A. Pikovsky, Reconstruction of a neural network from a time series of firing rates. Phys. Rev. E **93**, 062313 (2016)
3. J.D. Hart, K. Bansal, T.E. Murphy, R. Roy, Experimental observation of chimera and cluster states in a minimal globally coupled network. Chaos **26**, 094801 (2016)
4. A.B. Siddique, L. Pecora, J.D. Hart, F. Sorrentino, Symmetry- and input-cluster synchronization in networks. Phys. Rev. E **97**, 042217 (2018)
5. L.M. Pecora, F. Sorrentino, A.M. Hagerstrom, T.E. Murphy, R. Roy, Cluster synchronization and isolated desynchronization in complex networks with symmetries. Nat. Commun. **5**, 4079 (2014)

Chapter 15
Analysis of Synchronization of Mechanical Metronomes

Tohru Ikeguchi[1(✉)] and Yutaka Shimada[2]

[1] Department of Information and Computer Technology, Tokyo University of Science, 6-3-1 Niijuku, Katsushika, Tokyo 125-8585, Japan
tohru@rs.tus.ac.jp

[2] Department of Information and Computer Sciences, Saitama University, 255 Shimo-ohkubo, Sakura-ku, Saitama 338-8570, Japan
yshimada@mail.saitama-u.ac.jp

Abstract. Synchronization phenomena are ubiquitous around us and are observed in various real systems, for example, hands clapping (rhythmic applause) at the concert hall, light emission of fireflies, callings of frogs, circadian rhythms, pendulum clocks, mechanical metronomes placed on a plate, pedestrians on a suspension bridge, water flowing out of plastic bottles connected by hoses, candle flames fluctuation. In this paper, we focused on the synchronization phenomena observed in a mechanical system: mechanical metronomes on a plate. In particular, we discussed how to construct a mathematical model, or the equations of motion, which describe dynamical behavior of synchronization of mechanical metronomes put on a plate hung by strings. We also investigated their dynamical behavior by solving the equations of motions numerically. In the numerical experiments, parameter values of the equations of motion are experimentally obtained from the experimental equipment. We found that synchronization behavior of mechanical metronomes depends on the following two factors: relation between the frequencies of the metronomes and the plate, and initial angles of the metronomes. We also found the individual difference of the metronomes strongly affects the final behavior. In addition, the results also indicate that if the number of mechanical metronomes increases, it becomes extremely harder to observe the in-phase synchronization until energy applied to the metronomes through spiral springs is exhausted.

15.1 Introduction

If a dynamical system has a unique rhythm such as limit cycles, it is called an autonomous oscillator. In order to create a stable rhythm, the existence of nonlinear dynamics inherent in the system is essential. Therefore, these autonomous oscillators are sometimes referred to as nonlinear oscillators. When

V. In et al. (Eds.): *Proceedings of the 5th International Conference on Applications in Nonlinear Dynamics*, Understanding Complex Systems, https://doi.org/10.1007/978-3-030-10892-2_15

these autonomous nonlinear oscillators are weakly coupled, it is possible to observe an interesting phenomenon called synchronization [1]. For example, synchronized flashing of fireflies [2], frog calls [3], mechanical metronomes [4–7], pedestrians on suspension bridges [8,9], water flowing out of plastic bottles [10,11], flame of candles [12,13], inflow and outflow of salt water in a cup with a small hole at the bottom [14,15] and so on.

In this paper, we focused on the synchronization phenomena observed in mechanical systems, which is movements of mechanical metronomes put on a swinging plate. We analyzed nonlinear dynamical behavior of the mechanical metronomes and their synchronization behavior. In particular, we focused on how to construct a mathematical model, or the equations of motion, which describe dynamical behavior of synchronization of mechanical metronomes put on a plate hung by strings. We also investigated their dynamical behavior by solving the equations of motion numerically. We analyzed the synchronization phenomenon of mechanical metronomes by using a mathematical model of the motion equation. We put several mechanical metronomes on a plate hung by strings. When we conduct numerical simulations, we used the parameter values in the mathematical model estimated from the handmade experimental equipment. As a result, the time required for the in-phase synchronization becomes short when the frequency of the plate becomes large.

15.2 Synchronization of Mechanical Metronomes

The metronome is a typical example of a nonlinear oscillator. We can observe mutual coupling synchronization using the mechanical metronomes. For example, using a swinging plate hung by wires, we can observe synchronization phenomena [4,6]. Alternatively, using columnar objects like an empty can and laying a plate placed on them, we can observe synchronization phenomena [7].

We have already made an experimental equipment by the former method. Arranging mechanical metronomes on the plate, we conducted physical experiments to observe what kind of synchronization phenomenon occurs. One of the famous examples is the synchronization experiment with 32 metronomes [4]. In this experiment [4], a heat insulation board with vertical 600 [mm] × horizontal 1000 [mm] dimension is used as a plate to arrange the metronomes. A rectangular parallelepiped frame using a resin pipe (Vertical 800 [mm] × Wide 900 [mm] × Height 600 [mm]) We are hanging the platform with wire (Fig. 15.1a).

The metronome used for the experiment is called "Lupina," manufactured by Nikko Seiki Co., Ltd. The size of this mechanical metronome is 110 [mm] in height, 32 [mm] in width, 51 [mm] in depth, The mass is about 200 [g] (Fig. 15.1b). Then, mechanical metronomes are placed on the plate, then the initial position of the rod is randomly applied. As you can see from the movie [4] even if the number of mechanical metronomes is not so small, for example 32, it is possible to observe in-phase synchronization. Now, we have already succeeded in case that the number of metronomes is 100. In [5], successful observation of in-phase synchronization with 100 mechanical metronomes is published.

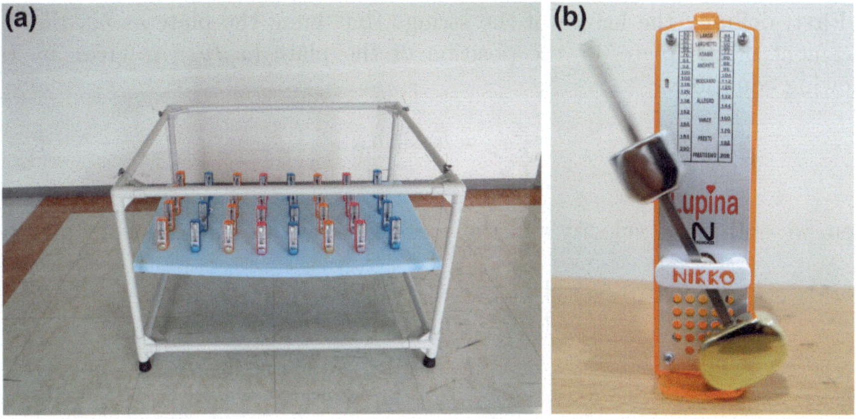

Fig. 15.1. **a** Experimental equipment with 32 mechanical metronomes, In this equipment, the plate is hung by four strings. **b** The mechanical metronome called "Lupina," which is used in the experiments of metronome synchronization

15.3 Mathematical Model of Metronome Synchronization

In this section, to analyze synchronous behavior using multiple metronomes, we derive a mathematical model. The results of numerical investigation will be introduced in the next section. In the following, taking multiple metronomes and the plate carrying them as a single system, we derive the equation of motion of the metronomes and the plate.

15.3.1 Experimental Setups and Introduction of Several Variables

The experimental equipment (Fig. 15.1a) consists of multiple metronomes and the plate hung by four strings. To derive the equations of motions of the metronomes and the plate, we define several variables. These are shown in Fig. 15.2.

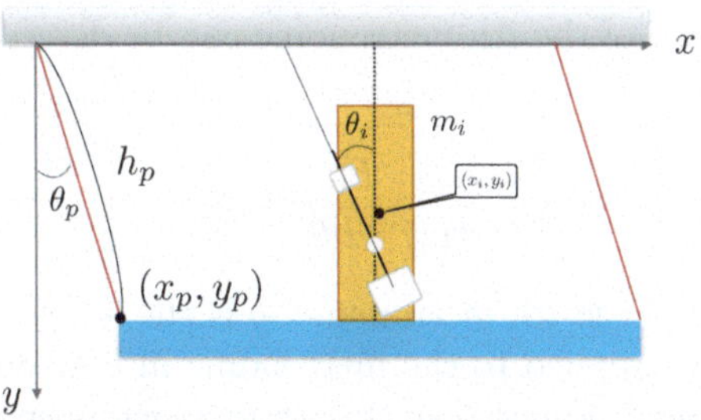

Fig. 15.2. Definition of variables to describe dynamical behavior of mechanical metronomes. In this figure, only one metronome is shown

First, defining the length of the strings that hung the plate as h_p, the displacement of the plate, or the position of the plate (x_p, y_p), is given by the following equations:

$$\begin{cases} x_p = h_p \sin\theta_p, \\ y_p = h_p \cos\theta_p. \end{cases} \tag{15.1}$$

Then, we obtain the velocity and the acceleration of the motion of the plate:

$$\begin{cases} \dot{x}_p = h_p\dot{\theta}_p \cos\theta_p, \\ \dot{y}_p = -h_p\dot{\theta}_p \sin\theta_p, \end{cases} \tag{15.2}$$

and

$$\begin{cases} \ddot{x}_p = -h_p\dot{\theta}_p^2 \sin\theta_p + h_p\ddot{\theta}_p \cos\theta_p, \\ \ddot{y}_p = -h_p\dot{\theta}_p^2 \cos\theta_p - h_p\ddot{\theta}_p \sin\theta_p. \end{cases} \tag{15.3}$$

In the similar way, we obtain the position of the ith metronome ($i = 1, 2, \ldots, n$), defining the length of the pendulum of the ith metronome h_i

$$\begin{cases} x_i = x_p + h_i \sin\theta_i, \\ y_i = y_p + h_i \cos\theta_i. \end{cases} \tag{15.4}$$

Then, the velocity and the acceleration can be obtained:

$$\begin{cases} \dot{x}_i = \dot{x}_p + h_i\dot{\theta}_i \cos\theta_i, \\ \dot{y}_i = \dot{y}_p - h_i\dot{\theta}_i \sin\theta_i, \end{cases} \tag{15.5}$$

and

$$\begin{cases} \ddot{x}_i = \ddot{x}_p - h_i\dot{\theta}_i^2 \sin\theta_i + h_i\ddot{\theta}_i \cos\theta_i, \\ \ddot{y}_i = \ddot{y}_p - h_i\dot{\theta}_i^2 \cos\theta_i - h_i\ddot{\theta}_i \sin\theta_i. \end{cases} \tag{15.6}$$

15.3.2 The Equation of Motion of the *i*th Metronome

The equation of motion of the ith metronome can be described by the following equations:

$$\begin{cases} m_i\ddot{x}_i = F_{x_i}, \\ m_i\ddot{y}_i = m_i g + F_{y_i}, \end{cases} \tag{15.7}$$

where m_i is the mass of the ith metronome, g is the acceleration of gravity, F_{x_i} and F_{y_i} are the forces applied to the metronome in horizontally and vertically. Let us define the moment of inertia of the ith metronome at its center of gravity, I_{G_i}, the equation of motion of the rotation can be described as

$$I_{G_i}\ddot{\theta}_i = \boldsymbol{x}'_i \times \boldsymbol{F}_i = x'_i F_{y_i} - y'_i F_{x_i}, \tag{15.8}$$

where $\boldsymbol{x}'_i = (x'_i, y'_i)^\mathsf{T} = (x_i - x_p, y_i - y_p)^\mathsf{T}$ and $\boldsymbol{F}_i = (F_{x_i}, F_{y_i})^\mathsf{T}$. Using Eq. (15.7),

$$\begin{aligned} I_{G_i}\ddot{\theta}_i &= m_i x'_i(\ddot{y}_i - g) - m_i y'_i \ddot{x}_i \\ &= m_i h_i \sin\theta_i(\ddot{y}_i - g) - m_i h_i \cos\theta_i \ddot{x}_i. \end{aligned}$$

In addition, using Eqs. (15.3) and (15.6), we obtain

$$\begin{aligned} I_{G_i}\ddot{\theta}_i &= -m_i h_i g \sin\theta_i + m_i h_i \sin\theta_i \ddot{y}_i - m_i h_i \cos\theta_i \ddot{x}_i \\ &= -m_i h_i g \sin\theta_i \\ &\quad + m_i h_i \sin\theta_i(-h_p\dot{\theta}_p^2 \cos\theta_p - h_p\ddot{\theta}_p \sin\theta_p - h_i\dot{\theta}_i^2 \cos\theta_i - h_i\ddot{\theta}_i \sin\theta_i) \\ &\quad - m_i h_i \cos\theta_i(-h_p\dot{\theta}_p^2 \sin\theta_p + h_p\ddot{\theta}_p \cos\theta_p - h_i\dot{\theta}_i^2 \sin\theta_i + h_i\ddot{\theta}_i \cos\theta_i) \\ &= -m_i h_i g \sin\theta_i \\ &\quad + m_i h_i h_p \sin(\theta_p - \theta_i)\dot{\theta}_p^2 - m_i h_i h_p \cos(\theta_p - \theta_i)\ddot{\theta}_p - m_i h_i^2 \ddot{\theta}_i. \end{aligned} \tag{15.9}$$

Then,

$$(I_{G_i} + m_i h_i^2)\ddot{\theta}_i = -m_i h_i g \sin\theta_i + m_i h_i h_p \left\{\sin(\theta_p - \theta_i)\dot{\theta}_p^2 - \cos(\theta_p - \theta_i)\ddot{\theta}_p\right\}. \tag{15.10}$$

Here, let us $\omega_i^2 = m_i g h_i/(I_{G_i} + m_i h_i^2)$,

$$\ddot{\theta}_i = -\omega_i^2 \sin\theta_i + \frac{\omega_i^2 h_p}{g}\left\{\sin(\theta_p - \theta_i)\dot{\theta}_p^2 - \cos(\theta_p - \theta_i)\ddot{\theta}_p\right\}. \tag{15.11}$$

Adding the viscosity term to the equation, we obtain

$$\ddot{\theta}_i = -\omega_i^2 \sin\theta_i + \frac{\omega_i^2 h_p}{g}\left\{\sin(\theta_p - \theta_i)\dot{\theta}_p^2 - \cos(\theta_p - \theta_i)\ddot{\theta}_p\right\} - 2\zeta_i\omega_i\dot{\theta}_i. \tag{15.12}$$

where ζ_i is the dumping ration of the ith metronome. To derive the dimensionless equation, we introduce $\mathrm{d}\tau = \omega_p \mathrm{d}t$; namely, we use $\varphi_i = \dfrac{\mathrm{d}\theta_i}{\mathrm{d}\tau}$, $\varphi_p = \dfrac{\mathrm{d}\theta_p}{\mathrm{d}\tau}$, $\dot{\varphi}_i = \dfrac{\mathrm{d}^2\theta_i}{\mathrm{d}\tau^2}$ and $\dot{\varphi}_p = \dfrac{\mathrm{d}^2\theta_p}{\mathrm{d}\tau^2}$. Then, Eq. (15.12) can be rewritten in the following style:

$$\begin{aligned} \dot{\varphi}_i &= -\left(\frac{\omega_i}{\omega_p}\right)^2 \sin\theta_i - 2\zeta_i\left(\frac{\omega_i}{\omega_p}\right)\varphi_i \\ &\quad + \left(\frac{\omega_i}{\omega_p}\right)^2 \sin(\theta_p - \theta_i)\varphi_p^2 - \left(\frac{\omega_i}{\omega_p}\right)^2 \cos(\theta_p - \theta_i)\dot{\varphi}_p. \end{aligned} \tag{15.13}$$

15.3.3 The Equation of Motion of the Plate

In this subsection, we will derive the equation of motion of the plate on which n metronomes are put. Let us define the mass of the plate m_p, and its displacement in the horizontal and vertical directions. Then, the equations of motion can be described:

$$\begin{cases} m_p\ddot{x}_p = -T_p\sin\theta_p - \sum_{j=1}^{n} F_{x_j}, \\ m_p\ddot{y}_p = m_p g - T_p\cos\theta_p - \sum_{j=1}^{n} F_{y_j}, \end{cases} \tag{15.14}$$

where T_p is the tension of the strings. Using these two equation to remove T_p, we obtain

$$\begin{aligned} m_p(\ddot{x}_p\cos\theta_p - \ddot{y}_p\sin\theta_p) = &-\cos\theta_p\sum_{j=1}^{n} F_{x_j} \\ &- m_p g\sin\theta_p + \sin\theta_p\sum_{j=1}^{n} F_{y_j}. \end{aligned} \tag{15.15}$$

From Eq. (15.7),

$$\begin{cases} F_{x_i} = m_i\ddot{x}_i = m_i(-h_p\dot{\theta}_p^2\sin\theta_p + h_p\ddot{\theta}_p\cos\theta_p - h_i\dot{\theta}_i^2\sin\theta_i + h_i\ddot{\theta}_i\cos\theta_i), \\ F_{y_i} = m_i\ddot{y}_i - m_i g = m_i(-h_p\dot{\theta}_p^2\cos\theta_p - h_p\ddot{\theta}_p\sin\theta_p - h_i\dot{\theta}_i^2\cos\theta_i - h_i\ddot{\theta}_i\sin\theta_i) - m_i g. \end{cases} \tag{15.16}$$

Substituting them into Eq. (15.15),

$$\begin{aligned} m_p h_p\ddot{\theta}_p = &\sum_{j=1}^{n} m_j\Big\{-\cos\theta_p(-h_p\dot{\theta}_p^2\sin\theta_p + h_p\ddot{\theta}_p\cos\theta_p - h_j\dot{\theta}_j^2\sin\theta_j + h_j\ddot{\theta}_j\cos\theta_j) \\ &+\sin\theta_p(-h_p\dot{\theta}_p^2\cos\theta_p - h_p\ddot{\theta}_p\sin\theta_p - h_j\dot{\theta}_j^2\cos\theta_j - h_j\ddot{\theta}_j\sin\theta_j)\Big\} \\ &-\left(m_p + \sum_{j=1}^{n} m_j\right) g\sin\theta_p \\ = &\sum_{j=1}^{n} m_j\left\{-h_p\ddot{\theta}_p - h_j\dot{\theta}_j^2\sin(\theta_p - \theta_j) - h_j\ddot{\theta}_j\cos(\theta_p - \theta_j)\right\} \\ &-\left(m_p + \sum_{j=1}^{n} m_j\right) g\sin\theta_p. \end{aligned} \tag{15.17}$$

Let us define $\gamma_i = m_i/m_p$ and $\eta_i = h_i/h_p$,

$$\begin{aligned} \left(1 + \sum_{j=1}^{n}\gamma_j\right)\ddot{\theta}_p = &-\sum_{j=1}^{n}\gamma_j\eta_j\left\{\sin(\theta_p - \theta_j)\dot{\theta}_j^2 + \cos(\theta_p - \theta_j)\ddot{\theta}_j\right\} \\ &-\omega_p^2\sin\theta_p\left(1 + \sum_{j=1}^{n}\gamma_j\right). \end{aligned} \tag{15.18}$$

because $\omega_p^2 = g/h_p$. Here, defining $\gamma_1 = \cdots = \gamma_n = \gamma$ and $\beta = \gamma/(1+n\gamma)$,

$$\ddot{\theta}_p = -\beta \sum_{j=1}^{n} \eta_j \left\{ \sin(\theta_p - \theta_j)\dot{\theta}_j^2 + \cos(\theta_p - \theta_j)\ddot{\theta}_j \right\} - \omega_p^2 \sin\theta_p. \tag{15.19}$$

Adding the viscosity term, we have

$$\ddot{\theta}_p = -\beta \sum_{j=1}^{n} \eta_j \left\{ \sin(\theta_p - \theta_j)\dot{\theta}_j^2 + \cos(\theta_p - \theta_j)\ddot{\theta}_j \right\} - \omega_p^2 \sin\theta_p - 2\zeta_p\omega_p\dot{\theta}_p \tag{15.20}$$

where ζ_p is the dumping ratio of the plate. Finally, we obtain the following dimensionless form

$$\dot{\varphi}_p = -\beta \sum_{j=1}^{n} \eta_j \left\{ \sin(\theta_p - \theta_j)\varphi_j^2 + \cos(\theta_p - \theta_j)\dot{\varphi}_j \right\} - \sin\theta_p - 2\zeta_p\varphi_p, \tag{15.21}$$

by $\mathrm{d}\tau = \omega_p \mathrm{d}t$. In Eq. (15.20), $\varphi_i = \dfrac{\mathrm{d}\theta_i}{\mathrm{d}\tau}$ and $\varphi_p = \dfrac{\mathrm{d}\theta_p}{\mathrm{d}\tau}$.

15.3.4 How to Solve the Equations of Motion Numerically

Substituting Eq. (15.21) into Eq. (15.13),

$$\begin{aligned}
\dot{\varphi}_i = &-\left(\frac{\omega_i}{\omega_p}\right)^2 \cos(\theta_p - \theta_i) \left\{ -\beta \sum_{j=1}^{n} \eta_j \sin(\theta_p - \theta_j)\varphi_j^2 - \beta \sum_{j=1}^{n} \eta_j \cos(\theta_p - \theta_j)\dot{\varphi}_j \right\} \\
&-\left(\frac{\omega_i}{\omega_p}\right)^2 \cos(\theta_p - \theta_i) \left\{ -\sin\theta_p - 2\zeta_p\varphi_p \right\} \\
&+\left(\frac{\omega_i}{\omega_p}\right)^2 \sin(\theta_p - \theta_i)\varphi_p^2 - \left(\frac{\omega_i}{\omega_p}\right)^2 \sin\theta_i - 2\zeta_i \left(\frac{\omega_i}{\omega_p}\right) \varphi_i.
\end{aligned} \tag{15.22}$$

Then,

$$\begin{aligned}
\dot{\varphi}_i - \beta \sum_{j=1}^{n} &\left(\frac{\omega_i}{\omega_p}\right)^2 \eta_j \cos(\theta_p - \theta_i)\cos(\theta_p - \theta_j)\dot{\varphi}_j \\
&= \left(\frac{\omega_i}{\omega_p}\right)^2 \cos(\theta_p - \theta_i) \left\{ \beta \sum_{j=1}^{n} \eta_j \sin(\theta_p - \theta_j)\varphi_j^2 + \sin\theta_p + 2\zeta_p\varphi_p \right\} \\
&\quad + \left(\frac{\omega_i}{\omega_p}\right)^2 \sin(\theta_p - \theta_i)\varphi_p^2 - \left(\frac{\omega_i}{\omega_p}\right)^2 \sin\theta_i - 2\zeta_i \left(\frac{\omega_i}{\omega_p}\right) \varphi_i.
\end{aligned} \tag{15.23}$$

Let us define the right hand side of Eq. (15.23) as b_i, and $\boldsymbol{b} = (b_1, \ldots, b_n)^\mathsf{T}$, $\boldsymbol{\varphi} = (\varphi_1, \ldots, \varphi_n)^\mathsf{T}$, $\dot{\boldsymbol{\varphi}} = (\dot{\varphi}_1, \ldots, \dot{\varphi}_n)^\mathsf{T}$, and introduce two matrices Z and Θ by the following definitions:

$$Z_{ij} = \left(\frac{\omega_i}{\omega_p}\right)^2 \eta_j \cos(\theta_p - \theta_i)\cos(\theta_p - \theta_j), \tag{15.24}$$

$$\Theta = E - \beta Z. \tag{15.25}$$

where E is an $n \times n$ unit matrix. Then, we have the following form of the equation,

$$\Theta \dot{\boldsymbol{\varphi}} = \boldsymbol{b}. \tag{15.26}$$

The determinant of Θ is calculated as

$$\det \Theta = 1 - \beta \sum_{j=1}^{n} \left(\frac{\omega_j}{\omega_p} \right)^2 \eta_j \cos^2(\theta_p - \theta_j). \tag{15.27}$$

If $\det \Theta \neq 0$, the inverse of Θ is

$$\Theta^{-1} = E + \frac{\beta}{\det \Theta} Z. \tag{15.28}$$

Then, we can express $\dot{\boldsymbol{\varphi}}$ as follows:

$$\dot{\boldsymbol{\varphi}} = \Theta^{-1} \boldsymbol{b} = \left(E + \frac{\beta}{\det \Theta} Z \right) \boldsymbol{b}. \tag{15.29}$$

In the similar way, substituting Eq. (15.13) into Eq. (15.21), we have

$$\begin{aligned} &\left(1 - \beta \sum_{j=1}^{n} \left(\frac{\omega_j}{\omega_p} \right)^2 \eta_j \cos^2(\theta_p - \theta_j) \right) \dot{\varphi}_p \\ &= \beta \sum_{j=1}^{n} \eta_j \left\{ - \sin(\theta_p - \theta_j) \varphi_j^2 - \left(\frac{\omega_j}{\omega_p} \right)^2 \cos(\theta_p - \theta_j) \sin(\theta_p - \theta_j) \varphi_p^2 \right. \\ &\quad \left. + \left(\frac{\omega_j}{\omega_p} \right)^2 \cos(\theta_p - \theta_j) \sin \theta_j + 2\zeta_j \left(\frac{\omega_j}{\omega_p} \right) \cos(\theta_p - \theta_j) \varphi_j \right\} \\ &\quad - \sin \theta_p - 2\zeta_p \varphi_p, \end{aligned}$$

which is reduced to the following equation:

$$\begin{aligned} \det \Theta \dot{\varphi}_p = \beta \sum_{j=1}^{n} \eta_j & \left\{ - \sin(\theta_p - \theta_j) \varphi_j^2 - \left(\frac{\omega_j}{\omega_p} \right)^2 \cos(\theta_p - \theta_j) \sin(\theta_p - \theta_j) \varphi_p^2 \right. \\ & \left. + \left(\frac{\omega_j}{\omega_p} \right)^2 \cos(\theta_p - \theta_j) \sin \theta_j + 2\zeta_j \left(\frac{\omega_j}{\omega_p} \right) \cos(\theta_p - \theta_j) \varphi_j \right\} \\ & - \sin \theta_p - 2\zeta_p \varphi_p. \end{aligned} \tag{15.30}$$

It is also essential how to decide the impulsive forces of the metronome. The power of the spring acts as the impulsive force twice per one period to keep a fixed amplitude with the real metronome. However, it is difficult to measure the impulsive force mechanically. Therefore, we estimated it from the damping ratio that we measured from the amplitude. Then, we set the power that absolute value of angular velocity 25.8 [deg/s] due to the impulsive force of the metronome when the angle θ_i becomes $\pm 10°$, and then the pendulum can continue to oscillate.

15.4 Results

In Fig. 15.3, we show the temporal changes of the angle values of the pendulum rods of each metronome. We can see that even if the number of metronomes is large, for example 100, we can observe in-phase synchronization after 60[s] in this numerical experiments.

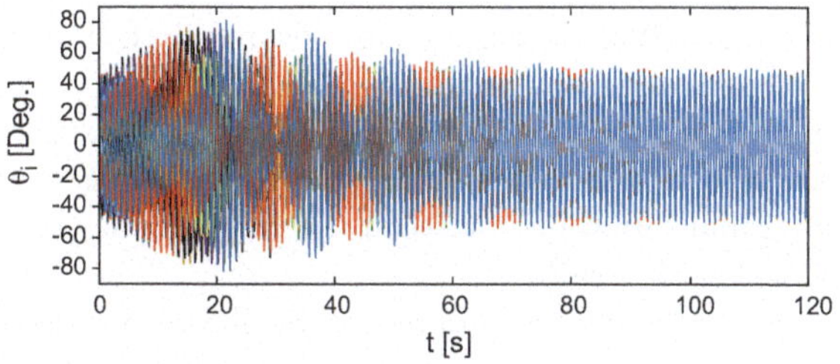

Fig. 15.3. The time series traces of angle values of the pendulum rods of 100 metronomes

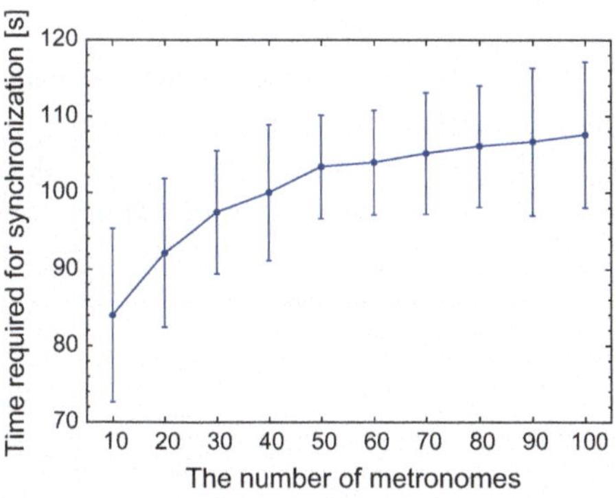

Fig. 15.4. The time required for achieving in-phase synchronization

In Fig. 15.4, the time required for in-phase synchronization was investigated in case of increasing the number of metronomes. To calculate the time required for in-phase synchronization, first, we used arbitrary two metronomes from n metronomes. The possible pair of two metronomes is ${}_nC_2$. Then, we calculated the correlation coefficients between two time series observed from these two metronomes. If the minimum value of the correlation coefficients is larger than 0.9, we defined that in-phase synchronization of metronomes is achieved. From Fig. 15.4, even if we increased the number of metronomes, the time for in-phase synchronization shows a tendency to converge. However, this tendency does not match to real physical experiments. In the physical experiments, if we increase the number of metronomes, it becomes very hard to achieve in-phase synchronization. The reason why all the metronomes exhibit in-phase synchronization relatively easily is that we did not introduce the small difference between metronomes. Namely, we assume that all the metronomes are homogeneous in this numerical experiments.

Thus, we conduct numerical experiments in case that metronomes have individual difference. In fact, even if we set the natural frequency of the metronomes to the same value, real oscillation frequencies become slightly different from each other. For example, if we set the frequency by adjusting the sliding weight of pendulum in Lupina, such as $f = 1.4$ [Hz] (168[bpm]), measured values become as follows: 1.385, 1.376, 1.389 and 1.382 [Hz]. Namely, it is very important to introduce the individual difference between the metronomes to discuss how the in-phase synchronization is achieved.

In Fig. 15.5, we show the probability of in-phase synchronization of two metronomes, if we changed the natural frequency of the plate f_p. To evaluate the synchronizability, we used the probability of achieving in-phase synchronization for 100 trials with different initial position of the pendulum rods of two metronomes. In Fig. 15.5, we expressed the individual difference by ε. Namely, the frequency of the first metronome f_1 is assumed to be 1.06 [Hz], and the frequency of the second metronome $f_2 = 1.06 + \varepsilon$.

If $\varepsilon = 0$ (inverted triangles), we can observe that the in-phase synchronization is achieved in case that f_p (the natural frequency of the plate) is smaller than the frequency of the metronomes. In addition, we confirmed that there exists relatively broader range of f_p (the natural frequency of the plate) for achieving in-phase synchronization. However, if we increased the value of ε, the regions for achieving the in-phase and anti-phase synchronization become smaller.

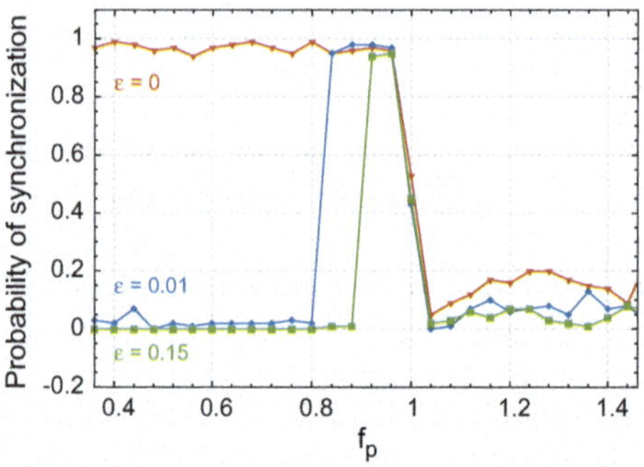

Fig. 15.5. Synchronizability measured by the probability of achieving in-phase synchronization for 100 trials with different initial conditions if the natural frequency of the plate f_p is changed

In Fig. 15.6, we evaluated how the individual differences affect the synchronizability. We defined the individual difference in the frequency of the metronome.

Using the nominal frequency $f = 1.06$ [Hz], we decided the frequency of each metronome by $f + \varepsilon \times u$, where $\varepsilon = 0.015$ and u is a uniformly distributed random numbers between -1 and 1.

In Fig. 15.6a, we show the distribution of the correlation coefficients (gray circles) and their averaged values (the solid blue line) in case of increasing the number of metronomes. In Fig. 15.6b, we derived probabilities of in-phase synchronization if we increased the number of metronomes. From these figures, we can see that if the number of metronomes is 30, the probability of in-phase synchronization becomes almost 0.5, and it becomes almost zero if the number of metronomes is 100, which indicates that it is almost impossible to achieve the in-phase synchronization with 100 metronomes until the energy source of the mechanical metronome is not exhausted.

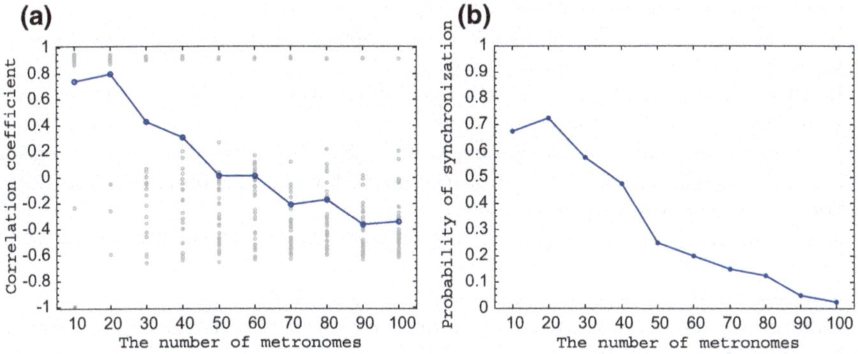

Fig. 15.6. How the individual difference between metronomes affect the synchironizability in case of increasing the number of metronomes. **a** The distribution of the correlation coefficients and **b** the probability of in-phase synchronization

15.5 Conclusion

In this paper, we discussed how to construct the equations of motion of synchronization of mechanical metronomes. We also investigated their dynamical behavior by solving the equations of motions numerically. In the numerical experiments, parameter values of the equations of motion are experimentally obtained from the experimental equipment.

We found that synchronization behavior of mechanical metronomes depends on the following two factors: the frequency of the metronomes and the natural frequency of the plate and initial angles of the metronomes. We also found the individual difference of the metronomes strongly affect the final behavior. In addition, if the number of mechanical metronomes increases, for example 100, it becomes extremely harder to observe the in-phase synchronization due to individual difference of metronomes until energy which is applied to the mechanical metronomes through spiral springs is consumed.

The research of TI was partially supported by JSPS KAKENHI Grant Numbers 15TK0112 and 17K00348, and the research of YS was supported by JSPS KAKENHI Grant Number 16K16126 and 18K18125.

References

1. A. Pikovsky, M. Rosenblum, J. Kurths, *Synchronization: A Universal Concept in Nonlinear Science*, Cambridge Nonlinear Science Series (Cambridge University Press, Cambridge, 2003)
2. F.E. Hanson, J.F. Case, E. Buck, J. Buck, Synchrony and flash entrainment in a new guinea firefly. Science **174**, 161–164 (1971)
3. I. Aihara, T. Mizumoto, T. Otsuka, H. Awano, K. Nagira, H.G. Okuno, K. Aihara, Spatio-temporal dynamics in collective frog choruses examined by mathematical modeling and field observations. Sci. Rep. **4**(3891) (2014)
4. Ikeguchi Laboratory, Synchnonization of 32 metronomes (2014), YouTube www.youtube.com/watch?v=JWToUATLGzs
5. Ikeguchi Laboratory, Synchnonization of 100 metronomes (2015), YouTube www.youtube.com/watch?v=suxu1bmPm2g
6. Y. Sato, K. Nagamoto, T. Nagamine, M. Fuse, Synchronized phenomena of oscillators (experimental and analytical investigation for two metronomes). Trans. Jpn. Soc. Mech. Eng. **66**(642), 9–15 (2000)
7. J. Pantaleone, Synchronization of metronomes. Am. J. Phys. **70**(10), 992–1000 (2002)
8. Y. Fujino, B.M. Pacheco, P. Warnitchai, S.I. Nakamura, Synchronization of human walking observed during lateral vibration of a congested pedstrian bridge. Earthq. Eng. Struct. Dyn. **22**(9), 741–758 (1993)
9. S.H. Strogatz, D.M. Abrams, A. McRobie, B. Eckhardt, E. Ott, Crowd synchrony on the milleninium bridge. Nature **438**, 43–44 (2005)
10. Ikeguchi Laboratory, Synchnonization of petbottle oscillators (in-phase) (2010), YouTube www.youtube.com/watch?v=Vvs14DUixIM
11. Ikeguchi Laboratory, Synchnonization of petbottle oscillators (out-of-phase) (2010), YouTube www.youtube.com/watch?v=k2aOIYhu9SQ
12. H. Kitahata, J. Taguchi, M. Nagayama, T. Sakurai, Y. Ikura, A. Osa, Y. Sumino, M. Tanaka, E. Yokoyama, H. Miike, Oscillation and synchronization in the combustion of candles. J. Phys. Chem. A **113**(29), 8164–8168 (2009)
13. Ikeguchi Laboratory, Synchnonization of candle oscillators (2009), YouTube www.youtube.com/watch?v=hUZbkZTD8jU
14. S. Nakata, T. Miyata, N. Ojima, K. Yoshikawa, Self-synchronization in coupled salt-water oscillators. Phys. D Nonlinear Phenom. **115**(3–4), 313–320 (1998)
15. Ikeguchi Laboratory, Synchnonization of salt-water oscillators (2011), YouTube www.youtube.com/watch?v=nXnAaIPJ4eE

Chapter 16
Hardware Implementation of Chaos Control Using a Proportional Feedback Controller

Benjamin K. Rhea, R. Chase Harrison, D. Aaron Whitney, Frank T. Werner, Andrew W. Muscha, and Robert N. Dean(✉)

Auburn University, Auburn, AL, USA
{bkr0001,rch0012,daw0043,ftw0001,azm0043,deanron}@auburn.edu

Abstract. This paper presents an electronic implementation of a controller for an exact solvable chaotic oscillator. The controller uses a proportional feedback control scheme to stabilize the chaotic oscillator, with both analog and digital components. The analog hardware implementation uses commercial-off-the-shelf (COTS) digital logic components and an analog feedback path in order to generate a control signal that produces small voltage perturbations in the oscillator's trajectory. The digital portion of the control effort is contained on a single microcontroller, which contains the information to be encoded into the chaotic oscillator. The perturbations are aperiodically applied using a clock that is generated from the oscillator's output signal so that the pulses can be applied to the oscillator at the correct times. In order to achieve different stabilized orbits, a variable gain stage was added to the controller, using an operational amplifier and a potentiometer, such that the magnitude of the control pulses can be adjusted. This controller is used to demonstrate stabilization of various periodic orbits in a double scroll exact solvable chaotic oscillator, which is shown in the time domain and in phase portraits. This type of controller could be used to encode information into chaotic waveforms for communication systems.

16.1 Introduction

Presented here is a mixed-signal proportional feedback controller implemented in hardware. This controller is demonstrated using a mixed-signal chaotic oscillator that is based on an exactly solvable piecewise-linear system. One potential application for chaos control is for encoding information into the resulting waveform for communication systems. Chaos based communications have been demonstrated using small voltage perturbations to control the symbolic dynamics of an electronic chaotic oscillator [6, 7].

The controller is comprised of two primary blocks, a digital portion and an analog portion. The digital portion of this controller contains information on

V. In et al. (Eds.): *Proceedings of the 5th International Conference on Applications in Nonlinear Dynamics*, Understanding Complex Systems, https://doi.org/10.1007/978-3-030-10892-2_16

two known trajectories' general time domain responses. These trajectories are stored in memory on a microcontroller. One of these trajectories is mapped to a logic level high, "1", and the other one is mapped to a logic level low, "0". The analog controller compares the desired trajectory with the current state of the oscillator, which determines how large of a voltage perturbation to apply. These voltage perturbations are applied at the local maxima and minima of the oscillator's trajectory. The analog controller determines when to apply this by tracking the derivative of the oscillator's output.

Using both digital and analog controller sections provides flexibility in how the controller can be used. Different trajectories can be saved in memory without any modification to the analog portion. This could allow for more complex encoding schemes. The analog portion is capable of fine-tuning the magnitude of the voltage perturbations across a wide range of values. This is important since both negative and positive voltage perturbations need to be applied to the oscillator in order to follow the desired trajectory. This is not possible using just a digital microcontroller, which typically operates on a single positive power supply only.

16.2 Background

16.2.1 Controlling Chaotic Oscillators

Chaotic oscillators present interesting challenges and opportunities for controlling them. Generally speaking, a chaotic oscillator is controlled by forcing it to maintain its trajectory in a periodic orbit. This is referred to as stabilizing the orbit of the oscillator. The inherent characteristics of a chaotic system, such as extreme sensitivity to initial conditions, an infinite variety of behaviors embedded in it, and an infinite number of unstable periodic orbits, can actually be exploited in controlling its behavior with a minimal amount of energy or effort, compared to controlling linear dynamic systems [5]. Several control techniques have been developed for controlling chaotic oscillators: OGY control, target-steering chaotic control, proportional feedback control, and delayed feedback control.

16.2.2 OGY Control

Edward Ott, Celso Grebogi and James Yorke developed a technique for controlling chaotic oscillators known as OGY control [9,12]. This technique takes advantage of the properties of chaotic systems in that trajectories eventually come very close to those of unstable periodic orbits. At that time, small perturbations can be applied to the system to nudge the trajectory along the desired periodic orbit. Additional small perturbations can be applied to keep the trajectory on that orbit, including in the presence of disturbances. Since an infinite number of unstable periodic orbits exist in the phase space of chaotic systems, very different controlled system responses can be obtained quite easily, compared to controlling linear systems.

16.2.3 Target-Steering Chaotic Control

Another technique for controlling chaotic systems is target steering [4]. In this chaos control technique, each state variable of the chaotic system is assigned a target value. At each time step, the values of the state variables are compared with their assigned target values. A ratio of them is then computed and used to proportionally adjust the states to steer the trajectory to the desired location.

16.2.4 Proportional Feedback Control

Proportional feedback control is another technique that has been used to control chaotic systems [8]. In this technique, a state or relevant signal in the system is sampled, either continuously or periodically. The sample is then compared with a desired level and its ratio is computed. The resulting ratio is then used to modulate a relevant control signal in the system to produce a favorable system response, thus completing the feedback path.

16.2.5 Delayed Feedback Control

Delayed feedback control is yet another technique that has been developed to control chaotic systems [10]. In this technique, the current state of the system is sampled and compared with the previously sampled state of the system, exactly one period in the past. The deviations between the states of the two sampling periods are computed and used to generate controlling perturbations that are applied to the system to steer it to the desired unstable periodic orbit.

16.3 Exact Solvable Chaotic Oscillator

The chaotic oscillator that has been chosen for this controller approach is based on the exactly solvable system previously developed by Saito and Fujita [11]. This system is a synthesis of a linear second order differential equation and non-linear discrete switching states. This is shown in Eqs. (16.1) and (16.2), where $\ln(\beta)$ is the positive dampening coefficient, ω is the fundamental frequency, u is the continuous time variable and $s(t)$ is the discrete state. This system is of particular interest because it has been shown to have an exact analytical solution, which was based on the summation of fixed linear basis pulses [2]. This system has been realized in electronic hardware [1]. From this analytical solution, a matched filter has been derived, which is the ideal filter for maximizing single-to-noise ratio (SNR) in the presence of additive white Gaussian noise (AWGN) [3]. This matched filter has been realized in hardware and used in a wireless communication system [13].

$$\ddot{u} - 2\beta\dot{u} + (\omega^2 + \beta^2)(u - s(t)) = 0 \tag{16.1}$$

$$s(t) = \begin{cases} +1 & u(t) \geq 0, \ \dot{u}(t) = 0 \\ -1 & u(t) < 0, \ \dot{u}(t) = 0 \end{cases} \tag{16.2}$$

16.4 Hybrid Controller

The hybrid controller that is based on proportional feedback consists of both an analog and a digital portion. The feedback structure has two separate paths, an inner control loop and an outer control loop, as shown in Fig. 16.1. The digital control portion was implemented using a microcontroller which was flashed with code that can be configured to send a square wave signal where the logic high maps to a specific trajectory and a logic low maps to another distinct trajectory. This signal was sent to the analog controller, which generated the appropriate magnitude of voltage perturbation to achieve the desired oscillator response. The microcontroller monitors the oscillator's analog output, V, and waits until one of the known trajectories has been completed.

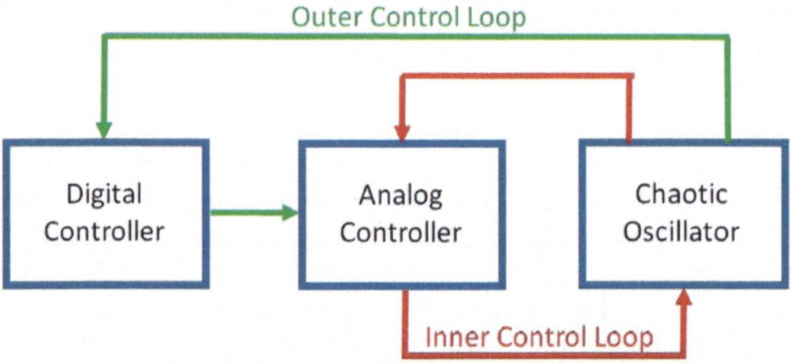

Fig. 16.1. Block diagram of the two feedback loops of the controller

The two feedback paths are physically connected using jumper wires that interconnect directly to the microcontroller and the analog controller header pins. The analog controller was implemented on a custom PCB using COTS parts, with various buffered testing points. The digital portion of the controller was programmed using C, compiled, and then flashed to an ARM NUCLEO-F446RE microcontroller using uVision IDE by Keil. The digital portion of the controller contains the information that is intended to be encoded into the oscillator. The analog portion of the controller scales the output from the digital controller to the appropriate voltage level and determines when to apply the voltage pulses by monitoring the output of the oscillator. The digital controller monitors the oscillator's feedback signal to make sure the correct trajectory is being encoded into the oscillator.

16.4.1 Analog Controller Section

The analog controller was realized using COTS op-amps, comparators, and a transmission-gate (T-gate) switch, as shown in Fig. 16.2. An analog difference amplifier compared a desired reference value that was stored on the digital microcontroller with the current output of the oscillator, V. This comparison generated

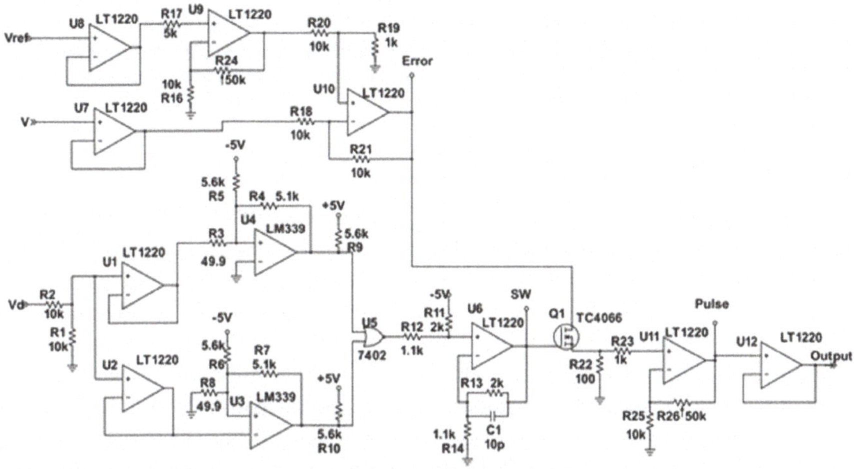

Fig. 16.2. Schematic of the analog controller

Fig. 16.3. Photograph of the analog controller PCB

an error signal that was aperiodically applied to the oscillator at the zero crossings of the derivative of the oscillator's output. A zero crossing detector circuit was designed using comparators, feedback resistors, and a NOR gate. The feedback resistors were configured in a Schmitt trigger topology with appropriate values for approximately 50 mV of hysteresis, in order to make the system more robust to false triggers in the presence of noise. At each of the zero crossings, a voltage pulse was applied to the oscillator node, V. Each of these pulses had a magnitude that was proportional to the magnitude of the computed error. The applied voltage perturbations had a fixed duration; however, the magnitude of each of these perturbations could vary in magnitude and direction. This voltage pulse was intended to push the oscillations to the desired known trajectory. The populated PCB is shown in Fig. 16.3.

16.4.2 Digital Controller Section

The digital portion of the controller was realized in software on an ARM microcontroller, a photograph of which is shown in Fig. 16.4. This software contained two arrays of data that corresponded to a known trajectory segment of the chaotic oscillator. These trajectory segments were previously determined from oscilloscope captures of the free running oscillator. One of these segments was mapped to a logic level high and the other one was mapped to a logic level low, to encode a "0" and "1" grammar. The microcontroller used a digital input capture to monitor the current state of the oscillator. This allowed for the controller to transition from a logic high to a logic low asynchronously. While this

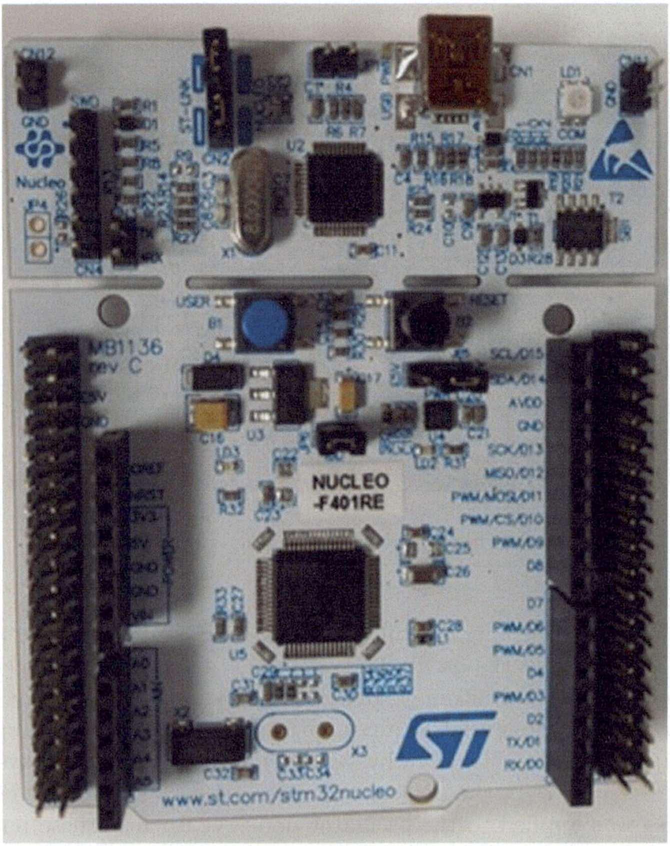

Fig. 16.4. Microcontroller PCB for the digital controller

additional feedback path appears unnecessary, it did improve performance of the system. This was due to the fact that the time of the transients between these logic switching events could vary widely, even when the same control effort was applied.

16.5 Controller Testing

In order to demonstrate how information could be encoded into a chaotic waveform, two distinct orbits from the free running chaotic oscillator were chosen. There were many different orbits that the controller can steer the free running oscillator into; however, switching between most of these orbits resulted in unpredictable transients in between each orbit. For this reason, two orbits were chosen that were very different from each other, while still attempting to maintain the shortest transient between the two orbits. The first orbit contained a trajectory that the output signal, V, circles both scrolls defined by the feedback state, $s(t)$. This is shown in the phase space in Fig. 16.5, and in the time domain response in Fig. 16.6. The other orbit was chosen where the output signal, V, oscillates around only one of the two orbits of the double scroll. The phase space and the time domain response for this orbit are shown in Figs. 16.7 and 16.8, respectively. Alternating between these two trajectories can be seen in the the phase space in Fig. 16.9, and in the time domain response in Fig. 16.10.

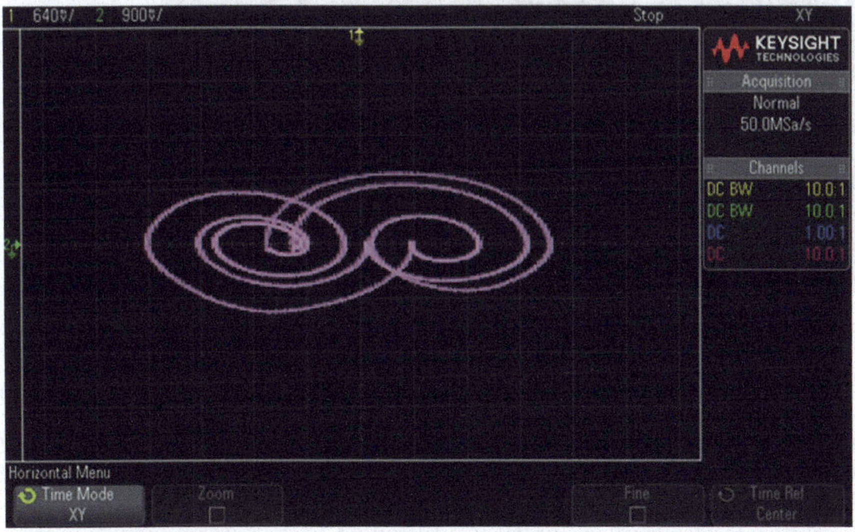

Fig. 16.5. Phase space oscilloscope image of 0-0-0 pattern

These two distinct orbits were chosen because it minimized transients of the oscillator's output from switching back and forth between these two orbits. This

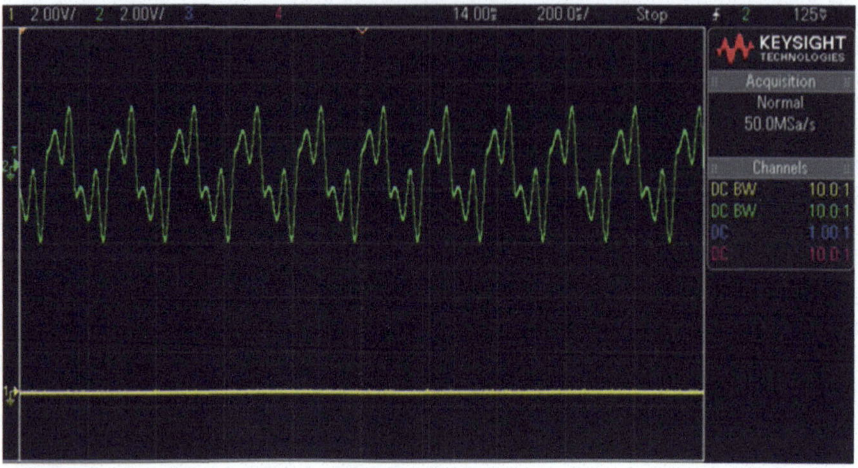

Fig. 16.6. Time domain oscilloscope image of 0-0-0 where green is the oscillator's output and yellow is the microcontroller's logic level

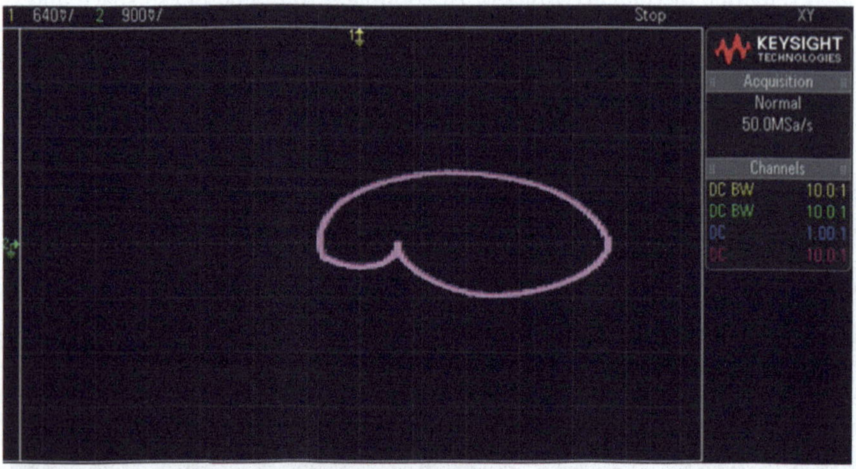

Fig. 16.7. Phase space oscilloscope image of 1-1-1 pattern

makes the amount of time between each of these two orbits be more predictable. This is important because the proposed encoding scheme relies on the two orbits being tuned to a similar length. Another measure to help ensure that these two orbits are of a similar duration is to repeat these patterns multiple times. While this sacrifices channel efficiency, it does result in a more reliable encoding and decoding scheme. Another observation, in particular, about the orbit that is mapped to a logic high, is that it may not completely be contained within the naturally allowable grammar of the free-running system. For this reason, the previously developed matched filter may not yield the optimum detection

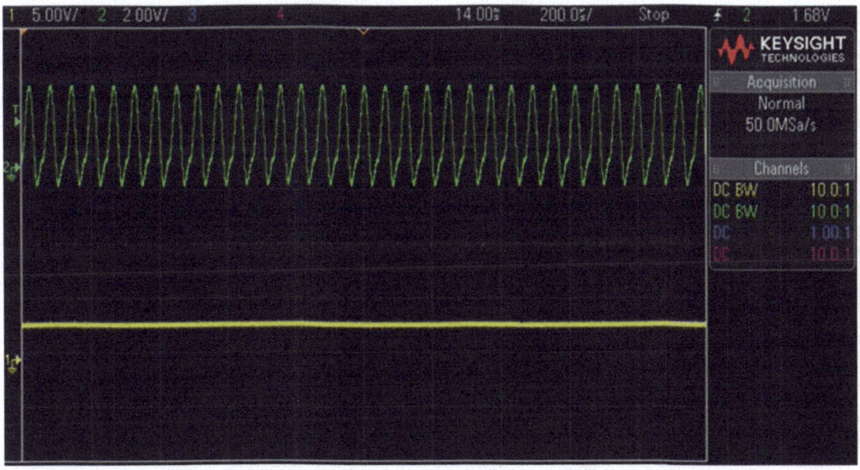

Fig. 16.8. Time domain oscilloscope image of 1-1-1 where green is the oscillator's output and yellow is the microcontroller's logic level

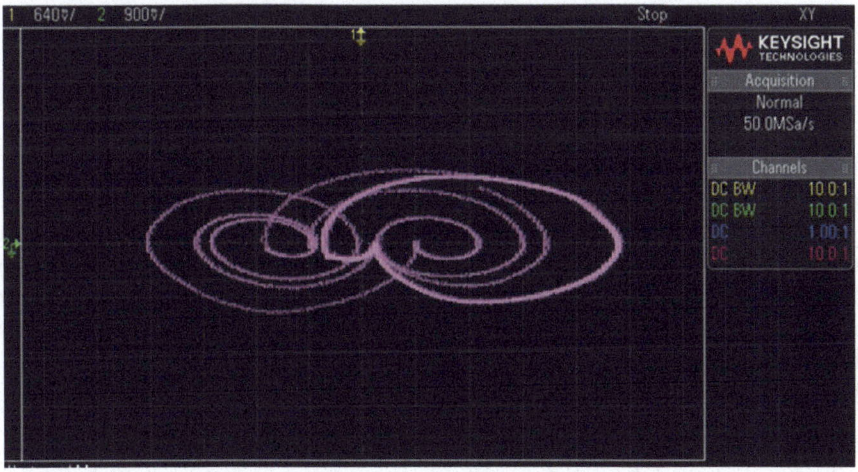

Fig. 16.9. Phase space oscilloscope image of 0-1-0 pattern

scheme. However, this design choice was exercised in order to realize a more practical and reliable hardware implementation.

The fundamental frequency of the analog chaotic oscillator used in this demonstration was approximately 18.4 kHz. Not factoring in the oscillator design, one of the limiting factors in increasing the frequency of operation of the overall system was the propagation delay of the feedback loop of the controller. While this was not an issue using a very low frequency chaotic oscillator, it could become a significant hurdle when looking to increase the bandwidth in

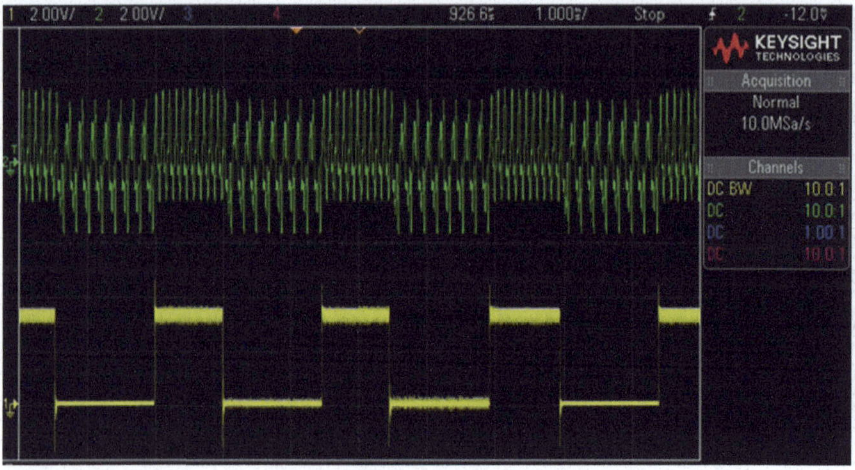

Fig. 16.10. Time domain oscilloscope image of 0-1-0 where green is the oscillator's output and yellow is the microcontroller's logic level

an application such as a communication system. These issues will be addressed in future development.

16.6 Conclusion

Presented here is a mixed signal controller implemented in hardware with potential applications in communication systems. The controller was composed of an analog portion and a digital portion. This controller was demonstrated by controlling an electronic chaotic oscillator to two distinct orbits. Using these two orbits, information can be embedded in the chaotic waveform by mapping one orbit to a logic high and one to a logic low. A demonstration of encoding information was presented using an alternating pattern of logic lows and highs. The two orbits were chosen in order to minimize transients between switching, which resulted in more reliable decoding.

References

1. A. Beal, J. Bailey, S. Hale, R. Dean, M. Hamilton, J. Tugnait, D. Hahs, N. Corron, Design and simulation of a high frequency exact solvable chaotic oscillator, in *MILCOM 2012* (IEEE, 2012), pp. 1–6
2. N.J. Corron, An exactly solvable chaotic differential equation. Dyn. Contin. Discret. Impuls. Syst. Ser. A Math. Anal. **16**, 777–788 (2009)
3. N.J. Corron, J.N. Blakely, M.T. Stahl, A matched filter for chaos. Chaos Interdiscip. J. Nonlinear Sci. **20**(2), 023,123 (2010)
4. J. Dattani, J.C. Blake, F.M. Hilker, Target-oriented chaos control. Phys. Lett. A **375**(45), 3986–3992 (2011)

5. C. Grebogi, Y.C. Lai, Controlling chaos. Handbook of chaos control (Wiley, New York, 1999), pp. 1–20
6. S. Hayes, C. Grebogi, E. Ott, Communicating with chaos. Phys. Rev. Lett. **70**(20), 3031 (1993)
7. S. Hayes, C. Grebogi, E. Ott, A. Mark, Experimental control of chaos for communication. Phys. Rev. Lett. **73**(13), 1781 (1994)
8. E.R. Hunt, Stabilizing high-period orbits in a chaotic system: the diode resonator. Phys. Rev. Lett. **67**(15), 1953 (1991)
9. E. Ott, C. Grebogi, J.A. Yorke, Controlling chaos. Phys. Rev. Lett. **64**(11), 1196 (1990)
10. K. Pyragas, Delayed feedback control of chaos. Philos. Trans. R. Soc. Lond. A Math. Phys. Eng. Sci. **364**(1846), 2309–2334 (2006)
11. T. Saito, H. Fujita, Chaos in a manifold piecewise linear system. Electron. Commun. Jpn. (Part I Commun.) **64**(10), 9–17 (1981)
12. T. Shinbrot, C. Grebogi, J.A. Yorke, E. Ott, Using small perturbations to control chaos. Nature **363**(6428), 411 (1993)
13. F.T. Werner, B.K. Rhea, R.C. Harrison, R.N. Dean, Electronic implementation of a practical matched filter for a chaos-based communication system. Chaos Solitons Fractals **104**, 461–467 (2017)

Chapter 17
Congestion Avoidance on Networks Using Independent Memory Information

Takayuki Kimura(✉)

Department of Electrical, Electronics, and Communication Engineering, Faculty of Fundamental Engineering, Nippon Institute of Technology, 4-1-1 Gakuendai, Miyashiro, Saitama 345-8501, Japan
tkimura@nit.ac.jp

Abstract. We propose in this paper a new routing method that utilizes hop distance information and transmitting history. Most of conventional routing methods use global and real-time information such as the number of waiting packets at nodes. In addition, they assumed that these real-time information can be accessed instantaneously at every node. These unrealistic network circumstances, however, limit applicability of routing methods. On the other hand, our proposed method in this paper uses transmitting histories held by each node to diversify routes of packets. In addition, any packets to exchange global information is not necessary. Numerical simulations indicate that our proposed method shows higher arrival rate of packets for various scale-free type communication network models.

17.1 Introduction

Congestion avoidance on networks such as the Internet, airplanes networks, vehicle traffic networks are inevitable of making future green society. Based on pioneering discoveries of small-world phenomena [1] and scale-free features [2], several studies for revealing congestion occurrence, or avoiding congestion on networks, are stimulated. Strategies for removing the congestion on the networks are categorized into two ways. The first one is to change underlying infrastructure or connections between nodes to alleviate congestion on networks [3]. The second one is to change routes on the networks by using sophisticated strategy. In fact, changing connections of networks is an effective way for removing congestion, however, it needs huge costs. Thus, many researchers have been developing better routing strategies for avoiding congestion, or enhancing network capacities so far.

As an example of network model to alleviate congestion, most of researchers uses communication network models [4]. We also employ the communication network models in this paper. A basic strategy that is commonly employed as the

V. In et al. (Eds.): *Proceedings of the 5th International Conference on Applications in Nonlinear Dynamics*, Understanding Complex Systems, https://doi.org/10.1007/978-3-030-10892-2_17

routing method in real computer networks is the shortest hop method (SP) [5], and this method has a significant problem. If the flowing of packets in the network increases, large volumes of packets are accumulated at nodes where many transmitting paths pass through. This causes delay of communication, or removal of packets in the worst case. To overcome this problem, Yan et al. developed a routing algorithm with local information such as degrees at each node [6]. Echenique et al. proposed a traffic aware method that uses hop distance and the number of waiting packets at nodes [7]. Tang et al. proposed a routing algorithm that employed a global and a local self-adjusting traffic awareness protocol [8]. Huang et al. proposed a probabilistic routing method that utilizes degree of nodes and hop distance information [9]. Wang et al. analyzed traffic dynamics for scale-free networks and proposed a routing algorithm using integrating local static and dynamic information [10,11]. A routing algorithm using link weight information and global dynamic information concerning network traffic was proposed in Ref. [12]. In addition, a routing method using artificial neural networks was proposed in Refs. [13–15] and this method was further improved using mutual connected chaotic neural networks [16–19].

As a routing method for alleviating congestion in communication network model, a memory routing method [20] has already been proposed. In this routing method, at each node selections of nodes for packet transmission are determined using hop distance information of networks, the number of waiting packets at adjacent nodes, and transmitting history. Although the memory routing method [20] avoids the congestion of packet effectively, each node always requires the number of waiting packets of the adjacent nodes at each time instant. The network model [20] is, therefore, assumed that each node instantaneously obtains the global and real-time information with any increase of packets in the whole networks. In addition, similar assumptions are seen in Refs. [6,7,10,12,21]. However, these unrealistic assumptions limit applicability of routing protocol for the real computer networks. In light of the above considerations, we propose a routing method that autonomously determines the transmitting nodes of packets at each node without global information in this paper. Numerical experiments show that our method keeps higher average arrival rate of packets than conventional routing methods for various scale-free type communication networks.

17.2 Communication Network Model

In this paper, an unweighted and undirected graph $G = (V, E)$ is used as a communication network model, where V is a set of nodes and E is a set of links. In this communication model, each node represents a host and a router in the network, and each link represents a connection between the nodes. A packet is then generated at a randomly selected node, and a destination of the packet that is different from the generated node is randomly assigned. Each node has a buffer for storing packets. If a packet is generated at a node, the packet is stored at the tail of the buffer of the node. In addition, a packet at the head of the buffer is transmitted to one of adjacent nodes. Here, adjacent means that nodes are directly connected by a link with each other. In other words, all packets are transmitted according to the First-In-First-Out principle. If the packet is transmitted to the node that has full volume of the packets in its buffer, the transmitted packet is removed from the network. Further, if a packet arrives at its destination, the packet is also removed from the network.

Similar to techniques to diversify node processing abilities introduced by [22], we assign to each node a packet storing capacity and transmitting performance. The packet storing capacity is the maximum number of packets each node can store in its buffer. In addition, the transmitting performance is the maximum number of packets the node can transmit the packets to its adjacent nodes at once time.

The packet storing capacity of the node i, B_i, is defined as

$$B_i = \mu k_i, \tag{17.1}$$

where $\mu > 0$ is a control parameter and k_i is the degree of the node i. By using Eq. (17.1), each node has the packet storing capacity that is proportional to its degree.

The transmitting capability of the node i, C_i, is defined as

$$C_i = 1 + \lfloor \lambda k_i + 0.5 \rfloor, \tag{17.2}$$

where $\lambda > 0$ is a tunable parameter.

If μ and λ are set to large values, congestion of packets hardly occurs because each node can store a large number of packets and transmits many packets to the adjacent nodes at once time. However, constructing such a communication network needs a huge cost. Thus, it is desirable to develop a packet routing strategy that works well with small values of μ and λ.

17.3 Realization of a Routing Method with Memory Information

In our routing method, an adjacent node j to which a packet will be transmitted from the node i at the $t+1$th time is determined using the following equation:

$$y_{ij}(t+1) = \xi_{ij}(t+1) + \zeta_{ij}(t+1), \tag{17.3}$$

where $y_{ij}(t+1)$ is an evaluation value of packet transmission from the node i to adjacent node j, $\xi_{ij}(t+1)$ is distance information from the node i to a destination through the adjacent node j, and $\zeta_{ij}(t+1)$ is memory information between the node i and j. ξ_{ij} and ζ_{ij} will be defined in Eqs. (17.4) and (17.5). If $y_{ij}(t+1)$ has the smallest value of the other nodes, the node i transmits a packet to the adjacent node j.

The distance information, $\xi_{ij}(t+1)$, is defined as follows:

$$\xi_{ij}(t+1) = \frac{d_{ij} + d_{jg(p_i(t))}}{\sum_{k \in N_i} (d_{ik} + d_{kg(p_i(t))})}, \tag{17.4}$$

where d_{ij} is the static hop distance between the node i and the adjacent node j, N_i is a set of adjacent nodes of the node i, $p_i(t)$ is a packet transmitted from the node i at the tth time, $g(p_i(t))$ is the destination of $p_i(t)$, $d_{jg(p_i(t))}$ is the shortest distance between the adjacent node j and $g(p_i(t))$, and this variable dynamically changes depending on $g(p_i(t))$.

The SP method that is commonly employed in the real communication networks is realized by using the distance information only: an adjacent node j to which the packet transmitted from the node i is determined by $\min y_{ij}(t+1) = \xi_{ij}(t+1)$.

The memory information of the node i, $\zeta_{ij}(t+1)$, is defined as follows:

$$\begin{aligned}\zeta_{ij}(t+1) &= \alpha \sum_{s=0}^{t} \gamma^s x_{ij}(t-s) \\ &= \alpha x_{ij}(t) + \gamma \zeta_{ij}(t),\end{aligned} \tag{17.5}$$

where $\alpha > 0$ is a control parameter that determines strength of the memory information, $0 < \gamma < 1$ is a decay parameter of the memory information. A memorizing variable of the packet transmission from the node i to the adjacent node j at the tth time, $x_{ij}(t)$, is defined as follows:

$$x_{ij} = \begin{cases} 1 & (\min y_{ij}(t+1)), \\ 0 & (\text{otherwise}). \end{cases} \tag{17.6}$$

By using memory information, each node successfully memorizes past transmitting history. If the node i frequently transmits the packets to the adjacent node j, ζ_{ij} increases. As a result, the node i avoids to transmit the next packet to the adjacent node j. We expect that this diversification of transmitting routes using the memory information expands network capacities effectively. In addition, any packets to exchange the dynamic information of networks such as the number of waiting packets at nodes is not necessary in our routing method because diversification of transmitting routes can be realized by the transmitting histories held by the nodes themselves. We consider this functionality increases applicability of our proposed method for the real-world systems.

17.4 Numerical Experiments

Since real communication networks are scale-free [23], we will adopt the scale-free topology as the communication network models. We compared performance of our proposed method with that of a SP and a SPr methods. The SP method transmits the packets on the fixed shortest paths between the sources and destinations once the network created and the shortest paths between any nodes are calculated by the Dijkstra algorithm. On the other hand, the SPr method randomly selects the transmitting routes of packets if two or more nodes have equal shortest distance to the destination.

First, we evaluate performance of the routing methods for the BA scale-free networks [2]. The BA scale-free networks are constructed by the following procedure. We begin with a complete graph with m_0 nodes. Then, we add a new node with m_0 links at every time step. Next, we connect m_0 links of the newly added node to the nodes that already exist in the network with probability $\Pi(k_i) = k_i / \sum_{j=1}^{N'} k_j$, where k_i is the degree of the ith node ($i = 1, \ldots, N'$), and N' is the number of the nodes at the current iteration.

Numerical simulations are conducted as follows. First, R packets with randomly selected sources and destinations are generated at each iteration. Here, we defined one iteration as at every node a selection of transmitting nodes among the adjacent nodes and transmissions of packets to the selected node. When a packet arrives at its destination the packet is removed from the networks. In addition, if a packet is transmitted to the adjacent node beyond its buffer size, the packet is removed from the networks. We used the number of iterations, T, for $T = 1,000$. In addition, α and γ in (17.5) are set to 0.01 and 0.99. We also set μ in Eq. (17.1) and λ in Eq. (17.2) to 1,000 and 0.4 respectively.

In these numerical experiments, we used the following four measures.

1. Average arrival rate of packets, $\bar{A}$:

$$\bar{A} = \frac{1}{RT} \sum_{t=1}^{T} a(t), \tag{17.7}$$

where R is the number of generating packets at each iteration, T is the number of iterations, and $a(t)$ is the number of arriving packets at the tth iteration. The average arrival rate is an important measure to evaluate the routing strategy. By reducing or inhibiting the packet congestion in the network, the routing strategy keeps higher arrival rate.

2. Average hop of arriving packets, $\bar{H}$:

$$\bar{H} = \frac{1}{|P_a|} \sum_{i \in P_a} h_i, \tag{17.8}$$

where P_a is a set of the arriving packets, $|P_a|$ is the number of elements of P_a, and h_i is the number of hops of the arriving packet i.

3. Average arrival time of arriving packets, $\bar{H}$:

$$\bar{T} = \frac{1}{|P_a|} \sum_{i \in P_a} t_i, \tag{17.9}$$

where P_a is a set of the arriving packets, $|P_a|$ is the number of elements of P_a, and t_i is the total arrival times of arriving packet i. The arrival time is a consumed time of routing path and the waiting time on nodes. By decreasing $\bar{T}$, the packets are quickly transmitted to their destinations.

4. Standard deviation of arrival times of arriving packets, $\mathrm{var}(T)$:

$$\mathrm{var}(T) = \sqrt{\frac{1}{|P_a|} \sum_{i \in P_a} (T_i - \bar{T})^2}, \tag{17.10}$$

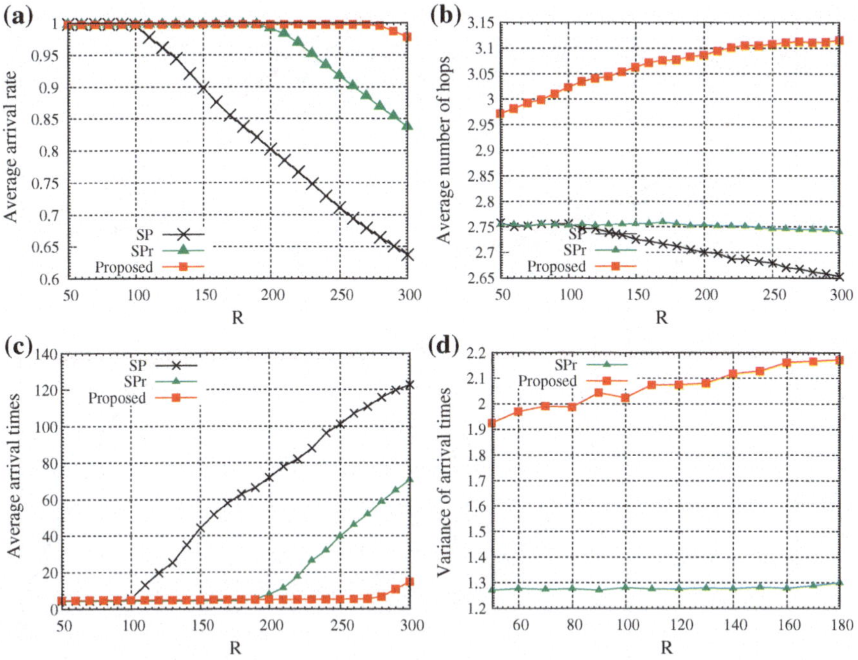

Fig. 17.1. Relationship between the number of generating packets (R) versus **a** an average arrival rate of packets ($\bar{A}$), **b** an average hop of arriving packets ($\bar{H}$), **c** an average arrival time of arriving packets ($\bar{T}$), and **d** a standard deviation of arrival time of arriving packets (var(T)) of SP, SPr, and proposed methods for the BA scale-free networks ($N = 300, m_0 = 4$). In all figures, the error bars are smaller than the symbol size

where P_a is a set of the arriving packets, $|P_a|$ is the number of elements of P_a, $\bar{T}$ is the total arrival time of packets defined by Eq. (17.9), and T_i is the number of arrival times of the arriving packet i.

Figure 17.1 shows the number of generating packets (R) versus an average arrival rate of packets ($\bar{A}$), an average hop of arriving packets ($\bar{H}$), an average arrival time of arriving packets ($\bar{T}$), and (d) a standard deviation of arrival time of arriving packets (var(T)) for the BA scale-free networks ($N = 300, m_0 = 4$).

In Fig. 17.1a, $\bar{A}$ according to the SP method decreases when R becomes 100, and that according to the SPr method start decreasing when R is larger than 190. On the other hand, the proposed method keeps 100% of arrival rate until R becomes 270. In Fig. 17.1b, $\bar{H}$ according to the proposed method is larger than that according to the SP and SPr methods. The proposed method decentralizes the transmitting nodes of packets using the memory information, however, $\bar{H}$ is slightly larger than that of SPr method. This result indicates that the proposed method transmits the packets using the paths that are slightly longer than the shortest paths. In Fig. 17.1c, $\bar{T}$ according to the SP method suddenly increases

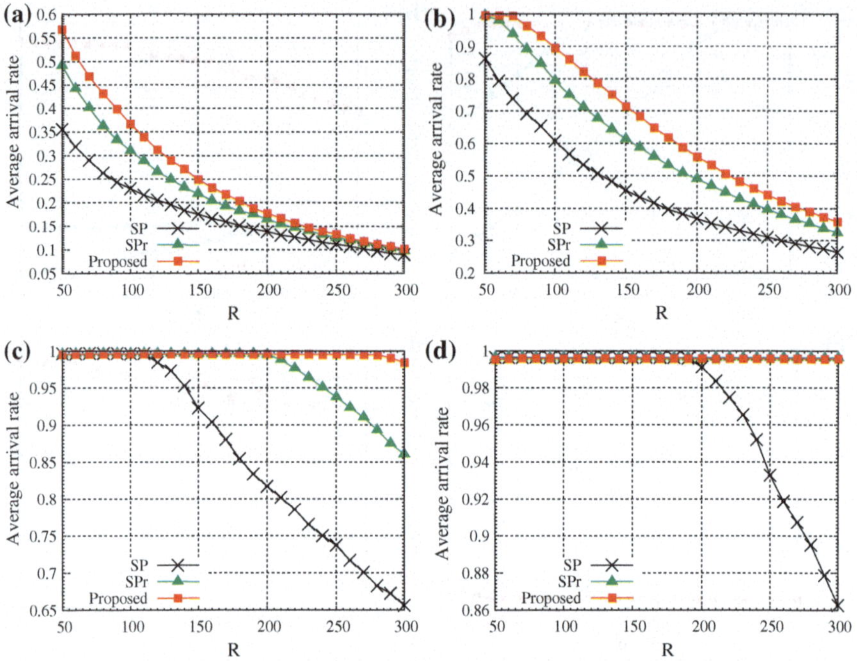

Fig. 17.2. Relationship between the number of generating packets (R) versus an average arrival rate of packets. The transmitting performance is set to **a** $\lambda = 0.0$, **b** 0.1, **c** 0.4, and **d** 0.7, respectively. The number of nodes is set to 300. In all figures, the error bars are smaller than the symbol size

when R is larger than 100. This indicates that the networks are congested if R is larger than 100. The packets are then trapped into the congested node, and they needs long iterations to be transmitted to their destinations. On the other hand, $\bar{T}$ according to the proposed method start increasing when R is larger than 270. In Fig. 17.1d, we compared var(T) of the proposed method with that of the SPr method on the condition that the networks has free-flow state. Figure 17.1d illustrates that var(T) according to the proposed method is larger than that according to the SPr method. These results indicate the proposed methods successfully diversifies the transmitting routes of the packets using the memory information because the proposed method keeps high $\bar{A}$ even if $\bar{H}$, $\bar{T}$, and var(T) increase.

$0 < \lambda \leq 1$ in Eq. (17.2) determines transmitting performance of each node: the number of packets a node can transmit the packets to its adjacent node at once time. If λ is set to a large value, the congestion hardly occurs, however, constructing such the communication networks needs much cost. To clarity the performance of the routing methods against rich or poor conditioned networks, we next evaluated the routing methods for the communication networks with different transmitting performance.

Figure 17.2 shows the number of generating packets (R) and an average arrival rate of packets ($\bar{A}$) for the BA scale-free networks with different transmitting performance. In Fig. 17.2a, each node transmits only one packet to its adjacent node at each iteration. Even if these poor conditioned networks, the proposed method keeps highest arrival rate of packets of the conventional routing methods. In Fig. 17.2b–d, $\bar{A}$ according to all the routing methods increases as λ becomes large. Especially, a point where $\bar{A}$ start decreasing of the proposed methods, i.e., the phase transition point from free-flow to congested state [24], drastically increases as λ becomes large. These results suggest that our method shows higher arrival rate both poor and rich conditioned communication networks.

Next, we evaluate the routing methods for the scale-free networks with different degree exponents. In scale-free networks, the degree distribution follows a power-law distribution, $P(k) \sim k^{-\gamma}$, where $P(k)$ is a probability that a node has degree k. In most of scale-free networks, γ is in the range of $[2, 3]$. We are then interested in how the performance of our method varies if the degree exponent changes. To construct the scale-free networks with different degree exponents, we adopted the network model proposed by Ref. [25]. The scale-free networks with adjustable degree exponent are generated by the following procedures. First, N isolated nodes to which positive integers ($i = 1, \ldots, N$) are indexed are generated in the network. Each node is then assigned an weight defined by $p_i = i^{-\eta}$ where $0 \leq \eta \leq 1$ is a tunable parameter. Next, nodes i and j are connected by a link using probability defined by the normalized weight $p_i / \sum_{k=1}^{N} p_k$ and $p_j / \sum_{k=1}^{N} p_k$ if there is no connection. This model [25] generates a scale-free network with the degree exponent following the power-law distribution $P(k) \sim k^{-\gamma}$, where γ is given by $\gamma = (1 + \eta)/\eta$. We added 10 N edges in these numerical simulations [9].

Figure 17.3 shows the number of generating packets (R) versus an average arrival rate of packets ($\bar{A}$) for the scale-free networks with different degree exponents. In Fig. 17.3, although the BA scale-free network has approximately 4 N links in the networks, this model has 10 N links. Thus, the transmitting routes of this network model are much larger than that by the BA scale-free networks. By using transmitting routes effectively, the proposed method keeps 100% of arrival rate even if R is over 700. In addition, the performance dependency against the different degree exponents cannot be seen in our proposed method. On the other hand, the $\bar{A}$ according to the SPr method becomes higher arrival rate as γ increases.

These numerical results indicate that our proposed method shows high performance for transmitting packets by using the memory information effectively. The memory information works to diversify the transmitting routes of packets and to prevent the communication network models from congestion. As a result, the packets successfully been transmitted to their destinations even if the number of packets in the networks increases.

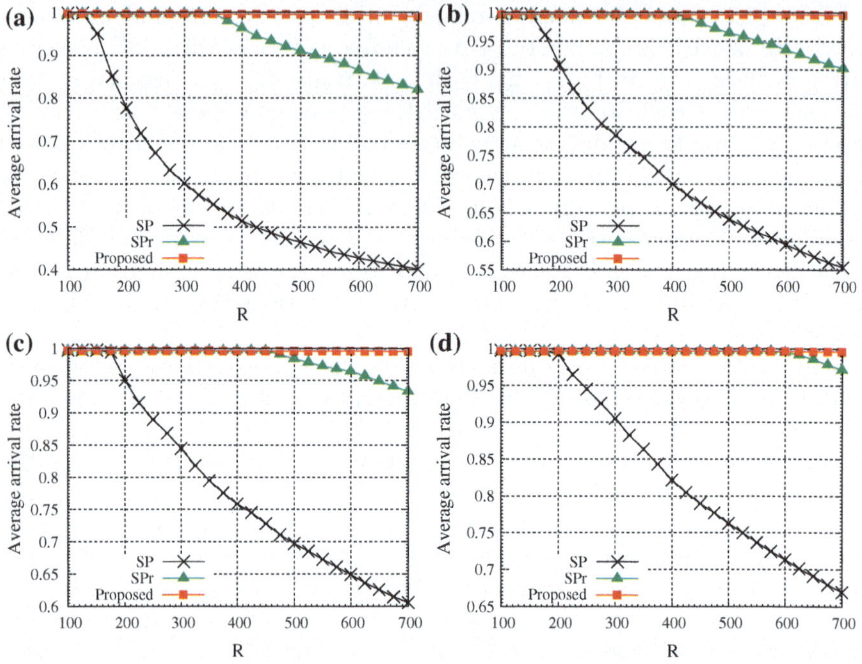

Fig. 17.3. Relationship between the number of generating packets (R) versus an average arrival rate of packets. Degree exponent γ of the scale-free networks is set to **a** $\gamma = 2.0$, **b** 2.5, **c** 2.75, and **d** 3.0. The number of nodes is set to 300. In all figures, the error bars are smaller than the symbol size

17.5 Conclusion

In this paper, we proposed a memory-based routing method that utilizes the hop distance information and the transmitting history for selecting routes of the packets. The key point of our method is that the global information of the networks such as the number of packets at the adjacent nodes is not necessary, however, our method shows higher arrival rate for various scale-free type communication network model even if the number of flowing packets increases. Autonomous selections of transmitting paths using the local information such as transmitting history held by each node has much possibility for real applications because no additional information exchange is required.

Acknowledgments. The research of T.K. was partially supported by a Grant-in-Aid for Young Scientists (B) from JSPS (No.16K21327).

References

1. D.J. Watts, D.J. Watts, S.H. Strogatz, S.H. Strogatz, Nature **393**, 440 (1998)
2. A.L. Barabási, R. Albert, Science **286**, 509 (1999)
3. R. Guimerà, A. Díaz-Guilera, F. Vega-Redondo, A. Cabrales, A. Arenas, Phys. Rev. Lett. **89**(24), 248701 (2002)
4. S. Chen, W. Huang, C. Cattani, G. Altieri, Math. Probl. Eng. **2012**(25), 1 (2012)
5. E.W. Dijkstra, Numer. Math. **1**(1), 269 (1959)
6. G. Yan, T. Zhou, B. Hu, Z.Q. Fu, B.H. Wang, Phys. Rev. E **73**(4), 046108 (2006)
7. P. Echenique, J. Gómez-Gardeñes, Y. Moreno, Europhys. Lett. **71**(2), 325 (2005)
8. M. Tang, Z. Liu, X. Liang, P.M. Hui, Phys. Rev. E **80**(2), 026114 (2009)
9. W. Huang, T.W.S. Chow, Chaos **19**(4), 043124 (2009)
10. W.X. Wang, C.Y. Yin, G. Yan, B.H. Wang, Phys. Rev. E **74**(1), 016101 (2006)
11. W.X. Wang, B.H. Wang, C.Y. Yin, Y.B. Xie, T. Zhou, Phys. Rev. E **73**(2), 026111 (2006)
12. C. Hong, Phys. A **424**, 242 (2015)
13. T. Kimura, H. Nakajima, T. Ikeguchi, Phys. A **376**, 658 (2007)
14. T. Kimura, T. Takamizawa, K. Kimura, K. Jin'no, Nonlinear Theory Appl. IEICE **6**(2), 263 (2015)
15. T. Kimura, T. Takamizawa, T. Matsuura, Am. J. Oper. Res. **06**(04), 343 (2016)
16. T. Kimura, T. Ikeguchi, Neural Comput. Appl. **16**(6), 519 (2007)
17. T. Kimura, T. Ikeguchi, Integr. Comput.-Aided Eng. **14**, 307 (2007)
18. T. Kimura, T. Hiraguri, T. Ikeguchi, Am. J. Oper. Res. **02**(03), 348 (2012)
19. Y. Morita, T. Kimura, Nonlinear Theory Appl. IEICE **9**(1), 95 (2018)
20. T. Kimura, T. Ikeguchi, C.K. Tse, Am. J. Oper. Res. **02**(01), 73 (2012)
21. X. Ling, M.B. Hu, R. Jiang, Q.S. Wu, Phys. Rev. E **81**(1), 016113 (2010)
22. M.B. Hu, W.X. Wang, R. Jiang, Q.S. Wu, Y.H. Wu, Phys. Rev. E **75**(3), 036102 (2007)
23. M. Faloutsos, P. Faloutsos, C. Faloutsos, in *SIGCOMM* (1999), pp. 251–262
24. A. Arenas, A. Díaz-Guilera, R. Guimerà, Phys. Rev. Lett. **86**(14), 3196 (2001)
25. K.I. Goh, B. Kahng, D. Kim, Phys. Rev. Lett. **87**(27), 278701 (2001)

Chapter 18
Opinion Network Modeling and Experiment

Michael Gabbay(✉)

Applied Physics Laboratory, University of Washington, 1013 NE 40th St, Seattle, WA 98105-6698, USA
gabbay@uw.edu

Abstract. We present a model describing the temporal evolution of opinions due to interactions among a network of individuals. This Accept-Shift-Constrict (ASC) model is formulated in terms of coupled nonlinear differential equations for opinions and uncertainties. The ASC model dynamics allows for the emergence and persistence of majority positions so that the mean opinion can shift even for a symmetric network. The model also formulates a distinction between opinion and rhetoric in accordance with a recently proposed theory of the group polarization effect. This enables the modeling of discussion-induced shifts toward the extreme without the typical modeling assumption of greater resistance to persuasion among extremists. An experiment is described in which triads engaged in online discussion. Simulations show that the ASC model is in qualitative and quantitative agreement with the experimental data.

18.1 Introduction

While the experimental study of social influence and opinion change in particular primarily remains the province of the social sciences, the modeling of social influence dynamics, however, has extended into other fields including physics, computer science, and electrical engineering [1–3]. The primary goal of opinion network models is to predict final opinions from initial ones typically via a process that updates node opinions over time. Continuous opinion models — the concern of this paper — allow for incremental shifts in opinion where the amount of change depends upon the distance between node opinions and the network of interpersonal influence that couples nodes. The DeGroot and Friedkin–Johnsen models, as well as the consensus protocol (a continuous time version of the DeGroot model), use a linear dependence in which the shift is proportional to the opinion difference [4–6]. Bounded confidence models posit a hard opinion difference threshold, within which nodes interact linearly, but beyond which the

V. In et al. (Eds.): *Proceedings of the 5th International Conference on Applications in Nonlinear Dynamics*, Understanding Complex Systems, https://doi.org/10.1007/978-3-030-10892-2_18

interaction vanishes [7]. The nonlinear model of [8] uses a soft threshold so that, rather than vanishing completely, the interaction decays smoothly with distance.

Modeling how opinions become more extreme has been of particular concern in the opinion network modeling literature. The primary contribution of this paper is to present an opinion network model, the Accept-Shift-Constrict (ASC) model, which provides an experimentally-supported depiction of group polarization, a classic social psychology effect in which discussion among like-minded group members tends to make groups more extreme. The ASC model describes opinion change processes over a network as group members exchange messages. These processes consist of, first, the acceptance of a persuasive message which can then lead to a shift in the receiver's opinion and also a constriction of the receiver's uncertainty level. In turn, this constriction narrows the extent to which subsequent messages advocating distant opinions are accepted.

This paper proceeds as follows. The next section discusses the group polarization effect along with its treatment in social psychology and the opinion network modeling literature. Section 18.3 describes a recent experiment involving discussion about betting on National Football League (NFL) games, the results of which challenge existing group polarization theory. In Sect. 18.4, an alternative *frame-induced* theory of group polarization is presented that can account for the experimental results. Sections 18.5 and 18.6 present the ASC model and experimentally-relevant simulation results.

18.2 Group Polarization Effect

In the group polarization effect, discussion among group members who are all on the same side of an issue induces more extreme decisions or opinions ("polarization" as used here connotes a group shifting further toward one pole of an issue rather than diverging toward opposite poles as in conventional usage) [9–11]. It was originally referred to as the "risky shift" as it was discovered in an experimental context involving small groups faced with choosing among options of varying risk levels; discussion tended to shift groups toward riskier options than the average of their pre-discussion preferences. Subsequent research observed systematic discussion-induced extremism in homogeneous groups in broader contexts including social and political attitudes and the severity of punishments in jury deliberations. A group is considered to be homogeneous with respect to an issue if all its members have initial preferences that lie on one side of the issue's neutral reference point. Group polarization is then said to occur if after the discussion the mean preference of the group shifts further away from the reference point compared with the mean prior to discussion. Polarization is typically observed for issues that have a substantial judgmental component as opposed to issues like math problems that have demonstrably correct solutions.

Two distinct processes, based on informational and normative influence respectively, are most commonly accepted in social psychology as causes of group polarization [9,10]. The informational influence explanation, known as persuasive arguments theory, focuses on the role of novel arguments. In essence, members of a homogeneous group, although inclined toward the same side of an issue, will typically possess different arguments in support of that side. The exchange of these arguments in discussion then exposes group members to even more information supporting their initial inclination and so shifts it further in the same direction. The normative influence explanation, social comparison theory, posits that the relationship of group member positions with respect to a culturally salient norm is critical rather than the information underlying those positions. The norm is taken to favor one pole of the issue. For example, a norm favoring risk-taking makes riskier positions more socially ideal than cautious ones. A major problem of the informational and normative influence theories is that they always predict polarization for an individual group whenever the polarization preconditions (homogeneous group and judgmental issue) are present, regardless of the distribution of initial opinions within the group. This problem stems from the fact that these theories were never reconciled with stronger, concurrent social influence phenomena such as majority influence and consensus pressure [12].

Within the opinion network modeling literature, extremism has been predominantly modeled by attributing higher network weights to nodes with more extreme initial opinions [13, 14]. This approach, which we refer to as "extremist-tilting," is necessitated by the property of most continuous opinion models that the mean opinion in networks with symmetric coupling remains constant at its initial value — a property that is at odds with the shift in mean exhibited in group polarization. Consequently, extremists must be assigned greater influence over moderates than vice versa in order to shift the mean. This explanation is different from the two more prominent theories above but shares their problem of uniformly predicting polarization for homogeneous groups.

18.3 Experiment

This section describes the group polarization experiment conducted in [12] in which three-person groups engaged in online discussion about wagering on National Football League (NFL) games. As is standard practice in NFL betting, spread betting was employed rather than wagering directly on which team will win the game. In spread betting, the terms "favorite" and "underdog" refer, respectively, to the likely winner and loser of the game itself. The point spread is the expected margin of victory of the favorite team as set by Las Vegas oddsmakers. A bet on the favorite is successful if its margin of victory exceeds the spread; otherwise a bet on the underdog is successful.[1] If Team A is the favorite by a spread of six points over the underdog Team B, then Team A has to win the game by more than six points in order for a bet on Team A to pay off. The objective of the spread is to endeavor to equalize the odds for either the favorite or underdog to win the bet.

In the experiment, an upcoming NFL game was chosen and a pre-survey then elicited subject initial preferences with respect to team choice and a wager

[1] In actual practice, bets are returned if the victory margin equals the spread.

amount on that team from \$0 to \$7 (in whole-dollar increments). On the basis of the pre-survey, discussion groups were constructed with respect to three dichotomous variables. The first is *policy side* of favorite or underdog corresponding to the team chosen as more likely to beat the spread. This variable imposes the polarization precondition of having like-minded group members as the groups are homogeneous with respect to the fundamental policy question of which team will win the bet. The second variable is *disagreement level* of "high" or "low" that depends upon the difference between the minimum and maximum wagers in the group. Each group consisted of low, intermediate, and high wager individuals with respective wagers w_1, w_2, and w_3. In all groups, the intermediate wager was set so that $w_2 \in \{\$3, \$4\}$. In the high disagreement condition, $w_1 = \$0$ and $w_3 = \$7$ giving a difference of \$7. In the low disagreement condition $w_1 \in \{\$1, \$2\}$ and $w_3 \in \{\$5, \$6\}$ so that the difference could be \$3, \$4, or \$5. The third variable is *network structure* of "complete" in which all members could communicate with each other or "chain" in which the intermediate wager member w_2 served as the center node connecting w_1 and w_3. After discussion, each member made their final wager. A group decision was not required but groups arrived at a consensus wager far more often than the alternative outcomes of a two-person majority or three different wagers. A winning (losing) bet resulted in a payoff of \$7 plus (minus) the wager, which was donated to a charity.

Polarization, or more specifically a risky shift, is observed for a group if its mean wager after discussion is greater than its initial mean wager. Most of the 198 groups reached a consensus wager. For these 169 consensus groups, statistically significant results were observed for all three of the manipulated variables. For policy side, only the favorite side exhibited a risky shift whereas the underdog side did not. For disagreement level, restricted to favorite groups (as underdog groups showed no systematic risky shift), high disagreement groups exhibited a greater risky shift than low disagreement groups. For network structure, similarly restricted to favorites, complete networks showed a greater risky shift than chains. All three of these behaviors can be seen in Fig. 18.1 in which substantial polarization is observed when the error interval is above the initial mean.

The above results are not readily explained by standard polarization theory. Particularly challenging is the policy side result as standard theory predicts that both policy sides should show a risky shift. For persuasive arguments theory, members of both the favorite and underdog groups presumably possess novel information in support of their team choice and should therefore increase their confidence and wager. For social comparison theory, a norm toward risk taking should cause both sides to increase their wager. The extremist-tilting explanation prevalent in opinion network modeling also fails to explain this differential polarization by policy side: if individuals with more extreme wagers are taken to be more confident and persuasive, then both favorite and underdog groups should display an equal tendency to increase their wagers.

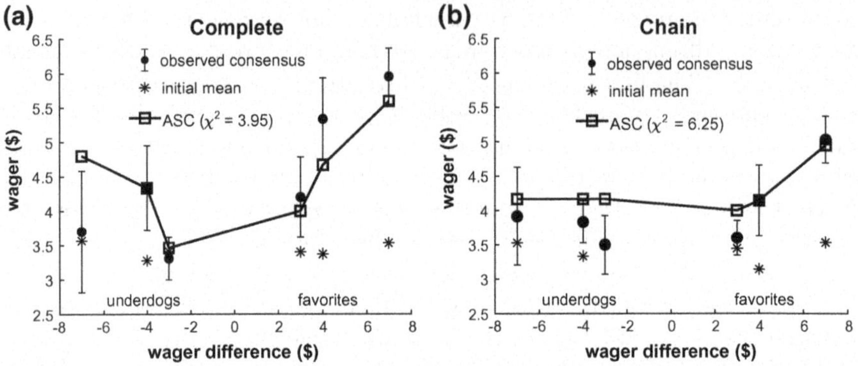

Fig. 18.1. Observed and simulated mean consensus wagers as function of initial wager difference, $w_3 - w_1$. **a** Complete network. **b** Chain network. Favorite groups shown on the positive x axis, underdogs on the negative. Observed consensus is average of final consensus wager (taken as positive for both favorites and underdogs) over groups at each difference value (no \$5 difference groups were used as there were only four total). Also shown is average of the group mean initial wager. Experimental data shown as circles. Error bars are standard errors. χ^2 value is the sum of the squared errors between the simulated and the experimental values normalized by the standard error at each data point. Simulation parameters: $\alpha = 0.034, \lambda(0) = 0.03, \lambda_{min} = 0.01$

18.4 Frame-Induced Polarization Theory

Reference [12] proposes a novel theoretical mechanism for group polarization that explains the results of the experiment. Central to the proposed mechanism is the distinction between the quantitative policy under debate and the *rhetorical frame* — the aspect of the policy upon which deliberations focus. The rhetorical frame will typically correspond to the dominant source of disagreement within the group due, for instance, to uncertainty as to the likelihood of an outcome. In a binary gamble such as in the experiment, the policy (e.g. wager amount) is linked to a given outcome (e.g. team) and so the rhetorical frame should be the subjective probability that that outcome will occur (e.g. win against the spread). The rhetorical frame position $\rho(x)$ is taken to be a function of the policy x. Groups will tend to shift toward the extreme if the functional relationship between the rhetorical position and the policy is concave ($\rho'' < 0$), that is, the rhetorical position increases more slowly as the policy becomes more extreme. For the experiment, such a concave relationship is expected between the subjective probability that a subject's chosen team will win against the spread and the wager amount (see Sect. 18.6).

The effect of concavity is to compress rhetorical distances toward the extreme relative to the distances between more moderate members, making it easier for majorities to form on the extreme side of the mean. Consequently, while the policy distribution may be symmetric so that no majority is favored on either side of the mean (as is approximately the case in our experiment), the distribution of rhetorical positions is skewed so that there is an initial majority on the extreme side of the rhetorical mean. This rhetorically-proximate majority (RPM) converges to a policy position more extreme than the mean to which the remaining minority of group members then concur, thereby resulting in a consensus policy that exhibits group polarization. The members of the F group (analogous to the favorite groups) in Fig. 18.2 provide an example of this mechanism. Although the intermediate member F_2 is equidistant in policy from the moderate F_1 and the extremist F_3, F_2 is rhetorically closer to F_3 and therefore (F_2, F_3) is the RPM pair. They agree on a policy halfway between them to which F_1 comes up due to majority influence. The RPM policy (b) is seen to be greater than the initial mean policy (a).

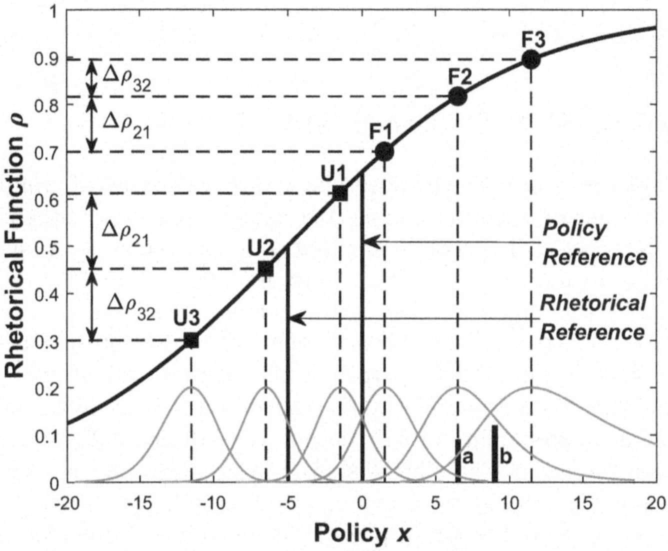

Fig. 18.2. Effect of rhetorical function concavity and offset reference. $\rho(x) = 1/(1 + e^{-\beta(x-x_0)})$ with $\beta = 0.13, x_0 = -5$. Short lines at bottom show alternative F group consensus policies: **a** mean policy $\bar{x} = x_2$; **b** RPM policy $\bar{x}_{23} = (x_2 + x_3)/2$. ASC model acceptance functions in gray at bottom

Although the concavity of the rhetorical function explains the basic group polarization effect, it cannot by itself account for unequal polarization on opposing policy sides as observed in the experiment. Capturing this differential polarization involves the freedom of the rhetorical function to have a different reference point than the policy. The policy reference is defined as the neutral point, taken to be $x = 0$, that demarcates opposing policy sides. The rhetorical reference is defined as the policy value that maps to the neutral point of the rhetorical frame. For a *proper* frame, the rhetorical reference is the same as the policy reference so that the pro and con policy sides coincide with the pro and con rhetorical sides. For an *improper* frame, the rhetorical and policy references are offset so that the rhetorical reference splits one of the policy sides. Figure 18.2 shows how an improper frame can lead to differential polarization by policy side. The rhetorical reference splits the con (negative) policy side, which results in the U group (analogous to underdog groups) being arrayed on the approximately linear part of the rhetorical function rather than on the shoulder as for the F group. Consequently, U_2 is roughly the same rhetorical distance from both U_1 and U_3. Considering the effects of uncertainty and noise, formation of the moderate (U_1, U_2) RPM pair is about as likely as the extreme (U_2, U_3) pair so that systematic group polarization is absent or much reduced as observed in the experiment for the underdog groups. An improper frame can result from the heuristic substitution of a simpler, intuitive frame in place of a more complex proper frame that directly corresponds to the policy [12]. In the experiment, the heuristic frame of which team will win the *game* replaces the proper frame of who will win against the *spread*.

18.5 Accept-Shift-Constrict Model

The ASC model evolves both the positions and uncertainties of group members in response to their dyadic interactions. We consider position first, which can be a policy or, more generally, an opinion about some matter. A persuasive message sent by one group member to another must first be accepted by the recipient in order to shift their policy. While a number of factors can affect whether a message is accepted, the distance between the message's rhetorical position and that of the receiver plays the key role in the ASC model: if the distance is within the *latitude of acceptance* (LOA), the message is likely to be accepted, but the acceptance probability rapidly decays beyond the LOA. If the message is accepted, then the receiver's policy is shifted in proportion to its distance from the sender's policy.

Formally, we encode the above process as an ordinary differential equation for $x_i(t)$, the policy position of the ith group member at time t. For a group with N members, the rate of change of x_i is given by

$$\frac{dx_i}{dt} = \sum_{j=1}^{N} \nu_{ij}(x_j - x_i) \exp\left\{-\frac{1}{2}\frac{(\rho(x_j) - \rho(x_i))^2}{\lambda_i^2}\right\}, \tag{18.1}$$

where ν_{ij} is the coupling strength from $j \to i$ and λ_i is i's LOA. The matrix formed by the coupling strengths defines a position-independent network of influence. In general, ν_{ij} depends on communication rate and other factors such as credibility and expertise ($\nu_{ii} = 0$).

The linear $x_j - x_i$ term in Eq. (18.1) represents the shift effect. The gaussian term represents the acceptance process and we refer to it as the acceptance function, $a(\Delta\rho, \lambda) = e^{-\Delta\rho^2/2\lambda^2}$. Although the acceptance function is always symmetric with respect to the sign of the rhetorical difference, $a(-\Delta\rho) = a(\Delta\rho)$, a concave $\rho(x)$ can causes it to appear asymmetric along the policy axis as clearly seen for F_2 and F_3 in Fig. 18.2.

In addition to position change, communication can also affect a person's uncertainty regarding their position. Group discussion has been observed to increase the level of certainty that members have in their quantitative judgments [15]. Accordingly, we introduce an uncertainty reduction mechanism in our model in which messages from those with similar positions constrict an individual's LOA so that they become more resistant to persuasion from distant positions. Messages originating within the LOA that are accepted decrease the LOA, but not beneath a certain minimum value λ_{min}. This yields for the LOA dynamics:

$$\frac{d\lambda_i}{dt} = \begin{cases} \sum_{j=1}^{N} \nu_{ij}(\lambda_{min} - \lambda_i)e^{-\Delta\rho_{ij}^2/2\lambda_i^2}, & |\Delta\rho_{ij}| \leq \lambda_i \\ 0, & |\Delta\rho_{ij}| > \lambda_i. \end{cases} \tag{18.2}$$

Equations (18.1) and (18.2) comprise the ASC model. Assuming no difference between rhetorical and policy positions, i.e, $\rho(x) = x$, Eq. (18.1) is equivalent to the model of [8] without the self-influence force that models a persistent effect of an individual's initial opinion. The uncertainty reduction dynamics represented by Eq. (18.2) is novel in opinion network modeling. The model of [13] includes a dyadic uncertainty interaction that results in uncertainty change only when dyad members have different uncertainties; this requires that uncertainty levels be visible to other group members, an assumption not present in Eq. (18.2), and does not allow equally uncertain individuals to mutually reinforce their opinions.

A crucial consequence of the uncertainty reduction dynamics in the ASC model is the ability for interim majorities to more effectively maintain their position in the face of minority influence. This effect is essential to the RPM process in the theoretical account of group polarization above (but it occurs regardless of whether or not the rhetorical function is different from the policy). Figure 18.3a illustrates the rough persistence of the majority position for a complete-network triad in which the intermediate member's position is taken to be halfway between the others. For sufficiently low initial disagreement, however, an interim majority will not form and the group equilibrium will be close to its initial mean (Fig. 18.3b).[2]

18.6 Simulation of Group Polarization

This section demonstrates the ability of the ASC model to produce the same qualitative effects as in the frame-induced polarization theory and as observed experimentally. Going beyond qualitative correspondence, its agreement with the data on a quantitative level is also shown. First, we discuss how the coupling strengths ν_{ij} are set. They are treated as dyadic communication rates as determined by simple topological considerations. For a complete network, on average,

[2] The persistence of majority positions on a continuous opinion axis is also found in the agent-based model of [16], which employs a confidence variable that must be transmitted between agents along with opinions, rather than the ASC model's use of an uncertainty interval not visible to others.

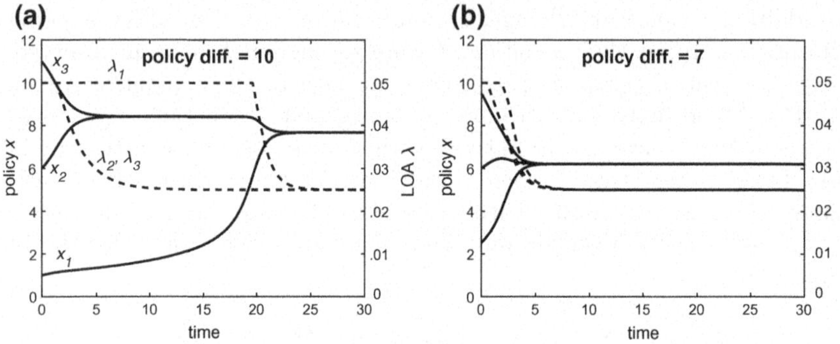

Fig. 18.3. Position and LOA trajectories in ASC model for a complete network. Solid curves show policy positions, dashed curves show LOAs. **a** High initial policy disagreement ($x_3 - x_1 = 10$) showing substantial shift between consensus and initial mean policy ($x_2(0)$). **b** Lower initial policy disagreement (7) results in near simultaneous convergence close to initial policy mean. $\lambda_{1,2,3}(0) = 0.05, \lambda_{min} = 0.025$; $\rho(x)$ as in Fig. 18.2

the communication rates are expected to be the same for all nodes, so we set $\nu_{ij} = 1/2$ for all three dyads. For the chain, if the sequence in which nodes send messages follows the chain path and the center node (node 2) predominantly opts to send its messages simultaneously to both outer nodes (rather than separately), then we expect node 2 to have about twice the communication rate with each of nodes 1 and 3. We therefore set $\nu_{12} = \nu_{32} = 1$ and $\nu_{21} = \nu_{23} = 1/2$.[3] These communication rate expectations are approximately borne out in the experiment [12].

Figure 18.4 displays simulation results for complete and chain network triads that are homogeneous with respect to policy side analogous to the experimental setup. The baseline case (dotted curve) consists of an intermediate node with an initial policy $x_2(0)$ halfway between the initial positions of the moderate $x_1(0)$ and the extremist $x_3(0)$. The other cases shown (light gray curves) account for position uncertainty by allowing $x_2(0)$ to deviate by various small amounts from the baseline case. The discussion-induced shift in the mean is plotted against the initial policy difference between the extremist and the moderate, where the opposing pro and con policy sides are shown on the positive and negative sides of the horizontal axis respectively. For the pro (con) side, a positive (negative) polarization shift indicates a shift toward the extreme — a higher wager in the case of the experiment. The mean over all the cases (solid dark curve) can be used to gauge the extent of systematic polarization.

The top row of Fig. 18.4 represents a proper rhetorical frame in which the policy and rhetorical references are coincident. In the experiment, the proper

[3] The sum of the communication weights is normalized to the same (arbitrary) value of 3 in both networks, a value that only affects the transient time and not the final equilibrium.

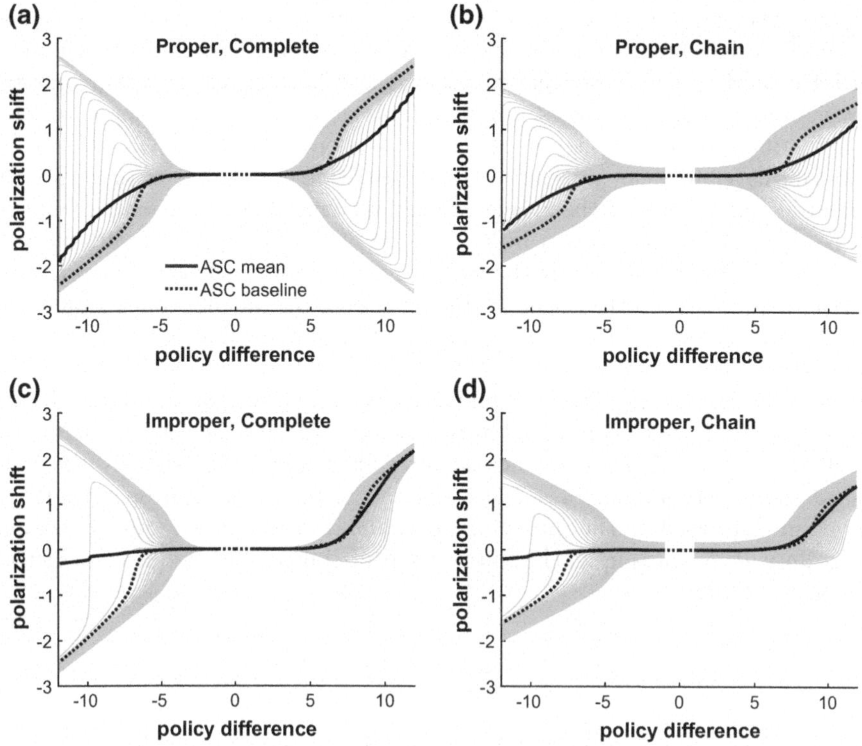

Fig. 18.4. ASC simulations for triad networks with variability in intermediate node policy. $\rho(x)$ taken as in Fig. 18.2. Top row shows proper rhetorical frame ($x_0 = 0$). Bottom row shows improper rhetorical frame ($x_0 = -5$). Positive and negative policy sides are on positive and negative horizontal axis respectively. Polarization shift, $\delta = \bar{x}(t_f) - \bar{x}(0)$, plotted as a function of the initial policy difference, $\Delta = x_3(0) - x_1(0)$. Shift toward the extreme corresponds to $\delta > 0$ for positive policy side and $\delta < 0$ for negative side. The position of the intermediate node was varied according to $x_2(0) = \pm(6+\epsilon)$ for the positive and negative policy sides, where ϵ takes on 41 uniformly-spaced values over the interval $[-1, 1]$. $x_1(0) = 6 - \Delta/2$ and $x_3(0) = 6 + \Delta/2$ for $\Delta > 0$ and analogously for $\Delta < 0$. ASC mean (black) taken over all ϵ values. Shifts for individual ϵ values shown as gray curves. Dotted curve shows $\epsilon = 0$ baseline case. Gap in the curves is the region where $x_2(0)$ would go beyond $x_1(0)$ or $x_3(0)$. ASC model parameters: $\lambda_{1,2,3}(0) = 0.05, \lambda_{min} = 0.025$

frame is the subjective probability of the favorite winning against the spread. The rhetorical function is taken to be concave with increasing policy extremity.[4] Regarding the mean, both policy sides exhibit equal polarization that increases with disagreement level and with the complete network showing more polariza-

[4] If the subjective probability of one of the binary outcomes is taken as the rhetorical frame and opposing policy sides have opposite signs, then concavity with increasing policy extremity yields an overall S-shaped rhetorical function as explained in [12].

tion than the chain. Considering higher disagreement levels, the mean polarizes less than the baseline case because some groups actually depolarize — those in which the moderate and intermediate node are sufficiently close to overcome the skewing effect of the rhetorical function. This ability to predict depolarization for individual groups despite the dominant tendency toward polarization is an important capability not present in the informational, normative, or extremist-tilting theories. Although the proper frame does exhibit polarization, accounting for the differential polarization by policy side observed experimentally requires use of an improper frame as is the subjective probability that the favorite will win the game. The bottom row of Fig. 18.4 employs an improper frame and indeed shows substantial systematic polarization for positive policies and little for negative ones.

The ASC model can also be quantitatively tested against the data. Groups can be simulated using their actual initial wagers and with the coupling strengths as set above. The rhetorical function $\rho(w)$ that maps the (signed) wager to the subjective probability of a favorite game victory (the improper frame) is derived in [12] based on the theory of individual decision making under risk and uncertainty. It depends upon the subjective probability $p(w)$ of a favorite victory (the proper frame)

$$p(w) = \frac{1}{2} - \frac{1}{8\alpha w} \pm \frac{1}{2}\sqrt{1 + \frac{1}{16\alpha^2 w^2}}, \tag{18.3}$$

where the $+$ $(-)$ sign implies bets on the favorite (underdog). The free parameter α is the risk aversion that quantifies how sensitive individuals are to variance around the expected value of the payoff. It is assumed to be identical for all subjects. The rhetorical function is then given by

$$\rho(w) = \frac{1}{2}\mathrm{erfc}\left\{\mathrm{erfc}^{-1}\left(2p(w)\right) - \frac{s_0}{\sigma\sqrt{2}}\right\}, \tag{18.4}$$

where $\mathrm{erfc}(u) = \frac{2}{\sqrt{\pi}}\int_u^\infty e^{-v^2} dv$. The parameter s_0 is the point spread for the game in question and $\sigma = 12.8$ is the empirical standard deviation for the margin of victory in NFL games. Both $p(w)$ and $\rho(w)$ are S-shaped implying a concave relationship between the subjective probability of the outcome estimated as more likely and the wager magnitude.

In addition to the risk aversion, there are two free parameters from the ASC model that need to be fit from the data, the initial LOA, $\lambda(0)$, and the minimum LOA, λ_{min}, both assumed identical for all subjects. The three parameters are estimated by minimizing the sum of χ^2 error values over both complete and chain networks. The simulation results are shown in Fig. 18.1. A three-parameter χ^2 goodness-of-fit test, which takes as its null hypothesis that the model is correct, yields a probability $Q = 0.33$ that χ^2 could have exceeded its observed value of 10.2 by chance. With a conservative threshold of $Q < 0.2$ for rejecting the null hypothesis, the ASC model is found to be consistent with the data.

18.7 Conclusion

The ASC model presented here describes a dual process of opinion and uncertainty change based on the greater acceptance rate of messages within one's LOA and the decrease in LOA due to exposure to similar views. A key dynamic in the model is the ability of proximate majorities to form and persist for symmetric networks, thereby enabling majorities to exert outsized influence and produce a consensus opinion different from the initial mean. Importantly, the ASC model does not involve the exchange of uncertainties over the network unlike other models in which uncertainties are directly coupled along with opinions [13,16]. Another important innovation of the ASC model is the conceptualization of distinct dimensions of opinion and rhetoric: opinion is an evaluation directly tied to a decision or other behavioral outcome of interest while rhetoric determines whether messages aimed at shifting opinions are found persuasive. If the rhetorical function mapping opinion to rhetorical position is concave, then proximate majority formation at the extreme is facilitated. Consequently, the ASC model can generate systematic group polarization due to the structure of the decision space rather than by assuming an asymmetric network structure in which influence is associated with extremity as typically done in opinion network modeling. The ASC model simulations shown here display the same qualitative phenomena as observed in the experiment: polarization on one policy side but not the other, increasing polarization with disagreement level, and greater polarization for complete networks than for chains. Furthermore, the ASC model is in quantitative agreement with the experimental data.

Acknowledgements. This work was supported by the Office of Naval Research under grant N00014–15–1–2549.

References

1. C. Castellano, S. Fortunato, V. Loreto, Rev. Mod. Phys. **81**(2), 591 (2009)
2. D. Kempe, J. Kleinberg, S. Oren, A. Slivkins, Netw. Sci. **4**(01), 1 (2016)
3. A.V. Proskurnikov, R. Tempo, Ann. Rev. Control **43**(Supplement C), 65 (2017)
4. M.H. DeGroot, J. Am. Stat. Assoc. **69**(345), 118 (1974)
5. N.E. Friedkin, E.C. Johnsen, *Social Influence Network Theory: A Sociological Examination of Small Group Dynamics* (Cambridge University Press, Cambridge, UK, 2011)
6. R. Olfati-Saber, J.A. Fax, R.M. Murray, Proc. IEEE **95**(1), 215 (2007)
7. J. Lorenz, Int. J. Mod. Phys. C **18**(12), 1819 (2007)
8. M. Gabbay, Phys. A **378**, 118 (2007)
9. D.G. Myers, H. Lamm, Psychol. Bull. **83**(4), 602 (1976)
10. D.J. Isenberg, J. Pers. Soc. Psychol. **50**(6), 1141 (1986)
11. C.R. Sunstein, J. Polit. Philos. **10**(2), 175 (2002)
12. M. Gabbay, Z. Kelly, J. Reedy, J. Gastil, Soc. Psychol. Q. **81**(3), 248 (2018)
13. G. Deffuant, F. Amblard, G. Weisbuch, T. Faure, J. Artif. Soc. Soc. Simul. **5**, 4 (2002)
14. N.E. Friedkin, IEEE Control Syst. **35**(3), 40 (2015)
15. J.A. Sniezek, Organ. Behav. Hum. Decis. Process. **52**(1), 124 (1992)
16. M. Moussaid, J.E. Kammer, P.P. Analytis, H. Neth, PLOS ONE **8**(11), 1 (2013)

Chapter 19
Analysis of Dynamics of Nonlinear Map Optimization

Kenya Jin'no(✉)

Faculty of Knowledge Engineering, Tokyo City University, Setagaya-ku, Tokyo 158-8857, Japan
jinno@nit.ac.jp

Abstract. We are developing a swarm intelligence optimization algorithm based on nonlinear dynamical system theory. In this article, we introduce Nonlinear Map Optimization (abbr. NMO) which we proposed. NMO is classified as swarm intelligence (abbr. SI) optimizer and consists of some search individuals whose dynamics is driven by a simple nonlinear map. The simple nonlinear map is regarded as a kind of circle map. For effective optimal solution search, the search point distribution of each search individual is important. The search point distribution is controlled by the simple nonlinear map. The parameters of the simple nonlinear map are controlled so that the search point distribution can be effectively searched for the optimal solution. Also, this map generates a chaotic search point time series while keeping the search range. Such a time series can efficiently search within the search range. As a result, NMO can search along the valley of the evaluation function. Namely, NMO is considered to have a rotation invariance and a scaling invariance. NMO can also be regarded as a system in which one-dimensional map oscillators move while being coupled with each other with a coupling strength according to distance. Therefore, the analysis of the dynamics of NMO gives new knowledge of the nonlinear coupled map.

19.1 Introduction

To search for an optimum value of a given objective function is a very important problem in various engineering fields. Optimization problems can be classified into two categories depending on whether the variables are continuous or discrete. We focus on the continuous optimization problem in this article. In order to solve the continuous optimization problem, gradient method is the most popular algorithm that the search direction is defined by the gradient of the objective function at the current search point. However, the gradient method cannot be utilized when the gradient information of the objective function cannot be obtained. Such problem is called as "Black-box problem".

V. In et al. (Eds.): *Proceedings of the 5th International Conference on Applications in Nonlinear Dynamics*, Understanding Complex Systems, https://doi.org/10.1007/978-3-030-10892-2_19

In order to solve the black-box problem, various kinds of solving algorithm are proposed [19]. Swarm Intelligence (abbr. SI) algorithm [2,3] is one of such solvers. SI contains ant colony optimization [6], artificial bee colony algorithm [14], firefly algorithm [21], cuckoo search [22], particle swarm optimization [5,15], and so on. SI systems consist typically of a population of simple plural agents interacting locally with one another and with their environment. The agents follow very simple rules. SI algorithms sample a set of solutions which is too large to be completely sampled, therefore the SI algorithm can find a feasible solution in a short time. Many SI algorithms implement some form of stochastic optimization, so that the solution found is dependent on the set of random variables generated. Also, the SI algorithms cannot guarantee to find a globally optimal solution. Therefore, we consider that the theoretical analysis of the dynamics of the agents is important.

In order to clarify the search mechanism of PSO, we proposed a canonical deterministic PSO (abbr. CD-PSO) [10,20] that the stochastic factors are removed, and we analyzed the behavior of each particle of PSO based on the dynamical system theory [7–10,20]. The particles of the PSO are scattered into the search space of the design variables, and the particles calculate an evaluation value corresponding to the design variable. And each particle shares its evaluation value and its parameter's information in a swarm.

What is important in the SI algorithm are "Exploration" and "Exploitation" [4] "Exploration" corresponds to a global solution search capability, and "Exploitation" corresponds to a local solution search capability. Based on the analysis result of the dynamics of the canonical deterministic PSO [7,10,20], the global search capability of the CD-PSO is related to sharing best solution information within the swarm. Also, we have clarified that a distribution of solution search points is very important for local solution searching capability of CD-PSO [11,16,17]. While various versions of stochastic PSOs have been proposed, standard PSO 2011 (abbr. SPSO2011) [23] has a superior solution search capability. The search point of SPSO2011 is shown in Fig. 19.1a that is similar to the normal distribution. The center of the horizontal axis denotes the the found best solution point. This distribution has a high center, and therefore has high local search capability [17]. On the other hand, Fig. 19.1b shows the distribution of the search points of the CD-PSO. The distribution at both ends is high and the distribution at the center is low. Due to such a distribution, the local search capability of CD-PSO is low. Therefore, we propose a new SI algorithm to improve the search point distribution [13]. The new SI algorithm consists of some search individuals driven by a simple nonlinear mapping which is classified into a kind of circle map. Since each search individual is driven by a nonlinear mapping, we call this system a nonlinear map model optimization method (abbr. NMO) [13]. The circle map which derives the dynamics of the search individuals can generate chaos. The chaotic motion leads diversity to the search point.

Comparing with other SI algorithms, NMO can search a feasible solution with a small number of search individuals. Therefore, the computational amount of NMO is smaller than other SI algorithms. Also, the dynamics of NMO is

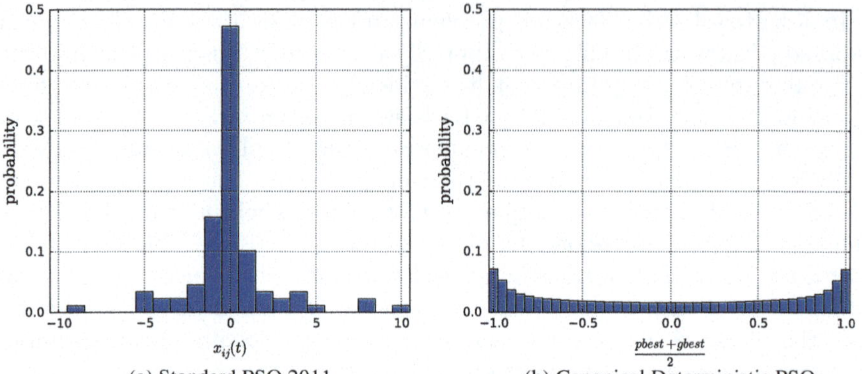

(a) Standard PSO 2011 (b) Canonical Deterministic PSO

Fig. 19.1. The search point distribution. The center of the horizontal axis denotes the found best solution point

described by a deterministic difference equation. For this reason, NMO is classified into deterministic systems.

19.2 Nonlinear Map Optimization

To improve the local search capability of SI algorithms, the search point distribution is important. The local search is carried out under the assumption that a better solution exists around the good solution found until the current iteration. Namely, we consider a search individual that can perform a local search while keeping a certain search range. In order to realize search individual having the above search properties, we consider a system which is consisted of some search individuals. The current jth dimensional position of the ith individual and the jth dimensional internal state variable of the jth individual are described by the following difference equation.

$$x_{ij}(t+1) = R_{ij}(t)\cos(\theta_{ij}(t)) + p_{ij}(t), \tag{19.1}$$

$$\theta_{ij}(t+1) = \begin{cases} \theta_{ij}(t) + \gamma \left| \frac{x_{ij}(t)-p_{ij}(t)}{R_{ij}(t)} + \frac{\pi}{2} - \arccos(\varepsilon_c) \right| + \frac{\pi}{2} - \arccos(\varepsilon_c) \\ \qquad \text{if } 0 < \sin\theta_{ij}(t)\cos\theta_{ij}(t) < \varepsilon_c, \\ \theta_{ij}(t) + \gamma \frac{|x_{ij}(t)-p_{ij}(t)|}{R_{ij}(t)} \qquad \text{otherwise.} \end{cases} \tag{19.2}$$

where $R_{ij}(t)$ denotes the jth dimensional search range of the ith search individual. The current iteration is t. $p_{ij}(t)$ is determined by the following equation.

$$p_{ij}(t) = \rho\, pbest_{ij}(t) + (1-\rho) gbest_j(t) \tag{19.3}$$

where $\boldsymbol{pbest}_i(t)$ denotes the personal best position of the ith search individual, and $\boldsymbol{gbest}(t)$ denotes the global best position in the swarm as follow.

$$\boldsymbol{pbest}_i(t) = \arg\min_{\boldsymbol{x}_i(\tau)} f(\boldsymbol{x}_i(\tau)), 0 \le \tau \le t \tag{19.4}$$

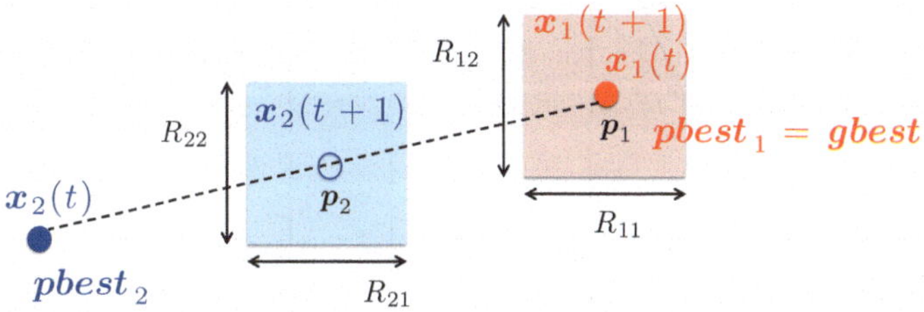

Fig. 19.2. The search search strategy of NMO. $\boldsymbol{p}_k$ denotes the current best position of the kth individual. The global search position is determined based on the best position $\boldsymbol{p}_k$. Also, each rectangle region around $\boldsymbol{x}_k$ denotes the local search range which is limited by the parameter $R_{ij}(t)$

$$\boldsymbol{gbest}(t) = \arg\min_i f(\boldsymbol{pbest}_i(\tau)), 0 \le \tau \le t \tag{19.5}$$

We assume 2 individuals are located in the search space as shown in Fig. 19.2. In this case, we suppose the 1st individual which denotes as $\boldsymbol{x}_1$(t) discovers the current global best position $\boldsymbol{gbest}$. $\boldsymbol{p}_k$ denotes the current best position of the kth individual which is calculated by Eq. (19.3). The best position of the 1st individual is not changed, the 1st individual searches within the rectangle region in the vicinity of $\boldsymbol{x}_1$(t). On the other hand, the current best position of the 2nd individual $\boldsymbol{p}_2$ is changed as shown in Fig. 19.2, the 2nd individual moves the new position which denotes as $\boldsymbol{x}_2(t+1)$. And the 2nd individual searches within the rectangle region in the vicinity of the renewal position $\boldsymbol{x}_2(t+1)$. Therefore the global search position is determined based on the best position $\boldsymbol{p}_k$. Also, each rectangle region around $\boldsymbol{x}_k$ denotes the local search range which is limited by the parameter $R_{ij}(t)$.

To analyze the dynamics of the local search capability, we consider a one-dimensional return map of the internal state variable $\theta_{ij}(t)$ as shown in Fig. 19.3. The parameters are set as $R(t) = 1\forall t$, $p_{ij}(t) = 0\forall t$, $\gamma = 0.795$, and $\varepsilon_c = 0.01$. The horizontal axis of Fig. 19.3 denotes the current internal state variable $\theta_{ij}(t)$, and the vertical axis denotes the next internal state variable $\theta_{ij}(t+1)$. Also, Figs. 19.3b, c show the enlargement figures of the return map in the vicinity of $\theta_{ij}(t) = 0$ and $\theta_{ij}(t) = \pi/2$.

Without loss of generality, we consider the case of $p_{ij}(t) = 0$ for simplicity. In this cases, Eq. (19.2) is rewritten as follows.

$$\theta_{ij}(t+1) = \theta_{ij}(t) + \gamma|\cos(\theta_{ij}(t))|. \tag{19.6}$$

The slope of the one-dimensional map around $\theta_{ij}(t) = \pi/2$ are given as

$$\frac{\mathrm{d}\theta_{ij}(t+1)}{\mathrm{d}\theta_{ij}(t)} = \begin{cases} 1+\gamma\sin(\theta_{ij}(t)) & \text{for } \frac{\pi}{2} \le \theta_{ij}(t), \\ 1+\gamma\sin(\theta_{ij}(t)+\arccos(\varepsilon_c)) & \text{for } \frac{\pi}{2} - \arccos(\varepsilon_c) < \theta_{ij}(t) < \frac{\pi}{2}, \\ 1-\gamma\sin(\theta_{ij}(t)) & \text{for } \theta_{ij}(t) \le \frac{\pi}{2} - \arccos(\varepsilon_c). \end{cases} \tag{19.7}$$

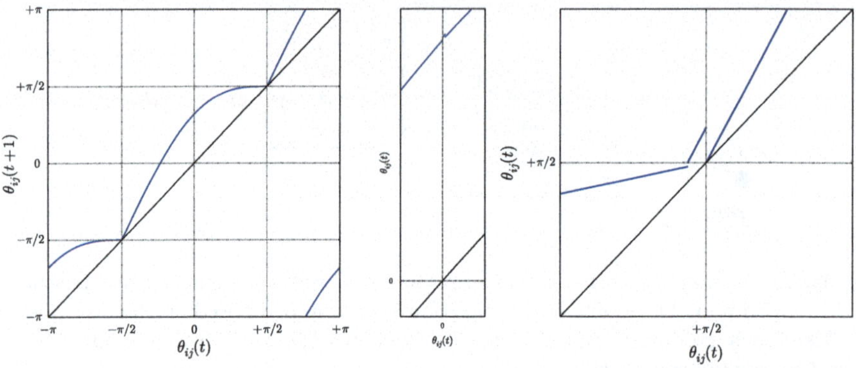

(a) Overview (b) vicinity of $\theta_{ij}(t) = 0$ (c) vicinity of $\theta_{ij}(t) = \pi/2$

Fig. 19.3. One-dimensional map of the internal state variable $\theta_{ij}(t)$ ($R(t) = 1$, $\gamma = 0.795$, $\varepsilon_c = 0.01$)

Here, we consider the stability in the vicinity of $\theta_{ij}(t) = \pi/2$. The gradient in $\theta_{ij}(t) \leq \pi/2 - \arccos(\varepsilon_c)$ is less than 1. On the other hand, the gradient in $\theta_{ij}(t) > \pi/2 \arccos(\varepsilon_c)$ is greater than 1. Therefore, the individual which is located within the region $\theta_{ij}(t) \leq \pi/2 - \arccos(\varepsilon_c)$ converges toward a point. However, the individual which is located within the region $\theta_{ij}(t) > \pi/2 - \arccos(\varepsilon_c)$ diverges from a point. Namely, the system does not have a stable point, therefore, the individual keeps to move.

Figure 19.4 shows the time evolution of the search point when the parameters set as $R(t) = 1$, and $\varepsilon_c = 0.01$. In Fig. 19.4, the vicinity of the center of the vertical axis corresponds to the best location found so far. The parameter γ controls the convergence speed. When γ is large, the convergence speed is fast

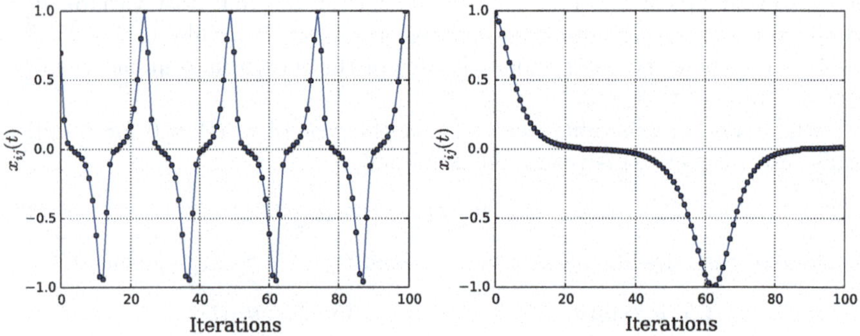

(a) $\gamma = 0.795$ (b) $\gamma = 0.200$

Fig. 19.4. The time evolution of the search point ($R(t) = 1$, and $\varepsilon_c = 0.01$). The convergence speed is controlled by the parameter γ

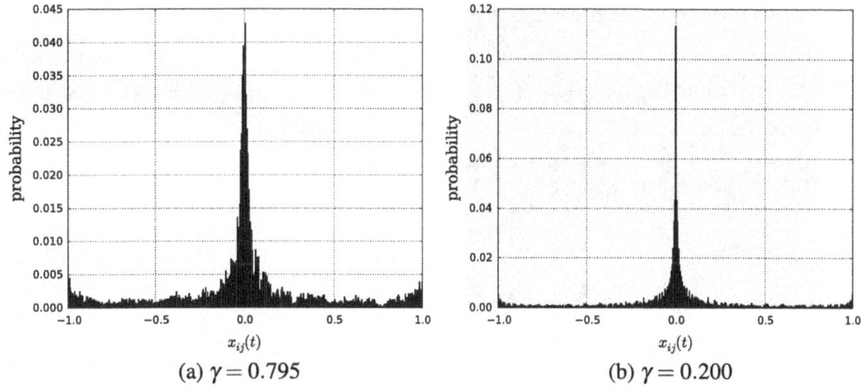

Fig. 19.5. The search point distributions of NMO ($R(t) = 1$, $\varepsilon_c = 0.01$). The distribution is controlled by the parameter γ

as shown in Fig. 19.4a. Conversely, the convergence speed is slow as shown in Fig. 19.4b. This time series indicates that each search individual keeps search range and intensively searches the central part. Since the search range can be limited by parameter $R_{ij}(t)$, the local search capability improves. Especially, the detailed search of the central part leads to the improvement of the search capability. Figure 19.5 illustrates the distribution of the search point of NMO. The search points are distributed within the range $[p_{ij}(t)-R_{ij}(t), p_{ij}(t)+R_{ij}(t)]$. Figure 19.5 indicates that the search point distribution in the vicinity of the center of the range is very high. This means that the search individual has a high local search capability.

The distribution is controlled by the parameter γ. Namely, the parameter γ is related to the variance of the distribution. If γ is small, the variance of the distribution becomes sharp.

In order to improve the local search capability, it is desirable to search for as many diverse points as possible. The system which is described by Eqs. (19.1) and (19.2) is regarded as a kind of circle map. On circle map, the parameter γ is the most important parameter. In order to investigate the influence of the parameter γ on search point distribution, we create a bifurcation diagram with the parameter γ. The bifurcation diagram is shown in Fig. 19.6a. The horizontal axis denotes the parameter γ, and the vertical axis denotes the search point in the search region. The γ on the horizontal axis is varied from 0.1 to 0.9. The bifurcation diagram indicates that the search point spreads throughout the search range at almost all parameter γ. Also, to confirm the property of the time series of the search point we calculate the Lyapunov exponent of the time series of the search point corresponding to Fig. 19.6a.

The one-dimensional map of the search point is given as

$$x_{ij}(t+1) = f(x_{ij}(t), \gamma). \tag{19.8}$$

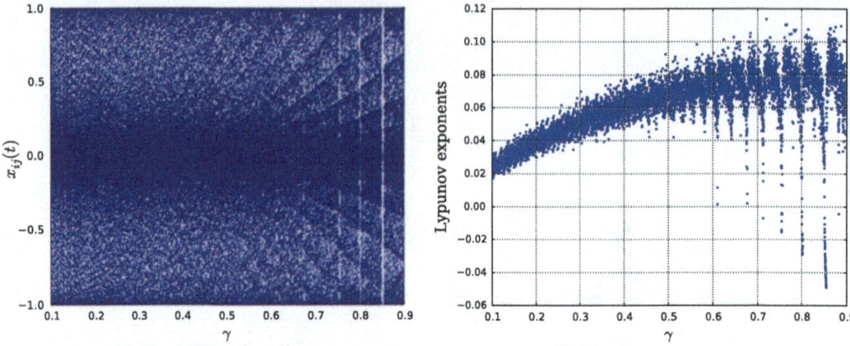

(a) The bifurcation diagram (b) The Lyapunov exponent

Fig. 19.6. The bifurcation diagram and the Lyapunov exponents of the search point time series. ($R(t) = 1$, $\varepsilon_c = 0.01$)

The Lyapunov exponent $\lambda(\gamma)$ of Eq. (19.8) is derived as follows.

$$\lambda(\gamma) = \lim_{N\to\infty} \frac{1}{N} \sum_{t=0}^{N} \ln \left| \frac{\mathrm{d}}{\mathrm{d}x_{ij}(t)} f(x_{ij}(t), \gamma) \right| \tag{19.9}$$

If the Lyapunov exponent of one-dimensional map system takes a positive value, the time series exhibits chaotic motion [1]. Figure 19.6b shows the Lyapunov exponent for each parameter γ. The horizontal axis denotes the parameter γ, and the vertical axis denotes the Lyapunov exponent. The result of Fig. 19.6b indicates that the Lyapunov exponent exhibits positive values at almost all parameter γ. Therefore, the corresponding time series of the search points exhibits a chaotic motion. Namely, since the time series of each dimension of each search individual is searched chaotically, it is possible to efficiently local search within the search range.

19.3 Search Ability

In order to confirm the search capability of NMO, we carry out some numerical simulations by using some well-known benchmark functions. Note that the search result of NMO depends only on the initial location of each search individual since NMO is a deterministic system. Therefore, the distribution of the initial arrangement of the search individuals is very important.

At first, we consider the case of '2D - Rotated Shift Ellipse function' (f_1) which is described in Eq. (19.10).

$$\begin{aligned} f_1(x,y) = 100 \left((x-4.3)\cos\tfrac{\pi}{6} - (y+0.6)\sin\tfrac{\pi}{6}\right)^2 \\ + \left((x-4.3)\sin\tfrac{\pi}{6} - (y+0.6)\cos\tfrac{\pi}{6}\right)^2 \end{aligned} \tag{19.10}$$

The global minimum value of f_1 is 0 at $(x, y) = (4.3, -0.6)$. This function has a dependency between variables by translating the coordinates of the original

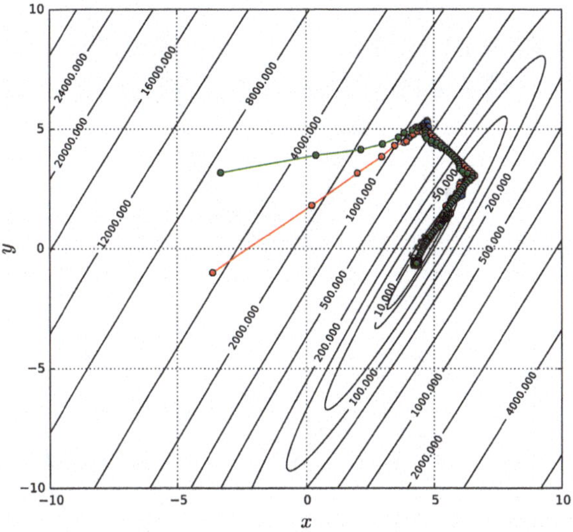

Fig. 19.7. The trajectories of three search individuals for Rotated Shift Ellipse function ($f_1(x, y)$). The search individuals move along the rotation angle

function and further rotating the axis. We consider the case where the NMO consists of only three search individuals. Figure 19.7 shows the contour map of Eq. (19.10), and the trajectories of three search individuals. We confirm that the search individuals move along the rotation angle and can reach the minimum point since the local search capability is improved.

Next, we consider the case of '2D - Rosenbrock function' which is described in Eq. (19.11).

$$f_2(x, y) = 100\left(y - x^2\right)^2 + (x - 1)^2. \tag{19.11}$$

The global minimum value of f_2 is 0 at $(x, y) = (1, 1)$ which is located inside a long, narrow, parabolic shaped flat valley. Therefore, the searching global minimum is difficult. The trajectories of three search individuals are shown in Fig. 19.8.

Also in Rosenbrock function, NMO can track the valley of the evaluation function. These results indicate that NMO is considered to have rotation invariance and scaling invariance [7–9]. However, the theoretical analysis on these invariants is insufficient at the present time.

The above two cases are unimodal functions. Next, we consider the case of multimodal functions. The following equation is '2D - Ackley function' which has many local minima.

$$f_3(x, y) = 20 - 20\exp\left(-0.2\sqrt{\frac{x^2 + y^2}{2}}\right) + e - \exp\left(\frac{\cos(2\pi x) + \cos(2\pi y)}{2}\right) \tag{19.12}$$

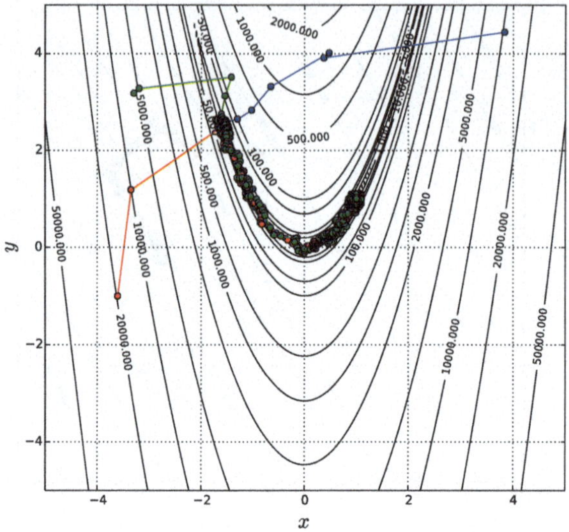

Fig. 19.8. The trajectories of three search individuals for Rosenbrock function ($f_2(x, y)$)

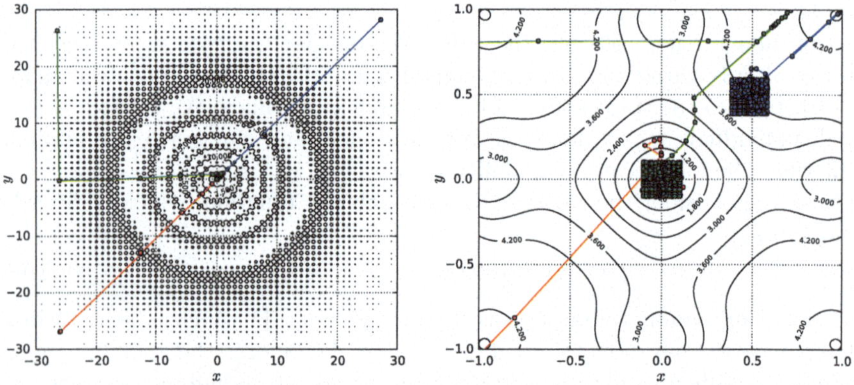

(a) Outline drawing (b) Enlargement around the global minimum

Fig. 19.9. The trajectories of three search individuals for Ackley function ($f_3(x, y)$)

The global minimum value of Ackley function is 0 at $(x, y) = (0, 0)$. Figure 19.9 shows the contour map of Eq. (19.12), and the trajectories of three search individuals. In this case, NMO finds the global minimum solution. However, depending on the search range $R_{ij}(t)$ and initial location of search individuals, NMO may not search the optimal solution.

Finally, we consider the case of '2D - Rastrigin function' which is described by the following equation.

$$f_4(x, y) = 20 + \left(x^2 - 10\cos(2\pi x)\right) + \left(y^2 - 10\cos(2\pi y)\right) \tag{19.13}$$

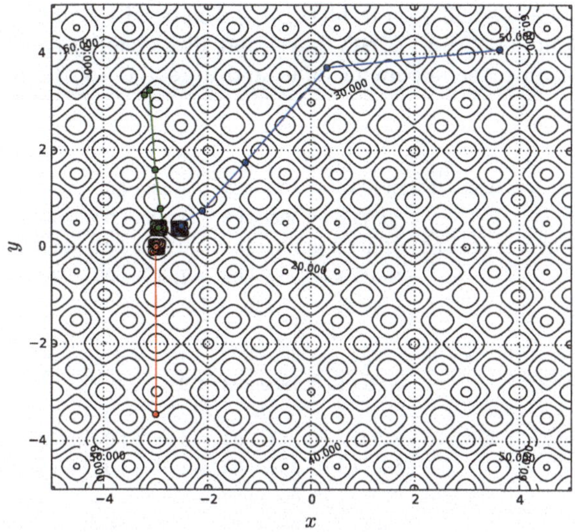

Fig. 19.10. The trajectories of three search individuals for Rastrigin function ($f_4(x, y)$)

The global minimum value of Rastrigin function is 0 at $(x, y) = (0, 0)$. Figure 19.10 shows the contour map of Eq. (19.13), and the trajectories of three search individuals. In this case, NMO traps the local minimum, and cannot find the optimum solution. The reason why the optimum value search fails in this manner that setting of the search range is inappropriate. The search range should be determined adaptively, but it has not been completed yet.

From the above results, the local search capability of NMO is improved, it is possible to search along the valley of the function, whereas the problem of global search capability remains.

19.4 Numerical Simulation

In order to confirm the fundamental solution search performance, we compare numerical simulation results of NMO with Standard PSO2011 [23]. We apply five 50 dimensional benchmark functions from the benchmark function set which was proposed in IEEE CEC2013 [18]. f_5 is Sphere function, f_6 is Rotated High Conditioned Elliptic function, f_7 is Rotated Bent Cigar function, f_8 is Rotated Discus function, and f_9 is Different Power function. Table 19.1 shows the numerical simulation results. The number of trials is 51. Table 19.1 shows the obtained minimum value, the median value, the maximum value, and the standard deviation. $f(*x)$ denotes the global minimum value of each benchmark function. The results indicate that the solution search performance of NMO is better than SPSO2011.

Table 19.1. Numerical simulation results. $f(*x)$ denotes the global minimum value of each benchmark function. 'Min', 'Med', 'Max', and 'Std' denote the obtained minimum value, the median value, the maximum value, and the standard deviation, respectively

Function	$f(*x)$		Min	Med	Max	Std
f_5 (Sphere)	−1.400e+03	SPSO	−1.400e+03	−1.400e+03	−1.400e+03	0.000e+00
		NMO	−1.400e+03	−1.400e+03	−1.400e+03	0.000e+00
f_6 (Rotated High Conditioned Elliptic)	−1.300e+03	SPSO	+3.776e+04	+6.785e+04	+1.126e+05	1.873e+04
		NMO	−1.299e+03	−1.297e+03	−1.290e+03	1.332e+00
f_7 (Rotated Bent Cigar)	−1.200e+03	SPSO	+1.995e+06	+4.365e+07	+5.711e+08	9.471e+07
		NMO	−1.180e+03	1.917e+03	2.514e+06	4.756e+05
f_8 (Rotated Discus)	−1.100e+03	SPSO	+3.113e+04	+4.987e+04	+7.704e+04	8.717e+03
		NMO	−1.098e+03	−1.097e+03	−1.092e+03	1.193e+00
f_9 (Different Power)	−1.000e+03	SPSO	−1.000e+03	−1.000e+03	−1.000e+03	0.000e+00
		NMO	−1.000e+03	−1.000e+03	−1.000e+03	0.000e+00

19.5 Conclusions

In this article, we analyzed the dynamics of our proposed Nonlinear Map Optimization. The NMO consists of some search individuals whose dynamics is driven by a simple circle map. The circle map generates a chaotic search point time series, and the distribution of the search points is a desirable distribution for the local search capability. As a result of improving the search capability in the vicinity of a good solution after guaranteeing the search range, NMO's solution searching ability has improved very much. It is insufficient to adaptively change the parameters, which is our future work. Also please note that while other SI algorithms are stochastic systems, NMO is a deterministic system.

Acknowledgements. This work was supported by JSPS KAKENHI Grant-in-Aid for Challenging Exploratory Research Number: 16K14271, and Grant-in-Aid for Scientific Research(C) Number: 15K06077.

References

1. K.T. Alligood, T.D. Sauer, J.A. Yorke, *Chaos: An Introduction to Dynamical Systems* (Springer, Berlin, 1996)
2. G. Beni, J. Wang, Swarm intelligence in cellular robotic systems, in *Proceeding of NATO Advanced Workshop on Robots and Biological Systems, Tuscany, Italy, 26–30 June 1989* (1989)
3. E. Bonabeau, M. Dorigo, G. Theraulaz, *Swarm Intelligence* (Oxford University Press, Oxford, 1999)
4. M. Clerc, J. Kennedy, The particle swarm explosion, stability and convergence in a multidimensional complex space. IEEE Trans. Evol. Comput. **6**, 58–73 (2002)

5. M. Clerc, *Particle Swarm Optimization* (Wiley-ISTE, New Jercy, 2006)
6. M. Dorigo, T. Stűtzle, *Ant Colony Optimization* (MIT Press, Cambridge, 2004)
7. Y. Hariya, T. Kurihara, T. Shindo, K. Jin'no, A study of robustness of PSO for non-separable evaluation functions, in *Proceeding of 2015 International Conference on Nonlinear Theory and its Applications (NOLTA2015), 1–4 Dec 2015* (2015), pp. 724–727
8. Y. Hariya, T. Shindo, K. Jin'no, An improved rotationally invariant PSO: a modified standard PSO-2011, in *Proceeding of 2016 IEEE World Congress on Computational Intelligence (IEEE WCCI 2016 (CEC2016)), Vancouver, Canada, 24–29 July 2016* (2016), pp. 1839–1844
9. Y. Hariya, T. Shindo, K. Jin'no, A novel particle swarm optimization for non-separable and ill-conditioned problems, in *Proceeding of IEEE 2016 International Symposium on Systems, Man, and Cybernetics (SMC2016), 9–12 Oct 2016* (2016), pp. 2110–2115
10. K. Jin'no, A novel deterministic particle swarm optimization system. J. Signal Process. **13**(6), 507–513 (2009)
11. K. Jin'no, T. Shindo, T. Kurihara, T. Hiraguri, H. Yoshino, Canonical deterministic particle swarm optimization to sustain global search, in *Proceeding of 2014 IEEE International Symposium on Systems, Man, and Cybernetics (SMC2014), San Diego, CA, USA, 5–8 Oct 2014* (2014), pp. 2470–2475
12. K. Jin'no, R. Sano, T. Saito, Particle swarm optimization with switched topology, nonlinear theory and its applications (NOLTA). IEICE **6**(2), 181–193 (2015)
13. K. Jin'no, Nonlinear map optimization, in *Proceeding of IEEE WCCI CEC2018* (2018), pp. 2082–2088
14. D. Karaboga, An idea based on honey bee swarm for numerical optimization. Technical Report TR06 (Erciyes University, Engineering Faculty, Computer Engineering Department, 2005)
15. J. Kennedy, R. Eberhart, Particle swarm optimization, in *Proceeding of IEEE 1995 International Conference on Neural Networks, 27 Nov–1 Dec 1995* (1995), pp. 1942–1948
16. K. Kohinata, T. Kurihara, T. Shindo, K. Jin'no, A novel deterministic multi-agent solving method, in *Proceeding of 2015 IEEE International Symposium on Systems, Man, and Cybernetics (SMC2015), HongKong, China, 9–12 Oct 2015* (2015), pp. 1758–1762
17. K. Kohinata, T. Kurihara, T. Shindo, K. Jin'no, Multi-agent search method with rotation angle dependent on the best position, in *Proceeding of 2016 International Conference on Nonlinear Theory and its Applications, 27–30 Nov 2016* (2016), pp. 463–466
18. J.J. Liang, B.Y. Qu, P.N. Suganthan, A.G. Hernández-Díaz, Problem definition and evaluation criteria for the CEC 2013 special session on real-parameter optimization (2018), http://alroomi.org/multimedia/CEC_Database/CEC2013/RealParameterOptimizationCEC2013_RealParameterOptimization_TechnicalReport.pdf. Accessed 25 May 2018
19. J.A. Nelder, R. Mead, A simplex method for function minimization. Comput. J. **7**(4), 308–313 (1965)
20. T. Shindo, K. Jin'no, Analysis of dynamics characteristic of deterministic PSO. Nonlinear Theory Appl. (NOLTA), IEICE **4**(4), 451–461 (2013). https://doi.org/10.1588/nolta.4.451
21. X. Yang, *Nature-Inspired Metaheuristic Algorithms* (Luniver Press, Frome, 2008)
22. X. Yang, S. Deb, Cuckoo search via Lévy flights, in *Proceeding of World Congress on Nature & Biologically Inspired Computing, 9–11 Dec 2009* (2009)

23. M. Zambrano-Bigiarini, M. Clerc, R. Rojas, Standard particle swarm optimisation 2011 at CEC-2013: a baseline for future PSO improvements, in *Proceedings of IEEE CEC 2013* (2013), pp. 2337–2344

Chapter 20 Analog-to-Digital Converters Employing Chaotic Internal Circuits to Maximize Resolution-Bandwidth Product - Turbo ADC

Zeljko Ignjatovic(✉) and Yiqiao Zhang

University of Rochester, Rochester, NY, USA
ignjatov@ece.rochester.edu, yzh187@ur.rochester.edu

Abstract. By applying information theoretic concepts to analog-to-digital convertor (ADC) design, we have created a mathematical framework from which the fundamental limit for the resolution-bandwidth product of any ADC may be derived. We found the surprising result that the limiting resolution of any ADC is proportional to oversampling-ratio (OSR), as opposed to widely-held belief that the resolution is proportional to $log_2(OSR)$, a dramatic increase in the achievable resolution. This result, which resembles Shannon's well-known result for the capacity of a communication channel, represents a paradigm shift in our understanding of data conversion methods and provides encouragement that new methods may be found. Furthermore, to achieve this theoretical limit, the internal analog modulator (or filter) of an ADC should be a chaotic system, so that both small as well as large changes in the input signal cause large (but bounded) deterministic changes at the output of the modulator - analogous to the "Butterfly effect". This led us to discover a new class of ADCs, which we call TurboADC's, that can trade off resolution for bandwidth on the fly, keeping their product equal to the fundamental information theoretic limit. These designs impose modest requirements on the analog front-end resources and power at the expense of greater complexity in the back-end decoder. A discrete-time TurboADC proposed here is a hybrid of a 1st order Delta-Sigma modulator and a Cyclic ADC, with the best features of both designs - oversampling, noise shaping, and simplicity from the Sigma-delta ADC approach and fast half-interval searching from Cyclic ADC's. Simulations of the proposed TurboADC confirm our finding that the resolution of an ADC may approach fundamental limit of OSR bits within the baseband.

20.1 Introduction

Analog-to-digital converters and conversion methods have experienced rapid growth in recent decades due to the steady development of CMOS technologies and the increasing demand for higher resolution and bandwidth. CMOS

V. In et al. (Eds.): *Proceedings of the 5th International Conference on Applications in Nonlinear Dynamics*, Understanding Complex Systems, https://doi.org/10.1007/978-3-030-10892-2_20

technologies have allowed more systems (both analog and digital) to be integrated into a single chip; thus, reducing manufacturing costs and allowing additional functions such as calibration techniques. There are two mainstream ADC techniques: Nyquist rate ADCs (e.g., Flash, Single-slope, Dual-slope, SAR, and Cyclic) and oversampling ADCs (e.g., oversampling PCM and $\Delta\Sigma$ converters). Nyquist rate ADCs are commonly used for low-to-moderate precision (resolution) and high bandwidth conversion applications, as seen in Fig. 20.1. Their resolution is limited by two fundamental sources of noise, thermal and flicker noise, as well as circuit imperfections such as DC offsets and non-linearity. A comprehensive review of current state-of-the-art Nyquist rate ADCs is provided in [1]. $\Delta\Sigma$ ADCs are used for high-precision low-to-moderate bandwidth applications. Their bandwidth is limited by the oversampling demands and the precision is limited by circuit noise and to a lesser extent by non-idealities such as DC offset, gain error, and non-linearity. In addition, $\Delta\Sigma$ ADCs are prone to instability due to the presence of a non-linear comparison operation within a feedback loop, which limits the order of $\Delta\Sigma$ ADCs in practical implementations. An overview $\Delta\Sigma$ ADC principles and state-of-the-art is provided in [2–4]. More recently, novel ADC methods have been introduced that rely heavily on joint-processing of digital samples to increase the RBW and decrease the complexity of analog components [5–7]. However, to the best of our knowledge, no single ADC method is able to cover the full breath of potential applications (starting from low-power conversions for bio-sensing and IoT applications to high-speed direct RF conversion in radar and communications).

By consolidating two distinct fields (Information theory on one side and the principles and methods of A/D conversion on the other), we established a mathematical framework that, among other things, helped us derive fundamental theoretical limit on the resolution-bandwidth product of Analog-to-digital converters (ADC) and also prove many essential and often unexpected results. For example, it is traditionally assumed that the quantization noise in ADCs is independent of the input analog signal. As a direct consequence to this assumption, the effective number of bits (ENOB) or resolution is always proportional to $log_2(OSR)$, where the OSR is the oversampling ratio. By using Information theory tools, we showed that this assumption is fundamentally flawed, and that the quantization noise is instead fully dependent on the input analog signal because its entropy is zero given the input analog signal. As a consequence, we show that the resolution is instead proportional to OSR. This represents a paradigm shift in understanding A/D conversion methods and allows us to discover novel methods of conversion. In this work, we introduce a novel class of ADC, termed TurboADC, that can trade resolution and bandwidth on the fly while preserving their product constant and equal to the fundamental theoretical limit with minimal use of analog front-end resources and power. Thanks to their simple front-end design (as simple as the 1st order $\Delta\Sigma$ modulator) and ease of integration, we envision that TurboADCs, will be able to replace most traditional ADC methods and even enable new applications such as software defined radio, direct RF signal conversion in communications, radar, ultrasound, and MRI imaging systems.

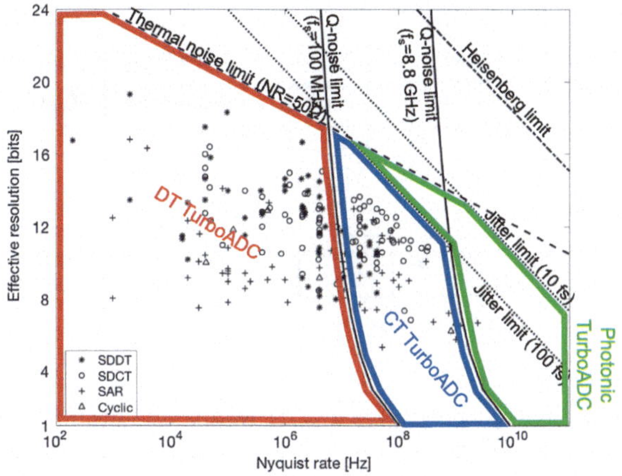

Fig. 20.1. TurboADC design space. Data points show effective resolution versus Nyquist rate of existing ADC methods including switched-capacitor $\Delta\Sigma$ (SDDT), continuous-time $\Delta\Sigma$ (SDCT), SAR and Cyclic ADC, [8]. We assume that the maximum sampling rate of CT TurboADC is close to 9 GHz, as demonstrated in [9] for CT $\Delta\Sigma$ ADCs

Figure 20.1 illustrates our understanding of the design space and how TurboADCs fit in. The data points represent effective resolution versus bandwidth for more than 200 traditional ADC designs including Successive Approximation (SA), Cyclic, Discrete-time (SDDT) and Continuous-time (SDCT) $\Delta\Sigma$ ADC, [8]. Designs that include pipelining, time-interleaving, and other means of parallelism (such as Flash ADC) are omitted for fair comparison due to their much-increased complexity, area, and power. This figure also shows upper bounds on resolution-bandwidth (RBW) product imposed by aperture jitter (for 100 and 10 fs RMS jitter), thermal noise corresponding to noise equivalent resistance of 50 Ω, and Heisenberg uncertainty principle as derived in [3]. It also shows upper bounds imposed by quantization noise (QN) according to our theory in [10]. Our vision is that the entire design space could be encompassed by our novel method that includes discrete-time (DT) TurboADC for high-resolution and medium speed, continuous-time (CT) TurboADC for high-speed conversion, and finally a Photonic TurboADC for ultra-high conversion speeds in hundreds of GS/sec, as shown in Fig. 20.1. The focus of this paper is on DT TurboADC's as described in Sect. 20.3.

20.2 ADC as Communication System and Conversion Capacity

We assume that each ADC can be described as a communication system, as shown in Fig. 20.2. The information source is analog, meaning it provides a continuous-time and continuous amplitude signal such as voltage, current, or

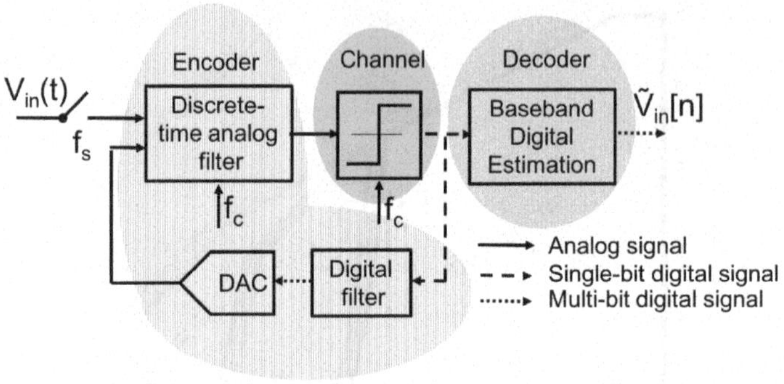

Fig. 20.2. Analog-to-digital converter as a communication system

charge to the ADC. The analog signal is sampled at a certain sampling rate f_s and the samples are fed to the analog filters and amplifiers of the ADC. The amplifier/filter structures process the analog signal by performing a form of encoding before the signal enters the comparator, which plays the role of the 'noisy' channel.

The output from the comparator generated at the rate of f_c is then decoded by digital circuitry and usually fed back to the encoder. In general, the rate f_c at which the comparator is operating is either equal to f_s (such as in $\Delta\Sigma$ ADCs) or larger (such as in SAR and Cyclic ADCs). In fact, for almost every ADC method, we were able to find a corresponding forward-error-correction (FEC) code. For example, the SAR ADC correspond to a form of block code with channel feedback. The $\Delta\Sigma$ ADCs correspond to convolutional codes, where the $\Delta\Sigma$ modulator acts as a convolutional encoder and its decimation filter plays the role of the maximum likelihood decoder. Also, the order of the $\Delta\Sigma$ modulator defines the memory length of the encoder. In describing an ADC as a communication system with FEC coding, we assume that the comparator(s) of an ADC is a 'channel' since it injects quantization noise into the transmitted signal even if the circuit components are otherwise noiseless. Once the comparator of the ADC is described as a communication channel, we can derive its intrinsic capacity (i.e., maximum number of information bits that can be digitized per second). Since the QN is neither Gaussian nor independent of the input signal to the comparator (in fact, the QN in an ideal ADC system is fully described given the input signal), the Shannon capacity formula $C = B * log_2(1 + SNR)$, where B is the channel bandwidth, as derived in [11], cannot be applied to calculate the capacity of such a system. A more general approach involving mutual information and entropy must be used.

Some important conclusions from our previous work about the ADC theory are listed below in the form of theorems (for proofs see [10]):

Theorem 1 (Capacity) *Maximum information rate at the output of an ADC employing M comparators operated at f_c comparisons per second is equal to $M *$*

f_c bits per second. We define this maximum information rate as a ***Conversion Capacity*** *C_{ADC}.*

Theorem 2 (Existence) *There exists at least one ADC that can operate at the C_{ADC} regardless of the input signal statistics. We term this type of ADC as* ***TurboADC*** *with reference to Turbo codes in communications that are able to approach Shannon's channel capacity.*

Theorem 3 (Necessary condition) *An ADC can achieve the conversion capacity if the autocorrelation function of the input to its internal comparator(s) is a delta function (i.e., white spectral properties) regardless of the input signal statistics.*

Consequently, two properties of a TurboADC can be derived.

Corollary 1 *The internal analog filter of a TurboADC that encodes the input signal before it is fed to the comparator, must be a* ***non-linear*** *filter (or a non-linear mapping).*

Corollary 2 *The output of the comparator in a TurboADC is a sequence of independent uniformly distributed bits.*

Theorem 4 (Oversampling) *If an ADC operates at its capacity and the input analog signal is oversampled by a factor of $OSR = f_c/2f_{in}$, the effective resolution in the baseband is equal to OSR bits.*

Perhaps the most interesting and unexpected property of a TurboADC is the one described in Theorem 4. It states that the resolution of a TurboADC is proportional to the OSR. In contrast, traditional oversampling $\Delta\Sigma$ ADCs achieve effective resolution that is proportional to $log_2(OSR)$. Clearly, for the same resolution, a TurboADC may operate at an exponentially lower sampling rate than the $\Delta\Sigma$ ADCs. Also, from Theorem 4 we conclude that resolution of a TurboADC trades linearly with its bandwidth such that the R-BW product is constant and equal to C_{ADC}. For example, to increase the resolution from 8 to 16 bits, a 2nd-order $\Delta\Sigma$ ADC would have to increase its sampling rate by a factor of 9.1 while a TurboADC would only have to double it (5 times reduction in power), which could prove crucial in battery-operated IoT devices. On the other hand, for the same technology node and power consumption, TurboADC may achieve data rates significantly higher than other ADC methods, which may enable new high-speed conversion applications. Finally, from Theorems 1 and 3 we prove the following theorem.

Theorem 5 (Chaotic encoder) *In order to achieve the theoretical limit to the R-BW product (the capacity) irrespective of the input signal statistics, the ADC's internal analog filter must be a deterministic system with aperiodic and bounded state trajectories for all input signal statistics – a chaotic system.*

Proof First, we prove deterministic property of the analog filter (or encoder). As in [10], mutual information between the comparator's 1-bit output $y[n]$ and the analog input $V_{in}[n]$ is defined as,

$$I(V_{in}[n], y[n]) = H(y[n] \mid y[n-1], \dots y[1]) - H(y[n] \mid y[n-1], \dots y[1], V_{in}[n], \dots V_{in}[1]), \tag{20.1}$$

Since the first term $H(y[n] \mid y[n-1], \dots y[1])$ can be at most equal to 1 bit, the mutual information term is maximized if and only if the second term is equal to zero. The second term is zero if and only if the state of the encoder is fully described given the input analog signal (i.e., it is not stochastic). Second, we prove the state boundedness by contradiction. If the state is unbounded it must grow to either positive infinite or negative infinite value (not both). Otherwise, its bandwidth would grow to infinity, which cannot be the case with discrete-time systems. Therefore, if the state becomes unbounded the output from the comparator $y[n]$ would be a constant value that carries no information (i.e., information rate falls below the capacity). Third, according to Theorem 3, since the state value over its trajectory must have a delta autocorrelation function it must follow aperiodic orbits (i.e., random-like nature). Finally, if an analog encoder is to produce an output that has white spectrum (aperiodic orbits) for any input signal statistics, it should do so even in the limiting case where the input signal is a delta function with the maximum bandwidth of $f_s/2$. In this case, the input signal affects only the initial state of the analog encoder and the subsequent state values continue to change on their own over aperiodic orbits. Therefore, it must be sensitive to initial conditions – a "Butterfly effect". An alternative limiting case, when the input analog signal is a DC signal, would lead to the same requirement about the analog encoder. □

20.3 Discrete-Time TurboADC

We first explore the use of a simple discrete-time dyadic transformation (or Bernoulli map) that can give rise to chaotic behavior. The phase space of this simple map is shown in Fig. 20.3a, which in its original form does not allow the use of an independent variable to affect the state's trajectory. There are many ways to ensure that an input analog signal is introduced into the chaotic map to affect the state's aperiodic trajectory. Figure 20.3b depicts the phase space of a modified Bernoulli chaotic map proposed in this work. This particular map is proposed for two reasons. First, it maximizes the dynamic range and signal-to-noise ratio (SNR) by allowing the amplitude of the input signal $V_{in}[n]$ to reach maximum level of V_{ref}. Second, it ensures a simple switched-capacitor circuit implementation, as shown in Fig. 20.4. Also, Eqs. (20.2a)–(20.2e) show the dynamical law of this chaotic system, where $s[n]$ is the internal state of the chaotic filter and V_{ref} is the reference analog voltage used by a TurboADC for

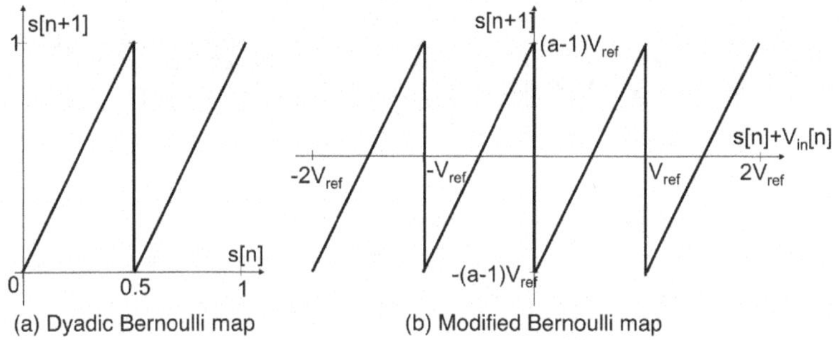

Fig. 20.3. 1-dimensional chaotic maps. **a** Traditional Bernoulli map. **b** Modified Bernoulli map as an analog encoder

digitization.

$$q[nT_s] = s[nT_s] + V_{in}[nT_s] \tag{20.2a}$$

$$y_0[nT_s] = \begin{cases} 1, & q[nT_s] \geq 0 \\ -1, & otherwise \end{cases} \tag{20.2b}$$

$$y_+[nT_s] = \begin{cases} 1, & q[nT_s] \geq V_{ref} \\ 0, & otherwise \end{cases} \tag{20.2c}$$

$$y_-[nT_s] = \begin{cases} -1, & q[nT_s] \leq -V_{ref} \\ 0, & otherwise \end{cases} \tag{20.2d}$$

$$s\big[(n+1)T_s\big] = a \cdot q[nT_s] - V_{ref}\big(y_0\,[nT_s] + 2 \cdot y_+\,[nT_s] + 2 \cdot y_-\,[nT_s]\big) \tag{20.2e}$$

The state $s[n]$ remains bounded in the $[-aVref, aVref)$ interval and trajectory is deterministic in absence of electronic noise. For certain values of the gain (e.g., $a = 2$) and initial state the map exhibits a true chaotic behavior with the Lyapunov exponent equal to $log(2)$. A block schematic of the described DT TurboADC based on the modified Bernoulli chaotic map in Eqs. (20.2) is shown in Fig. 20.4. Since the input to the internal quantizer is compared against three thresholds ($-V_{ref}$, 0, V_{ref}), a 2-bit quantizer and 2-bit feedback DAC are required in this implementation. In a way, DT TurboADC represents a hybrid between the $\Delta\Sigma$ modulator and the Cyclic ADC, with the best features of both designs - oversampling, noise shaping, and simplicity from the $\Delta\Sigma$ ADC and fast half-interval searching from Cyclic ADC. However, unlike the $\Delta\Sigma$ modulator that employs a DT integrator to shape the quantization noise outside the signal band, the TurboADC employs an unstable filter (pole $z_p = 2$ outside the unit circle), where both signal and quantization noise are shaped over aperiodic orbits. This allows it to achieve much higher R-BW products than the $\Delta\Sigma$ ADCs as demonstrated in Sect. 20.4. Also, unlike the Cyclic ADC, where each input signal sample is converted to digital independently of other input samples, the present state of the internal chaotic filter in TurboADC depends on the entire past of the analog input signal.

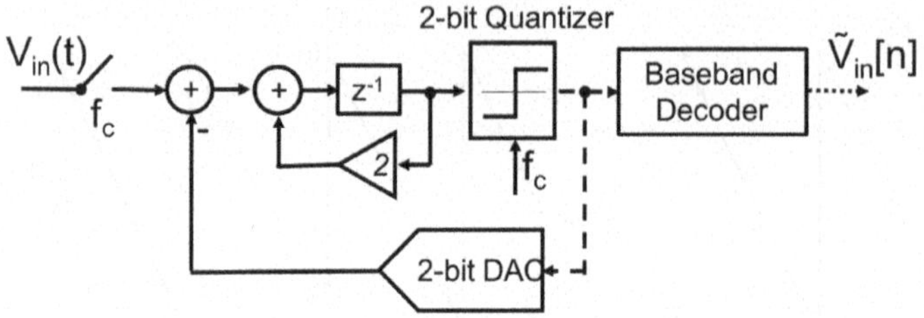

Fig. 20.4. Block schematic of the DT TurboADC employing modified Bernoulli chaotic map

The following example emphasizes the significance of this difference between the Cyclic ADC and TurboADC. Let us assume that a Cyclic ADC is designed for a sampling rate of $f_s = 8\,\mathrm{MHz}$ with 4-bit resolution. For each of the input signal samples, the Cyclic ADC produces 4 bits after cycling through four comparisons (i.e., comparator operates at $f_c = 32\,\mathrm{MHz}$), followed by resetting the internal state to a new input signal value. If we now assume that the actual analog signal applied to the Cyclic ADC is bandlimited to 1 MHz ($OSR = 4$), the best resolution that the Cyclic ADC can achieve in this case is 5 bits after averaging four original 4-bit samples. At the same time, if the TurboADC operates at the same speed ($f_c = 32\,\mathrm{MHz}$) and the input signal bandwidth is 1 MHz, the resolution will be 16 bits, which is an improvement of 11 bits over the Cyclic ADC. Additionally, the TurboADC would require much simpler anti-aliasing filter.

20.3.1 Baseband Decoding Method and Implementation

The single-bit stream produced by the comparator in TurboADC's must be decoded to produce a meaningful multi-bit representation of the input analog signal in baseband. Contrary to Cyclic ADC, where there is a one-to-one correspondence between the amplitude bits of the input signal samples and information bits at the output of the comparator, the TurboADC produces information bits that are affected by many past input signal samples. Therefore, input signal baseband samples must be estimated from the comparator's single-bit output stream. Unlike the $\Delta\Sigma$ ADC where the baseband multi-bit input signal samples are estimated with the help of a linear decimation filter, the TurboADC is a non-linear system, and so the baseband signal must be estimated with the help of non-linear estimation methods.

In the absence of electronic noise, the decoding method of the TurboADC (based on the modified Bernoulli map shown in Fig. 20.2 and Eqs. (20.2) can be implemented similarly to the non-linear decoder for $\Delta\Sigma$ ADC. The method is briefly described below (for more details see our work in [7,12]). We first assume that the output bit streams $y_0[n]$, $y_+[n]$, and $y_-[n]$ in Eqs. (20.2b)–(20.2d) are divided into non-overlapping frames of length N.

Equations (20.2b)–(20.2d) can be merged into a system of 3N inequalities and written in a matrix form as in (20.3)–(20.6).

$$\mathbf{s} = \mathbf{A} \cdot \mathbf{s} + \mathbf{B} \cdot \mathbf{v_{in}} - V_{ref} \cdot \mathbf{C} \cdot (\mathbf{y_0} + 2\mathbf{y_+} + 2\mathbf{y_-}) \quad (20.3)$$

$$\mathbf{y_0} \circ (\mathbf{s} + \mathbf{v_{in}}) \geq \mathcal{O} \quad (20.4)$$

$$\mathbf{y_+} \circ (\mathbf{s} + \mathbf{v_{in}} - V_{ref} \cdot \mathbf{1}) \geq \mathcal{O} \quad (20.5)$$

$$\mathbf{y_-} \circ (\mathbf{s} + \mathbf{v_{in}} + V_{ref} \cdot \mathbf{1}) \geq \mathcal{O} \quad (20.6)$$

where vector operation $\circ$ denotes Hadamard product, $\mathcal{O}$ is a zero-vector of length N, $\mathbf{1}$ is a one-vector of length N, and vectors

$$\mathbf{s} = \begin{bmatrix} s[1] \\ s[2] \\ \vdots \\ s[N] \end{bmatrix}; \quad \mathbf{v_{in}} = \begin{bmatrix} V_{in}[1] \\ V_{in}[2] \\ \vdots \\ V_{in}[N] \end{bmatrix}; \quad \mathbf{y_0} = \begin{bmatrix} y_0[1] \\ y_0[2] \\ \vdots \\ y_0[N] \end{bmatrix}; \quad \mathbf{y_+} = \begin{bmatrix} y_+[1] \\ y_+[2] \\ \vdots \\ y_+[N] \end{bmatrix}; \quad \mathbf{y_-} = \begin{bmatrix} y_-[1] \\ y_-[2] \\ \vdots \\ y_-[N] \end{bmatrix}$$

Matrix **A**, **B**, and **C** are the state transition matrices shown below for a gain of the internal chaotic circuit $a = 2$,

$$\mathbf{A} = \begin{bmatrix} 1\,0 \cdots 0\,0 \\ 2\,0 \cdots 0\,0 \\ 0\,2 \cdots 0\,0 \\ \vdots\;\vdots\;\ddots\;\vdots\;\vdots \\ 0\,0 \cdots 2\,0 \end{bmatrix}; \quad \mathbf{B} = \begin{bmatrix} 0\,0 \cdots 0\,0 \\ 2\,0 \cdots 0\,0 \\ 0\,2 \cdots 0\,0 \\ \vdots\;\vdots\;\ddots\;\vdots\;\vdots \\ 0\,0 \cdots 2\,0 \end{bmatrix}; \quad \mathbf{C} = \begin{bmatrix} 0\,0 \cdots 0\,0 \\ 1\,0 \cdots 0\,0 \\ 0\,1 \cdots 0\,0 \\ \vdots\;\vdots\;\ddots\;\vdots\;\vdots \\ 0\,0 \cdots 1\,0 \end{bmatrix}$$

We can further rearrange the matrix equation in (20.3) to express the state vector explicitly as in (20.7).

$$\mathbf{s} = (\mathbf{I} - \mathbf{A})^{-1}(\mathbf{B} \cdot \mathbf{v_{in}} - V_{ref} \cdot \mathbf{C} \cdot (\mathbf{y_0} + 2\mathbf{y_+} + 2\mathbf{y_-})) \quad (20.7)$$

Further constraints can be imposed on the input signal $\mathbf{v_{in}}$. For example, a band-limitation constraint can be introduced on the input analog signal as described in (20.8), where **W** is a band-limitation matrix of size $N \times \frac{N}{OSR}$ [12].

$$\mathbf{v_{in}} = \mathbf{W} \cdot \hat{\mathbf{v}}_{\mathbf{in}} \quad (20.8)$$

The vector $\hat{\mathbf{v}}_{\mathbf{in}}$ can be defined as a vector of discrete transform coefficients corresponding to the input signal band in which case the matrix **W** can be described as an inverse transform matrix. If the coefficients in $\hat{\mathbf{v}}_{\mathbf{in}}$ are amplitudes of orthonormal sinusoidal waves sampled at f_s, then each column of **W** is either a sine or cosine wave with frequencies varying from DC to $f_s/(2OSR)$.

Finally, we can combine Eqs. (20.5)–(20.8) into one matrix inequality as in (20.9), where $\Pi = (\mathbf{I} - \mathbf{A})^{-1}$.

$$\begin{bmatrix} \mathbf{y_0} \\ \mathbf{y_+} \\ \mathbf{y_-} \end{bmatrix} \circ \begin{bmatrix} \Pi \cdot (\mathbf{B} \cdot \mathbf{W} \cdot \hat{\mathbf{v}}_{\mathbf{in}} - V_{ref} \cdot \mathbf{C} \cdot (\mathbf{y_0} + 2\mathbf{y_+} + 2\mathbf{y_-})) + \mathbf{v_{in}} \\ \Pi \cdot (\mathbf{B} \cdot \mathbf{W} \cdot \hat{\mathbf{v}}_{\mathbf{in}} - V_{ref} \cdot \mathbf{C} \cdot (\mathbf{y_0} + 2\mathbf{y_+} + 2\mathbf{y_-})) + \mathbf{v_{in}} \\ \Pi \cdot (\mathbf{B} \cdot \mathbf{W} \cdot \hat{\mathbf{v}}_{\mathbf{in}} - V_{ref} \cdot \mathbf{C} \cdot (\mathbf{y_0} + 2\mathbf{y_+} + 2\mathbf{y_-})) + \mathbf{v_{in}} \end{bmatrix} \geq \begin{bmatrix} \mathcal{O} \\ \mathbf{y_+} + V_{ref} \cdot \mathbf{y_+} \\ \mathbf{y_-} - V_{ref} \cdot \mathbf{y_-} \end{bmatrix} \quad (20.9)$$

Computation of the baseband input signal can then be described as a linear feasibility problem (LFP) in (20.10) defined as finding a coefficient vector $\hat{\mathbf{v}}_{\mathbf{in}}$ that satisfies the inequality constraints in (20.9) and then transforming it to time domain as in (20.8) to obtain an input signal estimate $\hat{\mathbf{v}}_{\mathbf{in}}$.

$$\begin{aligned} &\textit{find} \quad \hat{\mathbf{v}}_{\mathbf{in}} \in R^{\frac{N}{OSR}} \\ &s.t.\ \hat{\mathbf{v}}_{\mathbf{in}}\ \textit{satisfies Eq. } 20.9 \end{aligned} \tag{20.10}$$

This LFP problem can be solved by using sequential projection algorithms such as the Kaczmarz method [13] and the Agmon, Motzkin, and Schoenberg (AMS) method, [14,15]. Let $\mathbf{Ax} = \mathbf{b}$ be a linear system and N be the number of rows of $\mathbf{A}$. Each row $\boldsymbol{a}_n$ of matrix $\mathbf{A}$ together with corresponding element b_n of vector $\mathbf{b}$ define the hyperplane $H_n = \{\mathbf{x} : \boldsymbol{a}_n\mathbf{x} = b_n, 1 \leq n \leq N\}$. Thus, a solution to the linear system can be obtained by sequentially projecting onto the hyperplanes H_n, as described in [13]. A system of linear inequalities, $\mathbf{Ax} \leq \mathbf{b}$, can be solved in a similar manner. The AMS method treats the system of linear inequalities as a set of half-spaces $S_n = \{\mathbf{x} : \boldsymbol{a}_n\mathbf{x} \leq b_n, 1 \leq n \leq N\}$, where the projection onto a half-space only occurs if the current inequality is violated, [14,15]. Given an arbitrary initial approximation $\mathbf{x}(0)$, the $(i+1)$th estimate of the solution is calculated as

$$\mathbf{x}(i+1) = \mathbf{x}(i) + \min\left\{0, \frac{b_n - \boldsymbol{a}_n\mathbf{x}(i)}{\| \boldsymbol{a}_n \|_2^2}\right\} \boldsymbol{a}_n^T \tag{20.11}$$

where $n = i \bmod N + 1$ and $\| \boldsymbol{a}_n \|_2^2$ is the Euclidean norm. From Eq. (20.11) the solution $\mathbf{x}$ at the $(i+1)$th iteration step is changed only if the nth inequality is violated. Otherwise, the current estimate remains unchanged (i.e., $\mathbf{x}(i+1) = \mathbf{x}(i)$).

20.4 Simulation Results

To demonstrate the DT TurboADC method proposed in Sect. 20.3, we set up simulations in MATLAB environment. First, a bandlimited input signal $\mathbf{v}_{\mathbf{in}}$ of length $N = 64$ is created by transforming a set of N/OSR coefficients $\hat{\mathbf{v}}_{\mathbf{in}}$ to time domain with the use of band-limitation matrix $\mathbf{W}$ as shown in Eq. (20.8). The coefficients $\hat{\mathbf{v}}_{\mathbf{in}}$ are chosen at random and independently from a normal distribution. The input signal $\mathbf{v}_{\mathbf{in}}$ is then normalized so its amplitude does not exceed maximum value of $V_{ref} = 1$. The output bit stream $\mathbf{y_0}$ and auxiliary outputs $\mathbf{y}_+$ and $\mathbf{y}_-$ from the DT TurboADC are then fed to the decoder as described in Sect. 20.3.1. The estimated input signal $\hat{\mathbf{v}}_{\mathbf{in}}$ is then compared to the actual input $\mathbf{v}_{\mathbf{in}}$ and the mean-squared-error (MSE) is calculated for various OSR values. SNR is then calculated as $SNR = 10log_{10}(V_{ref}^2/MSE)$. The results are compared against 1st order $\Delta\Sigma$ ADC with non-linear decoder as in [12], as shown in Fig. 20.5. The resulting SNR of the DT TurboADC indicates the effective resolution close to OSR (where the effective resolution is calculated as

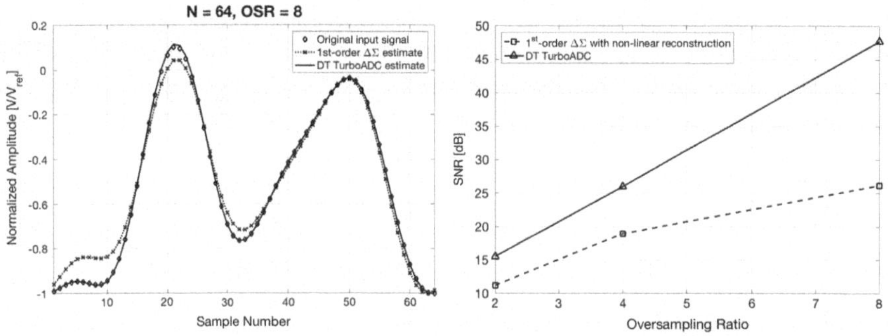

(a) Waveforms (b) SNR vs. OSR

Fig. 20.5. Simulations results. **a** shows time domain waveforms of original input signal, input signal estimated by a 1st-order $\Delta\Sigma$ ADC with a non-linear input signal reconstruction as in [12], and DT Turbo ADC for $OSR = 8$. **b** shows SNR as a function OSR for 1st-order $\Delta\Sigma$ ADC with a non-linear input signal reconstruction and DT TurboADC

$SNR/6.01$), while the 1st order $\Delta\Sigma$ ADC's resolution follows $log_2(OSR)$ trend (see Fig. 20.5b). For example, at $OSR = 8$, the DT TurboADC achieves an SNR of 47 dB (or 7.82 bits of effective resolution), while the conversion capacity limit is 8 *bits*. The 1st order $\Delta\Sigma$ ADC achieves only 4.3 *bits* for the same OSR. It should be noted that the DT TurboADC decoder as described in Sect. 20.3.1 did not always converge occasionally producing a high MSE. The convergence of the decoder was highly sensitive to the initial input signal estimate indicating that the solution set to the (20.10) may not be convex. Further research is needed to understand the convexity of the solution set to this estimation problem. If the solution set proves to be non-convex, further efforts will be made to introduce additional constraints to guarantee convergence of the decoder (e.g., an additional comparator with a direct access to the input analog signal might be used to provide the polarity of the input samples to the decoder). In addition, the decoder in Sect. 20.3.1 is deterministic assuming no electronic noise is affecting the state of the encoder. Thus, further research is needed to understand the effects of electronic noise on the choice for a specific decoder type and its design.

20.5 Conclusion

A new class of ADCs, termed TurboADC, capable of achieving fundamental theoretical limit to the resolution-bandwidth product (or conversion capacity) is presented. We prove that a TurboADC must employ a deterministic chaotic circuit to achieve the capacity. A discrete-time implementation of TurboADC with the front-end circuit complexity similar to a simple 1st order $\Delta\Sigma$ modulator is also proposed and its capacity achieving capabilities are demonstrated through simulations. The simulation results show that the resolution in the baseband is proportional to the OSR (defined as the ratio between one half the sampling frequency and the input analog signal's bandwidth) surpassing all existing ADC methods, whose resolution is proportional to $log_2(OSR)$, opening up possibilities for new data conversion applications such as high-speed direct RF signal conversion in radar, high-speed communications, and medical imaging.

References

1. F. Maloberti, *Data Converters* (Springer, Berlin, 2007), pp. 141–208
2. P.M. Aziz, H.V. Sorensen, J. vn der Spiegel, An overview of sigma-delta converters. IEEE Signal Process. Mag. **13**(1), 61–84 (1996)
3. R.H. Walden, Analog-to-digital converter survey and analysis. IEEE J. Sel. Areas Commun. **17**(4), 539–550 (1999)
4. S.R. Norsworthy, R. Schreier, G.C. Temes, IEEE Circuit & Systems Society, *Delta-Sigma Data Converters: Theory, Design, and Simulation* (IEEE Press, 1997)
5. I. Galton, H.T. Jensen, Delta-Sigma modulator based A/D conversion without oversampling. IEEE Trans. Circuits Syst. II Analog. Digit. Signal Process. **42**(12), 773–784 (1995)
6. M. Marijan, Z. Ignjatovic, Code division parallel delta-sigma A/D converter with probabilistic iterative decoding, in *Proceedings of 2010 IEEE International Symposium on Circuits and Systems, Paris* (2010), pp. 4025–4028
7. M. Marijan, Z. Ignjatovic, Reconstruction of oversampled signals from the solution space of delta-sigma modulated sequences, in *IEEE 54th International Midwest Symposium on Circuits and Systems (MWSCAS), Seoul* (2011), pp. 1–4
8. B. Murmann, ADC Performance Survey 1997–2017, http://web.stanford.edu/~murmann/adcsurvey.html
9. E. Martens et al., A 48-dB DR 80-MHz BW 8.88-GS, s bandpass $\Delta\Sigma$ ADC for RF digitization with integrated PLL and polyphase decimation filter in 40 nm CMOS, in *Symposium on VLSI Circuits - Digest of Technical Papers. Honolulu, HI* (2011), pp. 40–41
10. Z. Ignjatovic, M. Sterling, Information-theoretic approach to A/D conversion. IEEE Trans. Circuits Syst. I Regul. Pap. **60**(9), 2249–2262 (2013)
11. C. Shannon, *The Mathematical Theory of Communication* (University of Illinois Press, Urbana, 1998)
12. M. Marijan, Z. Ignjatovic, Non-linear reconstruction of delta-sigma modulated signals: randomized surrogate constraint decoding algorithm. IEEE Trans. Signal Process. **61**(21), 5361–5373 (2013)
13. S. Kaczmarz, Angenaherte auflosung von systemen linearer gleichungen. Bull. Int. Lacademie Pol. Sci. Lett. **35**, 355–357 (1937)
14. S. Agmon, The relaxation method for linear inequalities. Can. J. Math. **6**, 382–392 (1954)
15. T. Motzkin, I. Schoenberg, The relaxation method for linear inequalities. Can. J. Math. **6**, 393–404 (1954)
16. M. Mishali, Y.C. Eldar, O. Dounaevsky, E. Shoshan, Xampling: analog to digital at sub-nyquist rates. IET Circuits Devices Syst. **5**(1), 8–20 (2011)

Chapter 21
Calculating Embedding Dimension with Confidence Estimates

T. L. Carroll[1(✉)] and J. M. Byers[2]

[1] US Naval Research Lab, Code 6392, Washington, DC, MD 20375, USA
thomas.carroll@nrl.navy.mil
[2] US Naval Research Lab, Code 6395, Washington, DC, MD 20375, USA
jeff.byers@nrl.navy.mil

Abstract. We describe a method to estimate embedding dimension from a time series. This method includes an estimate of the probability that the dimension estimate is valid. Such validity estimates are not common in algorithms for calculating the properties of dynamical systems. The algorithm described here compares the eigenvalues of covariance matrices created from an embedded signal to the eigenvalues for a covariance matrix of a Gaussian random process with the same dimension and number of points. A statistical test gives the probability that the eigenvalues for the embedded signal did not come from the Gaussian random process.

21.1 Introduction

When analyzing a dynamical system based on a single variable time series, the first step is to embed the time series in a phase space to obtain a representation, or embedding, of the trajectory of the dynamical system [1]. Beginning with a digitized time series $s(i), i = 1 \ldots N$, the method of delays [2] is used to create a series of vectors,

$$\mathbf{s}(i) = [s(i), s(i+\tau), \ldots s(i+(d-1)\tau)], \tag{21.1}$$

where d is the embedding dimension and τ is the embedding delay. If the original dynamical system had a dimension of k, then the largest value of d necessary for $\mathbf{s}$ to be an embedding of the original dynamics is $2k + 1$, although in many cases $d = k$ is sufficient [3].

There are a number of methods for obtaining the d and τ: correlation integrals [4], false nearest neighbors [5], singular value decomposition [6], nonlinear modeling methods [7,8] and others [9–11]. If one has a large amount of low noise data from a low dimensional dynamical system, these methods can work well; in practice, however, we frequently have to make do with smaller amounts of noisy

V. In et al. (Eds.): *Proceedings of the 5th International Conference on Applications in Nonlinear Dynamics*, Understanding Complex Systems, https://doi.org/10.1007/978-3-030-10892-2_21

data. The previously mentioned dimension estimation methods require the user to estimate some parameter of the algorithm, which then affects the estimated dimension.

21.1.1 Error Estimates

Very few dimension estimation methods include a way to estimate confidence in the estimated dimension. One recent method does allow one to estimate the effect of noise on the dimension calculation [12], possibly aiding in determining the reliability of the result, although filtered noise is not discussed.

If one is 95% confident that the embedding dimension is 3, that is good; if one is only 5% confident, that is not so good. It is necessary to develop a method of dimension estimation that also allows the user to calculate the possible confidence in the final number. In this work, we show how the properties of random matrices may be used to put bounds on the eigenvalue spectrum of covariance matrices calculated from a finite dimensional attractor.

An additional feature of this method is that there is only 1 adjustable parameter, and the value for that parameter is based on a reasonable physical argument and is chosen before the calculation commences.

21.2 Covariance Matrices

The algorithm presented here has much in common with the singular value decomposition method already mentioned. Both methods create matrices from the data and seek to detect anisotropy in these matrices. The singular value decomposition method does a singular value decomposition on the data and looks at how many singular values are above the noise floor. There is no rigorous way to determine this threshold. The current method calculates the eigenvalues of the covariance matrix for the data and compares these eigenvalues to those expected for a Gaussian random signal. For the method described in this paper, the embedded time series is simply considered as a point cloud in phase space. No assumptions are made about dynamics. The null hypothesis is that this point cloud is drawn from a random process, and our goal is to disprove this hypothesis.

A chaotic attractor that can be embedded in a d dimensional phase space lies on an invariant manifold with a dimension of d or less. The manifold may have a local dimension $<d$. Curvature of the manifold is why the attractor itself requires d dimensions for embedding. The local dimension can be estimated by finding the eigenvalues of the covariance matrix for small regions on the attractor. These eigenvalues can be used to estimate the probability that the embedded signal is nonisotropic in a d dimensional phase space. Anisotropy is taken as an indication that the signal can be embedded in d dimensions.

Why find the covariance for local regions and not just the entire attractor? Chaotic attractors have a distinctive structure in phase space, as plots of attractors shown in this paper reveal. This structure means that the distribution of points on the attractor is anisotropic even when signal is embedded in too few dimensions to form an embedding of the dynamical system. It is necessary is divide the attractor into small regions for which the attractor density is approximately constant.

21.2.1 Clustering

If the local region on the attractor is too small, noise and digitization errors will obscure the local dimensionality. If the local region is too large, the curvature of the manifold and variations in density will cause errors in dimension estimation. In order to stay between these 2 size limits, local regions are found using a clustering algorithm. A small region on the attractor is divided into K equal size bins, and the number of points in each bin, m_k is counted. The empirical probability of finding a point in each bin is $\tilde{\pi}_k = m_k/M$, where M is the sum of the points in all K bins. The model probability is a constant over al K bins. Both sets of probabilities are used to update a prior containing the least information, and the posterior probabilities are compared using a Kullback–Leibler divergence, [13], a commonly used measure of the difference between probability distributions. An analytic formula for this Kullback–Leibler divergence was derived in [14]. An penalty function of $K\log_2(K)$ must be subtracted from this divergence function, as creating more bins is the equivalent of overfitting the data. The final formula for measuring how different the posterior probability distribution inferred from the $\tilde{\pi}_k$'s from the posterior model distribution is

$$R\left(m_k, K\right) = \frac{\frac{1}{\ln 2}\sum_{k=1}^{K}\left[\left(m_k - \rho_0 V\right)\cdot\psi\left(m_k + \frac{1}{2}\right) - \ln\Gamma(m_k + \frac{1}{2}) + \ln\Gamma\left(\rho_0 V + \frac{1}{2}\right)\right] - K\log_2\left(K\right)}{K} \tag{21.2}$$

where $\rho_0 = \sum_{k=1}^{K} m_k \Big/ (KV)$, where V is the volume of an individual bin, the function ψ is the digamma function and Γ is the gamma function. The units of $R(m_k, K)$ are bits/bin. A reasonable minimum threshold for $R(m_k, K)$ is 1 bit/bin. For this threshold, the attractor density is approximately constant over the K bins.

The embedded time series vector $\mathbf{s}(i)$ (Eq. 21.1) is clustered by first picking a random index point on the attractor. A set of $d+1$ nearest neighbors to the index point is found and partitioned into $K = 2^d$ bins. The Kullback–Leibler divergence $R(m_k, K)$ is then found for this set of neighbors. If $R(m_k, K)$ <1 bit/bin, more near neighbors are included. The region around the index point is expanded until $R(m_k, K)$ >1 bit/bin. The binning and expansion process is then started again with a new randomly chosen index point. The clustering process is continued until at least 90% of the points in $\mathbf{s}(i)$ have been included in a cluster. The different clusters may overlap.

Points on the same trajectory may be included in the group of neighbors, possibly introducing spurious correlations caused by time correlation in the 1-d signal $s(i)$. These correlations will be suppressed using a Theiler exclusion [15], in which points within a certain number of time steps along the same trajectory are excluded. In addition, the method of surrogates [16] was used to below account of the effects of time correlations.

From the M_l points in the l'th cluster, a $M \times d$ dimensional vector $\mathbf{x}$ is created. The vector $\mathbf{x}$ is normalized by subtracting the mean from each component and dividing by the standard deviation

$$\mathbf{y}_j = \frac{\mathbf{x}_j - \bar{\mathbf{x}}_j}{\sqrt{\left[\sum_{j=1}^{d}\sum_{i=1}^{M}\left(\mathbf{x}_j\left(i\right) - \bar{\mathbf{x}}_j\right)^2\right]}}. \tag{21.3}$$

where the overbar operator indicates a mean, and the subscript j indicates one of the d components of the vector $\mathbf{x}$. Next, the $d \times d$ covariance matrix is formed:

$$\mathbf{C} = \frac{\mathbf{y}^T\mathbf{y}}{M}. \tag{21.4}$$

A d dimensional Gaussian random process will be isotropic in a d dimensional space, and the mean covariance matrix for this process will be proportional to the identity matrix. The possible covariance matrices $\mathbf{C}$ for the Gaussian random process may be drawn from a Wishart distribution [17] with a mean covariance proportional to the identity matrix, and it is possible to place some limits on the eigenvalues of $\mathbf{C}$. If the eigenvalues of $\mathbf{C}$ do not fall within the limits for a Gaussian random process, then we reject the null hypothesis that $\mathbf{s}(i)$ was obtained from a Gaussian random process. The covariance eigenvalues may fall outside the limits for a Gaussian random process if the signal is not isotropic when embedded in a d dimensional space.

The probability distribution of random matrices $\mathbf{X}$ with covariance Σ is the Wishart distribution, [17]

$$\begin{aligned} f\left(\mathbf{X}, \Sigma, n\right) &= \frac{|\mathbf{X}|^{((n-d-1)/2)} e^{\left(-\frac{1}{2}\mathrm{trace}\left(\Sigma^{-1}\mathbf{x}\right)\right)}}{2^{nd/2}\pi^{(d(d-1))/4}|\Sigma|^{n/2}\Gamma_d(n/2)} \\ \Gamma_d\left(\tfrac{n}{2}\right) &= \pi^{\frac{d(d-1)}{4}} \prod_{j=1}^{d} \Gamma\left(\tfrac{n}{2} + \tfrac{1-j}{2}\right) \end{aligned} \tag{21.5}$$

where n is the number of degrees of freedom (number of points in the time series), $\mathbf{X}$ and Σ are $d \times d$ matrices where $n \geq d$, and $||$ indicates a determinant.

For n and d approaching ∞, the probability distribution for the eigenvalues of a random matrix converges to the Marchenko–Pastur distribution [18]. For low dimensional attractors, the Marchenko–Pastur distribution is not a good approximation, so the range of possible eigenvalues for a Gaussian random process must be estimated from a Monte-Carlo process by drawing random $n \times d$ matrices from the Wishart distribution.

The function wishrnd() in MATLAB was used to create covariance matrices drawn from a Wishart distribution with a mean covariance matrix equal to the identity.

21.2.2 Time Series Correlations

The Wishart distribution that was used to find limiting values of eigenvalues for the covariance matrix is a distribution for covariance matrices of Gaussian random processes having a flat power spectrum. Such a random process is isotropic in space. The power spectrum of a measurement of an actual physical system is not flat, but is limited in frequency. These frequency limits cause a time series from an experiment to have some correlation in time. When the data is embedded in a phase space, this time correlation can make the embedded signal non-isotropic in phase space, so the eigenvalues for a covariance matrix from such a signal may be outside the limits for the eigenvalues of a Gaussian random process, even if the embedded signal is just filtered noise. It is well known that filtered noise signal can cause problems for dimension estimation algorithms [15, 19], and there have always been concerns that using embedding delays that are too short can lead to false correlations.

To detect this time correlation, we create a surrogate signal [16] from our original time series. The original time series is Fourier transformed, the Fourier components are multiplied by random phase factors, and then the phase randomized Fourier signal is inverse transformed. The result is a random surrogate signal with the same power spectrum as the original signal. Because the power spectrum is the same as the original signal, the correlation properties of the surrogate are the same as the original signal. If the eigenvalues for the covariance matrices from the surrogate signal embedded in d dimensions are outside the limits for a Gaussian random signal, then the anisotropy could have been caused by time correlation. It is still possible that the signal is a deterministic signal that can be embedded in d dimensions; it just isn't possible to tell if the anisotropy is caused by determinism or by time correlation.

21.3 Dimension Estimates

The example attractor here came from the Rossler system [20]. A time series of 20,000 points was generated from the Rossler equations

$$\begin{aligned} \tfrac{dx}{dt} &= -y - z \\ \tfrac{dy}{dt} &= x + 0.2y \\ \tfrac{dz}{dt} &= 0.2 + z\,(x - 5.7)\,. \end{aligned} \tag{21.6}$$

The Rossler equations were integrated using a 4th order Runge-Kutta integration routine with a time step of 0.1 s. Figure 21.1 is a plot of the attractor created by embedding the x signal with a delay of 2. Figure ?? is the autocorrelation of the Rossler signal.

The embedded Rossler signal was clustered according to the methods of Sect. 21.2.1, with a threshold of $R(m_k, K) > 1$ bit/bin. The l'th cluster contained M_l points. These M_l points were used to create a d dimensional covariance matrix, as described in Eqs. (21.3)–(21.4), and the d eigenvalues of the covariance matrix were calculated.

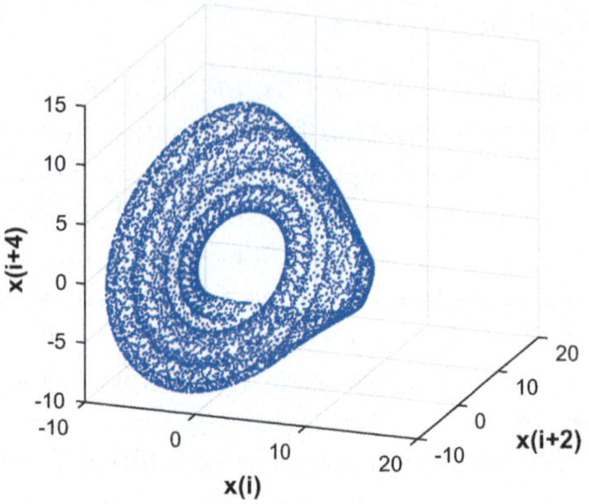

Fig. 21.1. Rossler attractor obtained by embedding the x signal from Eq. 21.6 with a delay of 2

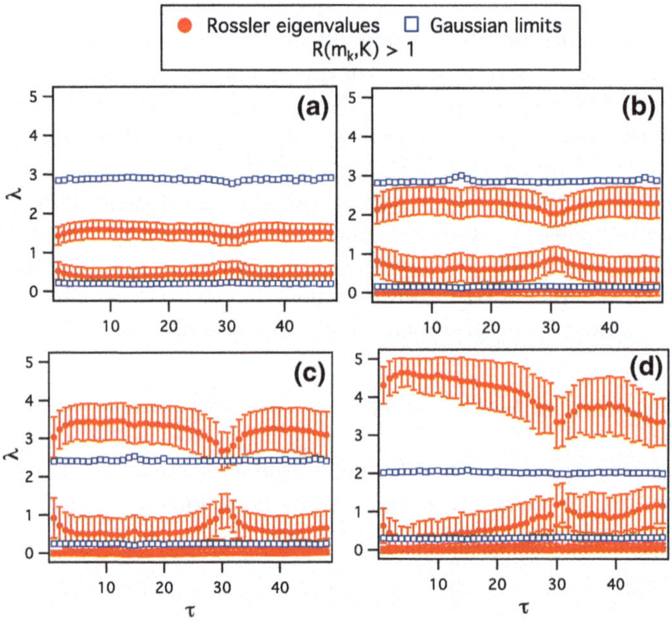

Fig. 21.2. Filled circles with error bars are the mean eigenvalues $\lambda_j(d, \tau)$ $j = 1 \ldots d$ as a function of embedding delay τ for covariance matrices for the clustered Rossler attractor. The open squares are the mean limiting values for covariance matrices of Gaussian random d dimensional signals, based on a Monte Carlo simulation. **a** corresponds to d=2, **b** is d=3, **c** is $d = 4$ and **d** is $d = 5$

For a given dimension d and embedding delay τ, the Rossler attractor was clustered into N_c clusters, resulting in N_c sets of d eigenvalues $\lambda_{l,j}(d, \tau)$ $l = 1 \ldots N_c, j = 1 \ldots d$.

21.3.1 Limiting Eigenvalues

For each value of d and τ, the eigenvalues $\lambda_{l,j}(d,\tau)$ for the Rossler system must be compared to the limiting values for the eigenvalues λ_{max} and λ_{min} of the covariance matrix of a d dimensional Gaussian random signal, as found by the Monte Carlo process above. The limiting eigenvalues λ_{max} and λ_{min} depend on the number of points in the cluster, M_l (as shown in Fig. ??). Each of the clusters for a given d and τ may contain a different number of points. For the l'th cluster, containing M_l points, the limiting eigenvalues are $\lambda_{max}(d, M_l)$ and $\lambda_{min}(d, M_l)$.

Figure 21.2 shows the mean eigenvalues for the covariance matrices from the Rossler system as a function of τ for $d = 2, 3, 4, 5$. In Fig. 21.2a, the mean eigenvalues $\lambda_j(d,\tau)$ for covariance matrices for a 2-d embedding of the Rossler system are well within the limits for the eigenvalues for the covariance matrix for a Gaussian random process. In 2 dimensions, the null hypothesis can't be disproved- the covariance matrices for the clusters on the Rossler attractor could come from a Gaussian random process. In Fig. 21.2b, the 3-d embedding, one of the mean Rossler eigenvalues lies outside the range of eigenvalues for a Gaussian random process for $\tau < 50$. From Fig. 21.2b we can say that for $\tau < 50$, the Rossler signal is not isotropic in $d = 3$. The conclusion is that the Rossler system can be embedded in 3 dimensions. Figure 21.2c, d, in 4 and 5 dimensions, also reject the null hypothesis for those dimensions.

21.3.2 Filtered Noise

Filtered noise can be a difficult test for dimension estimation algorithms. Because filtered noise is correlated in time, it can appear to have anisotropy in phase space. A 20,000 point filtered noise signal was clustered as in the previous examples, with an information threshold of $R(m_k, K) > 1$ bit. As before, the filtered noise signal was embedded in different dimensions with different delays. As with the Rossler system, the eigenvalues for covariance matrices from the filtered noise signals were calculated. The mean eigenvalues and the mean limiting values of eigenvalues for random covariance matrices are plotted in Fig. 21.3.

The filtered noise signal appears to have some anisotropy when embedded in 4 or 5 dimensions, as seen in Fig. 21.3c, d. This apparent anisotropy is a consequence of the time correlation of the filtered noise signal. To detect when anisotropy could be caused by time correlation, a surrogate signal is required.

21.4 Surrogate Signals

Time correlation can produce the appearance of anisotropy in the embedded signal, particularly for small values of the embedding delay τ. To discover this spurious anisotropy, a surrogate signal is created as described above [16]. The surrogate signal will have the same power spectrum as the Rossler signal, so linear correlations will be preserved. Because the surrogate signal is otherwise

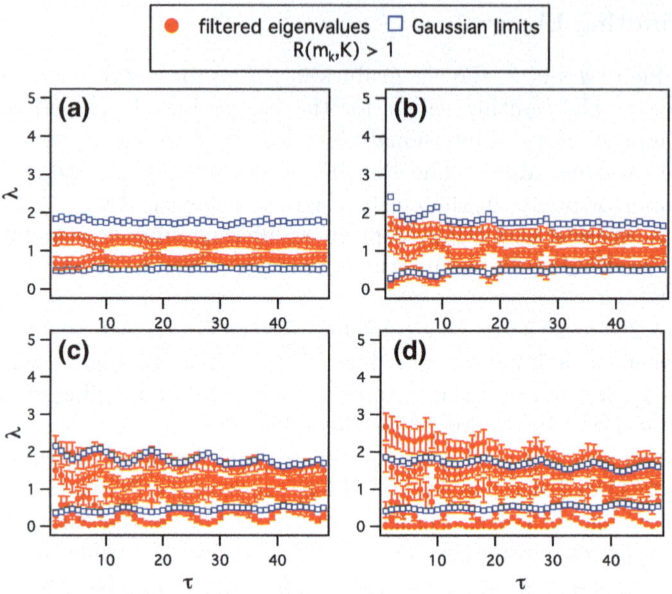

Fig. 21.3. Filled circles with error bars are the mean eigenvalues $\lambda_j(d,\tau)$ $j = 1 \ldots d$ as a function of embedding delay τ for covariance matrices for the clustered filtered random noise signal. The open squares are the mean limiting values for covariance matrices of Gaussian random d dimensional signals, based on a Monte Carlo simulation. **a** corresponds to $d = 2$, **b** is $d = 3$, **c** is $d = 4$ and **d** is $d = 5$

random, any anisotropy seen in the covariance matrices for the surrogate signal could is caused by the time correlations in the original Rossler signal.

The Rossler signal is Fourier transformed, the Fourier components are each multiplied by a random phase factor, and the randomized Fourier is inverse transformed to yield the surrogate signal. The surrogate signal is embedded in the phase space, and the eigenvalues of the covariance matrices are plotted in Fig. 21.4.

In Fig. 21.4, the eigenvalues for the surrogate signal from the Rossler signal are outside the bounds of the eigenvalues for the covariance matrix of a Gaussian random system for low values of the embedding delay τ for embedding dimensions 4 and 5. Figure 21.4c, d would appear to show that the embedded Rossler surrogate signal is anisotropic in dimensions 4 and 5, but this apparent anisotropy results from the time correlation in the Rossler x signal.

In the next section, information from Figs. 21.2 and 21.4 is combined to give a probability that the covariance matrix algorithm can determine that the Rossler x signal is not isotropic when embedded in d dimensions.

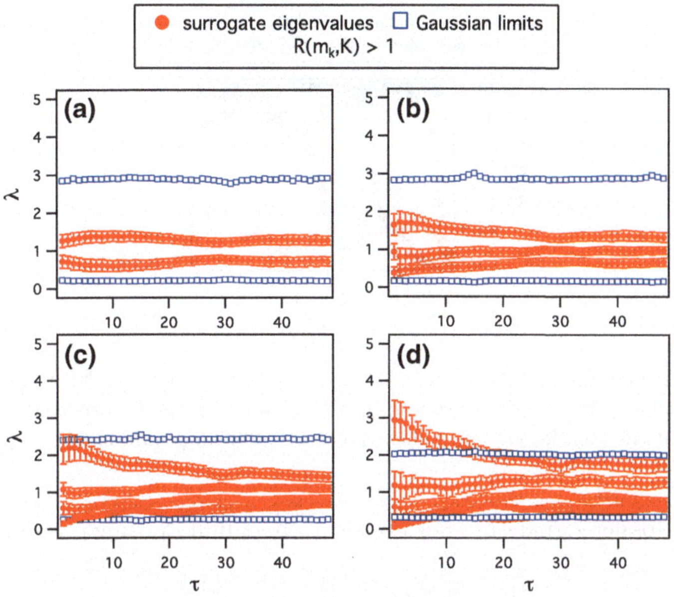

Fig. 21.4. Filled circles with error bars are the mean eigenvalues $\lambda_j(d, \tau)$ $j = 1 \ldots d$ as a function of embedding delay τ for covariance matrices for phase randomized surrogate signal created from the Rossler x signal. The open squares are the mean limiting values for covariance matrices of Gaussian random d dimensional signals, based on a Monte Carlo simulation. **a** corresponds to $d = 2$, **b** is $d = 3$, **c** is $d = 4$ and **d** is $d = 5$

21.5 Surrogate Signals and Probabilities

The simplest way to use the eigenvalue spectrum to estimate embedding dimension is to estimate the probability that the embedded signal is nonisotropic in d or fewer dimensions. The ability to estimate probabilities is the major difference between this dimension algorithm and other algorithms.

Because there is some deviation in the values of the eigenvalues of the covariance matrices for a signal, the question of whether or not an eigenvalue $\lambda_j(d, \tau)$ is outside the limits specified by the upper and lower limiting eigenvalues for a Gaussian random process, $\lambda_{max}(d, \tau)$ and $\lambda_{min}(d, \tau)$ is not a yes or no question. To estimate the probability $\rho(d, \tau)$ that at least one of the eigenvalues for the covariance matrices is outside the limits for a Gaussian random process, we count the fraction of times that this occurs for each combination of d and τ.

The probability $\rho_n(d, \tau)$ is the probability that the Rossler x signal does not have the eigenvalues of a Gaussian random signal when embedded in d dimensions or fewer using a delay of τ. The chance that the anisotropy in the covariance matrices comes from time correlation alone, and not determinism, must be accounted for. The probability $\rho_{surr}(d, \tau)$ is calculated from the surrogate Rossler x signal. The probability that this algorithm can indicate that the

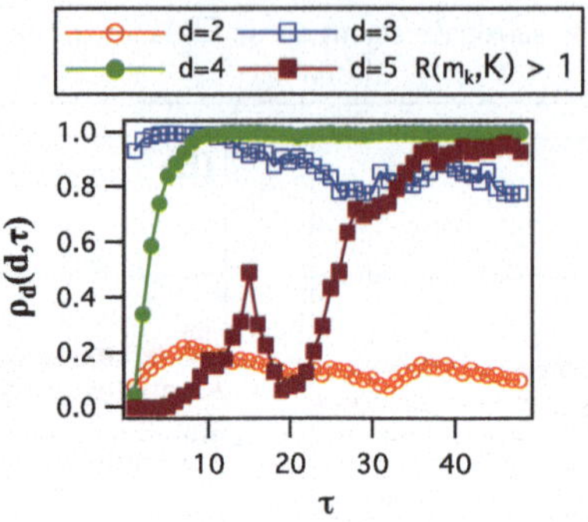

Fig. 21.5. Probability $\rho_d(d, \tau)$ (Eq. 21.7) that the embedded Rossler x signal did not come from a Gaussian random process (i. e. it is not isotropic), corrected for the time correlation of the Rossler signal. The x signal was 20,000 points long

embedded Rossler signal does not appear to be a uniform random signal is then

$$\rho_d(d, \tau) = \rho_n(d, \tau) - \rho_{surr}(d, \tau). \tag{21.7}$$

The value of $\rho_d(d, \tau)$ is plotted in Fig. 21.5. The probability plotted in Fig. 21.5 only shows the probability that this algorithm could determine that the embedded signal was not a random signal when embedded in d dimensions with an embedding delay of τ. It is possible that time delays that show a low probability could still be legitimate embedding delays, but this algorithm couldn't distinguish between a low dimensional signal and filtered noise for those delays.

The plots in Fig. 21.5 do show that for delays between 2 and 18, there is a better than 90% probability that the embedded signal is anisotropic when embedded in $d = 3$. The drop in probability for the 3-d embedding in Fig. 21.5 most likely occurs because for larger delays, the effects of curvature of the manifold occupied by the Rossler attractor become large enough to affect the covariance matrix.

21.5.1 Filtered Noise

The surrogate signal method was also applied to the filtered noise signal. Figure 21.6 shows the probability that the dimension algorithm could determine that the structure seen in the filtered noise signal was due to its finite dimension and not caused by time correlation. A probability of <0 means that the surrogate probability had a higher probability of being anisotropic than the regular signal; we take negative probabilities to be the same as 0.

In Fig. 21.6, the dimension algorithm detects no anisotropy in the filtered random noise signal.

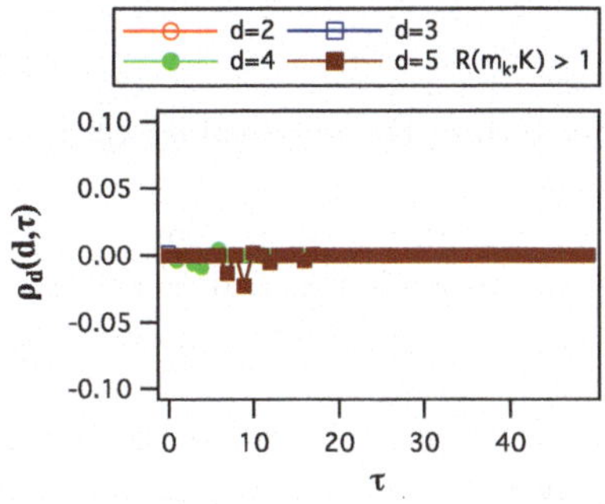

Fig. 21.6. Probability $\rho_d(d, \tau)$ that the embedded filtered random noise signal did not come from a Gaussian random process, corrected for the time correlation of the filtered signal. A probability of <0 means that the surrogate probability had a higher probability of being anisotropic than the regular signal; we take negative probabilities to be the same as 0

21.6 Algorithm Summary

Combining all the sections above, the dimension estimation algorithm is:

1. Before starting, estimate limits on eigenvalues for covariance matrices of a Gaussian random process for different dimensions and numbers of points, using a Monte Carlo simulation (Eq. 21.5). Store these values in a lookup table.
2. Embed a time series $s(i), i = 1 \ldots N$ in d dimensions with an embedding delay of τ: $\mathbf{s}(i) = [s(i), s(i+\tau), \ldots s(i+(d-1)\tau)]$.
3. Apply clustering algorithm using probability threshold of Eq. (21.2).
4. For each of N_c clusters, calculate normalized covariance matrix of Eq. (21.3)–(21.4).
5. Find the eigenvalues of the normalized covariance matrix for each cluster.
6. The l'th cluster contains M_l points. From the lookup table containing the limiting eigenvalues for a random Gaussian process containing M_l points and embedded in d dimensions, retrieve the maximum and minimum possible eigenvalues $\lambda_{max}(d, M_l)$ and $\lambda_{min}(d, M_l)$.
7. Find the probability that one of the eigenvalues $\lambda_{l,j}(d, \tau), l = 1 \ldots N_c, j = 1 \ldots d$ is outside the limits $\lambda_{max}(d, M_l)$ and $\lambda_{min}(d, M_l)$. This probability is $\rho_n(d, \tau)$.
8. Create a phase randomized surrogate signal from $s(i), i = 1 \ldots N$.
9. Repeat steps 1–9 for the phase randomized signal to get a probability $\rho_{surr}(d, \tau)$.
10. Calculate the probability $\rho_d(d, \tau) = \rho_n(d, \tau) - \rho_{surr}(d, \tau)$. Plot $\rho_d(d, \tau)$ as a function of d and τ.

21.7 Discussion

This algorithm, based on finding the eigenvalues of a covariance matrix for an embedded signal, estimates the probability that the embedded signal is not isotropic when embedded in d dimensions. Anisotropy is taken as an indication that the signal can be embedded in d dimensions. The covariance matrix eigenvalues for filtered random signals may also lie outside the range expected for a Gaussian (isotropic) d-dimensional process, so to eliminate this possibility, it's necessary to generate a phase randomized surrogate of the signal to be embedded, and calculate the eigenvalues for covariance matrices for an embedded version of this signal.

The algorithm described here doesn't solve the problem of finding the best embedding delay τ. The algorithm gives the probability for some value of embedding delay τ that an unknown signal is not isotropic when embedded in d dimensions, and that the anisotropy could not have been a result of time correlation. There may be legitimate values of the embedding delay τ that give a low probability in this algorithm; the low probability simply means that this algorithm can't tell unambiguously that a signal is nonisotropic when embedded in d dimensions. There has been work that shows that different values of the embedding delay τ are useful for different applications [1].

This dimension estimation method is not as computationally efficient as some methods, but it outputs a confidence level, so that the user can understand how reliable the dimension measurement is.

References

1. E. Bradley, H. Kantz, Chaos **25**, 097610 (2015)
2. H.D.I. Abarbanel, R. Brown, J.J. Sidorowich, L.S. Tsmring, Rev. Mod. Phy. **65**, 1331–1392 (1993)
3. F. Takens, *Detecting Strange Attractors in Turbulence* (Springer, New York, 1980)
4. P. Grassberger, I. Procaccia, Phys. D Nonlinear Phenom. **9**, 189–208 (1983)
5. M.B. Kennel, R. Brown, H.D.I.A. Abarbanel, Phys. Rev. A **45**, 3403–3411 (1992)
6. G.P. King, R. Jones, D.S. Broomhead, Nucl. Phys. B- Proc. Suppl. **2**, 379–390 (1987)
7. A. Maus, J.C. Sprott, Commun. Nonlinear Sci. Numer. Simul. **16**, 3294–3302 (2011)
8. Y.I. Molkov, D.N. Mukhin, E.M. Loskutov, A.M. Feigin, G.A. Fidelin, Phys. Rev. E **80**, 046207 (2009)
9. L. Cao, Phys. D Nonlinear Phenom. **110**, 43–50 (1997)
10. T. Buzug, G. Pfister, Phys. Rev. A **45**, 7073–7084 (1992)
11. L.M. Pecora, L. Moniz, J. Nichols, T.L. Carroll, Chaos: Interdiscip. J. Nonlinear Sci. **17**, 013110–013119 (2007)
12. J.F. Restrepo, G. Schlotthauer, Phys. Rev. E **94**, 012212 (2016)
13. S. Kullback, R.A. Leibler, Ann. Math. Stat. **22**, 79–86 (1951)
14. T.L. Carroll, J.M. Byers, Phys. Rev. E **93**, 042206 (2016)
15. J. Theiler, Phys. Rev. A **34**, 2427–2432 (1986)

16. J. Theiler, S. Eubank, A. Longtin, B. Galdrikian, J.D. Farmer, Phys. D **58**, 77–94 (1992)
17. J. Wishart, Biometrika **20A**, 32–52 (1928)
18. V.A. Marchenko, L.A. Pastur, Math. USSR-Sb. **1**, 457–483 (1967)
19. A. Provenzale, A.R. Osborne, in *Dynamics and Stochastic Processes: Theory and Applications*, ed. by R. Lima, L. Streit, R.V. Mendes, vol 355 (1990), pp. 260–275
20. O.E. Rossler, Z. Naturforsch **38a**, 788–801 (1983)
21. T.L. Carroll, J.M. Byers, Chaos: Interdiscip. J. Nonlinear Sci. **27**, 023101 (2017)

Chapter 22
Bio-Inspired Approach to Quantify Nonlinearities in Time-Series Measurements Using the Nuttall-Wiener-Volterra (NWV) Method

Derke R. Hughes[1(✉)], Richard A. Katz[1], Robert M. Koch[1], and Albert H. Nuttall[2]

[1] Naval Undersea Warfare Center, Newport, RI 02841, USA
derkehughes@navy.mil
[2] Nuttall Analysis, Old Lyme, CT, USA

Abstract. This research offers an additional approach to the increased interest in information theoretic techniques utilized in the Theory of Communications, Electrical Engineering and Signal Processing disciplines for extracting nonlinear behavior in dynamical systems. This new approach was, in part, motivated by a diligent effort to create a man-made system that mimics the sound generation of a cicada. This insect has tremendous sound production capacity for its size. For the Okanagana and Magicicada species studied in this research, these cicadae ranged in size from five to six centimeters and produce sounds that are heard several hundred meters away. The evolution of this new signal processing algorithm from this bio-inspired research is explained in this article. This investigation initially examined the cicada hypothesized nonlinear system, by employing a number of numerical techniques in which to identify nonlinearity in a measurement times series. One such technique, the Nuttall modified-Volterra approach would serve as the validation and verification process for confirming that the inherent artificiality introduced by converting the sound production system of the biologic system to a man-made device did not corrupt the inherent dynamics of the cicada mating call. The technical advantage gained from quantification of the expansion kernels using the Nuttall approach, is the creation of more characterization clues by extending beyond the linear kernel response. This unique method is based on an extension of earlier developments of Vito Volterra and Norbert Wiener. The new Nuttall-Wiener-Volterra (NWV) method identifies the existence of nonlinearity in a measurement time series and determines the power distribution of individual nonlinear components. Moreover, the NWV method, unlike other methods that are likely less computationally efficient due to the Curse of Dimensionality (COD), significantly reduces the computational workload, thereby making characterizations of nonlinear systems with memory at higher orders possible. The nonlinear system kernel responses reveal identification and characterization of linear and nonlinear dynamics contained within the system under investigation. Thus, the nonlinear kernel responses computed for the cicada exposed a critical development for the NWV technique, namely that in order to obtain meaningful NWV kernel responses (i.e., to have physically and mathematically sound

V. In et al. (Eds.): *Proceedings of the 5th International Conference on Applications in Nonlinear Dynamics*, Understanding Complex Systems, https://doi.org/10.1007/978-3-030-10892-2_22

computational results), there are restrictive requirements for the system input excitation to be (a) band-limited, (b) white Gaussian and (c) zero mean. By studying the anatomical structures in the cicada sound production system and developing the wave propagation and finite element (FE) models this effort also then attempted an approach to confirm the accuracy of these models by employing the NWV nonlinear (and linear) analysis method.

22.1 Introduction to an Investigation of Bio-Inspired Sound Generation

The initial step in the generation of this bio-inspired source is to understand the in-air cicada transmission capability. The effort focused on characterizing the cicada sound production system by measuring the sound produced using microphones and a laser-Doppler vibrometry. These measurements led to an analysis and comparison of the sound production capabilities of the cicada as compared to traditional transducers operating in air in order to develop a figure-of- merit relating to the sound production efficiency gain that is produced. A second-order signal processing model using the Volterra method was developed to verify the presence of nonlinear behavior in the cicada mating call, leading to the development of new fundamental design equations that replicate the cicada mating call and are able to produce accurate representations of both the temporal and spectral signal structure. The cicada appears to down-convert multiple higher frequency components in order to provide an inter-modulated band pass signal structure in the 3–14 kHz region of the audio band by a technique that does not exist in man-made systems to date. A finite element analysis model for the cicada was created to simulate the cicada's sound production system in-air which could help explain the structural acoustics generated by the cicada's anatomy.

22.2 Development of a Finite Element Analysis (FEA) Model of Cicada Sound System

Micro-computer tomography (micro CT) images of the cicada were scanned to the appropriate resolution to develop a meshed computer aided design (CAD) model. The meshed CAD model is analyzed with ABAQUS Finite Element (FE) analysis software which generates pressure values at a prescribed distance from the structures analyzed. In Fig. 22.1, the FE model shows the essential anatomical structures that produce the cicada mating call as well as the synthesized tymbal modeled structure.

The tymbals provide the clicking noise from the buckling of the tymbal ribs, and the abdomen acts as an amplifier for the impulse train caused from the snap of the tymbal ribs. A simulated tymbal muscle response is utilized as the forcing function for the FE model. This FE model illustrated in lower part of Fig. 22.1 has air elements placed around the structure and the pressure history is computed as a function of time as shown in Fig. 22.2.

The simulated results depicted in Fig. 22.2 are the initially computed output for the in-air FEA model used to compare to experimental data; however, there are several

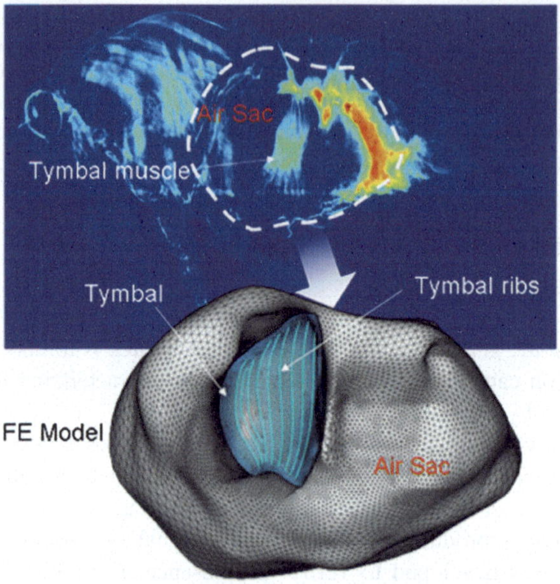

Fig. 22.1. The essential anatomical sound production system

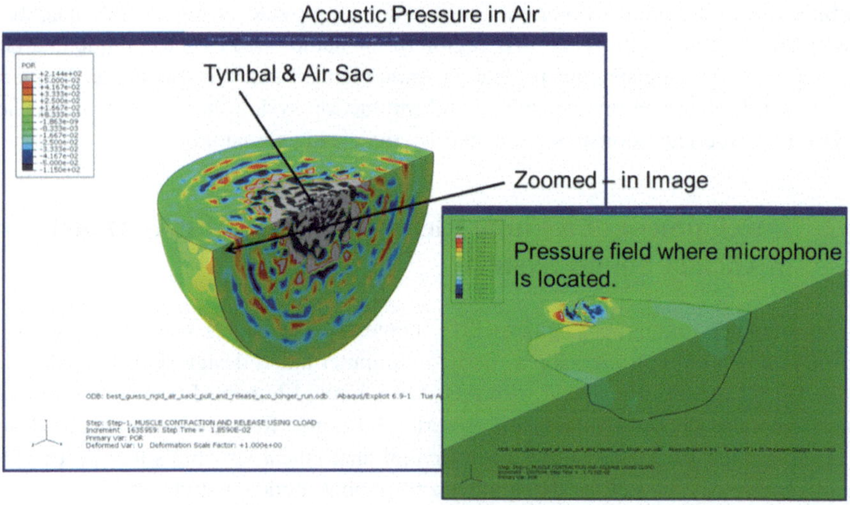

Fig. 22.2. ABAQUS pressure field results

difficult finite element analysis (FEA) modeling problems with the sound system of the cicada to be overcome and addressed such as the rib-stiffened buckling transduction. Previous FEA models were built to address the step-by-step development required to simulate the tymbal vibration. Current commercial software packages cannot address the parameters which govern the challenges created by the cicada's unique acoustic transduction system. The material properties of the tymbal, frequencies generated, and the difficulties of modeling buckling acoustics have not been fully examined using current FE software. Non-commercial software may possess the mathematical fidelity to compute the structural acoustic interface between the tymbal and air with cicada material properties and its operating frequencies such as hybrid fluid-structural code (DYSMAS).

22.3 Development of a Wave Propagation Model for Cicada Mating Call

The complexities associated with a rib-stiffened buckling transduction system are obvious and thus a systematic methodology is being considered in this article. The tymbals in the cicada not only are ribbed stiffened, which present modeling and computational issues – the ribs also buckle to produce the mating call. Determining the potential frequency ranges that are capable with a multimodal buckling structure is a revolutionary concept in acoustic sound production. Since air does not impose a significant surface load on an object creating sound, the opportunity to analyze multimodal structural acoustics is possible with the in-air FE model. The in-air FE model simplification is to verify that the FE modeling software can maintain fluid-to-structure contact accurately through a simulation. Consequently, the FE software requires a qualification process, namely that the FE results match known theoretical solution. To this end the development of an analytical solution is necessary. For example, in Fig. 22.3, there is an end-cap portion of a sphere which vibrates at a given velocity and frequency. The derivation to obtain the radiation pattern is as follows:

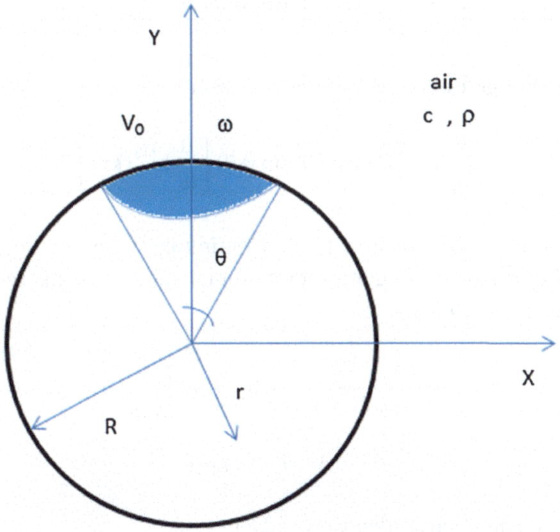

Fig. 22.3. End-cap portion of a sphere vibrating

the velocity v,

$$\begin{aligned} v_r &= Re\{V_0 e^{-i\omega t}\} & & 0<\theta<\theta_o \\ & & & r=R \\ &= 0 & \theta_o<\theta<\tfrac{\pi}{2} & \quad r=R \end{aligned} \tag{22.1}$$

is based on r the radius for the real portion of the velocity. Outside of the sphere the wave equation is given by the following expression for the pressure field:

$$p = Re\left\{\sum_{l=0}^{\infty} A_l P_l(cos\theta) h_l(kr) e^{-i\omega t}\right\}. \tag{22.2}$$

This equation contains the Hankel functions h_l with A_l coefficients and Legendre P_l polynomials where l represents the indices. The Hankel function is shown as an outward propagating wave function. Consider Euler's equation where the velocity of the end-cap portion of the sphere equals pressure generated by the motion the wave number k and the radius r of the source, as follows:

$$\begin{aligned} \hat{v}_r &= \frac{1}{i\omega\rho}\sum_l A_l P_l(\cos\theta)\left\{\frac{d}{dr}h_l(kr)\right\} \\ -i\omega\rho\hat{v}_r &= -\frac{\partial}{\partial r}\hat{p}. \end{aligned} \tag{22.3}$$

Using orthogonality with Legendre polynomials and an initial velocity V_o

$$\int_0^{\theta_o} V_0 P_l(cos\theta) sin\theta d\theta \tag{22.4}$$

this expression is plugged into the Euler's equation

$$p = \frac{1}{i\omega\rho} A_l \int_0^{\pi} P_l(\cos\theta)^2 \sin\theta d\theta \left\{\frac{d}{dr} h_l^{(1)}(kr)\right\}_{r=R} \tag{22.5}$$

for the Hankel function of the first kind. Continuing with the solution to Hankel function of first kind and the Legendre polynomial indices being set to 0 and 1 the expressions are given as follows

$$= \frac{A_l}{i\omega\rho}\frac{2}{2l+1}\left\{\frac{d}{dr}h_l^{(1)}(kr)\right\}_{r=R} \tag{22.6}$$

$$P_0 = 1 \quad P_1 = cos\theta \tag{22.7}$$

along with a few trigonometric identities (22.8)

$$\begin{aligned} &\int_0^{\theta_o} sin\theta\, d\theta = 1 - cos\theta_0 \\ \int_0^{\theta_o} cos\theta sin\theta\, d\theta &= \int_0^{\theta_o} \frac{1}{2}\frac{d}{d\theta} sin^2\theta\, d\theta \\ &= \frac{1}{2} sin^2\theta_0. \end{aligned} \tag{22.8}$$

The Hankel solutions at equilibrium on the surface of the end cap of the sphere are given in the following equations:

$$\begin{aligned} V_0(1 - cos\theta_0) &= \frac{A_0}{i\omega\rho} 2\left\{\frac{\mathrm{d}}{\mathrm{dr}} \mathrm{h}_0^{(1)}(\mathrm{kr})\right\}_{\mathrm{r=R}} \\ V_0\left(\frac{1}{2} sin^2\theta_0\right) &= \frac{A_0}{i\omega\rho}\frac{2}{3}\left\{\frac{\mathrm{d}}{\mathrm{dr}} \mathrm{h}_1^{(1)}(\mathrm{kr})\right\}_{\mathrm{r=R}} \\ h_0^{(1)}(kr) &= -i\frac{e^{ikr}}{kr} \quad h_1^{(1)}(kr) = -i\left(\frac{i}{kr} - \frac{1}{(kr)^2}\right)e^{ikr} \end{aligned} \tag{22.9}$$

Now, the Hankel functions for small kr at the edge of sphere at equilibrium are

$$\begin{aligned} h_0^{(1)}(kr) &\cong -\frac{i}{kr} + 1 \\ h_1^{(1)}(kr) &\cong -i\left(\frac{i}{kr} - 1 - \frac{1}{(kr)^2} - \frac{i}{kr} + \frac{1}{2}\right). \end{aligned} \tag{22.10}$$

Then, the Hankel coefficients A_l at the edge of the end cap using small angle assumption are solved in the next set of equations:

$$\begin{aligned} V_0(1 - cos\theta_0) &= \frac{A_0}{i\omega\rho} 2\frac{i}{kR^2} \\ V_0\frac{1}{2} sin^2\theta_0 &= \frac{A_1}{i\omega\rho}\frac{2}{3}\frac{-2\mathrm{i}}{\mathrm{k}^2\mathrm{R}^3} \end{aligned}$$

$$A_0 = kR^2\frac{\omega\rho}{2} V_0(1 - cos\theta_0) \tag{22.11}$$

$$A_1 = -k^2R^3\frac{3}{8}\omega\rho V_0 sin^2\theta_0. \tag{22.12}$$

In order to determine a radiation pattern or acoustic pressure field the derivation continues as follows. The acoustic pressure for large kr is derived in the following manner by the Hankel coefficient plugged into the Euler equation:

$$h_0^{(1)}(kr) \rightarrow \frac{-i}{kr}e^{ikr} \quad \text{and} \quad h_1^{(1)}(kr) \rightarrow \frac{1}{kr}e^{ikr} \tag{22.13}$$

$$\begin{aligned}\hat{p} &\approx kR^2\frac{\omega\rho}{2}V_0(1 - cos\theta_0)\left(\frac{-i}{kr}e^{ikr}\right) - k^2R^3\frac{3}{8}\omega\rho V_0 sin^2\theta_0 cos\theta_0\frac{1}{kr}e^{ikr}\\ \hat{p} &\approx \omega\rho V_0R^2\left\{\frac{-i}{2}(1 - cos\theta_0) - \frac{3}{8}kRsin^2\theta_0 cos\theta_0\right\}\frac{e^{ikr}}{r}.\end{aligned} \tag{22.14}$$

Also, the angle θ_0 is assumed small to form the following first approximations $1 - cos\theta_0 \approx \frac{1}{2}\theta_0^2$ and $sin^2\theta_0 \approx \theta_0^2$, which form this pressure

$$\hat{p} \approx \omega\rho V_0R^2\frac{1}{4}\theta_0^2\left\{-i - \frac{3}{2}kRcos\theta_0\right\}\frac{e^{ikr}}{r} \tag{22.15}$$

approximation. Consequently, the radiation pattern is proportional to the magnitude of bracketed expression $\left\{-i - \frac{3}{2}kRcos\theta_0\right\}$. So, the radiation pattern for the acoustics is equal to $1 + \frac{9}{4}(kR)^2cos^2\theta_0$ for a given initial velocity V_o, which generates a pressure of around 248 kPa at 10 m/s for a 10 kHz end-cap source at standard environmental conditions in air. The graph for the radiation pattern is shown in Fig. 22.4.

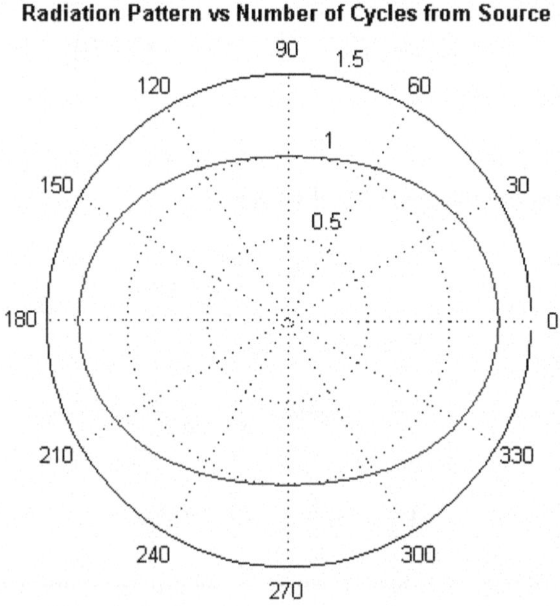

Fig. 22.4. Radiation pattern for end-cap portion of the sphere

This radiation pattern is a key step in helping to understand the acoustic pressure created by such a surface vibrating on a sphere.

22.4 Understanding Wave Propagation of Acoustic Signals for Use in Cicada Generated Sound

Previous work by Young [1] and Bennet-Clark [2, 3] studied a linear model of sound propagation in cicadas, however, linear models fail to adequately explain how the signals interact and propagate over long distances. On the other hand, non-linear models are more difficult to analyze. The findings presented in Edoh, Hughes, and Katz [4] suggest that a linear model does not fully capture the acoustic waveform distortion and proposes a nonlinear model instead, studying Burgers' equation [5]. Burgers' equation is a simplification of the Westervelt equation that incorporates nonlinear effects into a forward-propagating plane-like wave, and is stated as

$$\frac{\partial v}{\partial x} - \frac{\beta}{c^2} v \frac{\partial v}{\partial \tau} = \frac{\delta}{c^3} \frac{\partial^2 v}{\partial \tau^2} \tag{22.16}$$

where $\tau = t - \frac{x}{c}$, x is the spatial variable measured in meters, t is time in seconds, c is the speed of sound (c = 343.2 meters per second in air at 20 °C), and we take $\beta = 1.2$ as the parameter of nonlinearity, and $\delta = 1.9 \times 10^{-6}$, for propagation in air.

In this simulation, the study refines this wave propagation model to determine whether the significant features observed in the recorded cicada data are present. This is a semi-empirical approach in the sense that the theory guides the determination of whether the propagation can be simulated using a simplified model like Burgers' equation, or if higher order nonlinearities must be included in the model equation. Numerical challenges are due to nonlinearity and non-smooth nature of the signals along with high frequency content. However, Burgers' equation solutions are known to form multiple simultaneous solutions with smooth initial data. Therefore, the Burgers' solution should adequately handle the frequency component interaction.

In [4], the authors implemented a Fourier spectral solver and low order time integration method to solve Burgers' equation. However, the performance of spectral methods depends on the smoothness of the solution (and initial data); yet the recorded data shows that the signals are not necessarily smooth, see Fig. 22.3 for example.

In order to gain more confidence in the results, the solver was re-implemented using a Weighted Essentially Non-Oscillatory (WENO) finite volume approach [6] coupled with a Runge–Kutta solver. These methods are designed specifically for functions that have multiple simultaneous solutions (for mathematicians this is one form of shock) and are known to perform well. Preliminary results seem to indicate that this approach is well-suited to the problem, but the computed results at 15 inches from the insect appear less attenuated than the recorded data. Figure 22.6 below displays the result computed with WENO in Fig. 22.6a using the signal in Fig. 22.5 as a source, while the image in the Fig. 22.6b right displays the recorded data, both at 15 inches from thc cicada. It is apparent that the major features of the signal are preserved, however further analysis is necessary to study the errors and determine whether the lack of attenuation

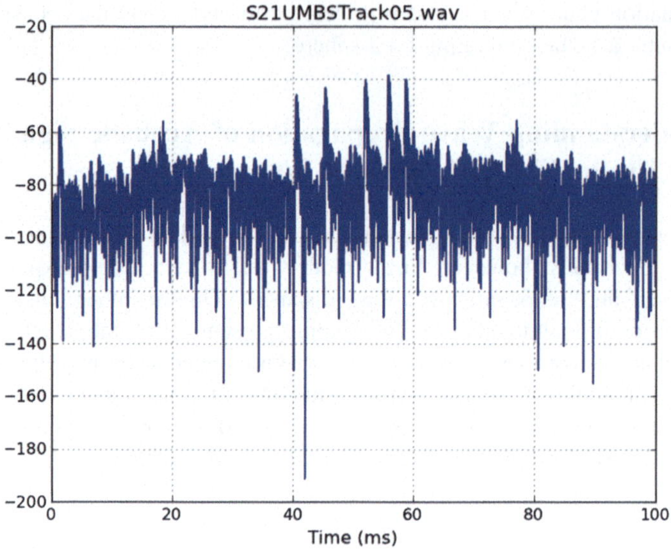

Fig. 22.5. Recorded signal level (dB) 5 inches from the cicada

in computed results is due to experimental conditions or inaccuracies in the Burgers' equation dissipation.

The cicada research has two identifiable parts; the acoustic generation and the sound propagation. The majority of the research is performed on the creation mechanism since this project has always considered the sound generation to be nonlinear. Also, the project maintains that tymbal buckling is a nonlinear process. Therefore, measurements were devised and obtained to quantify the cicada performance using the following: from micro-computer tomography (micro CT) scans, acoustic beamformed data, input-to-output data comparisons based on laser Doppler vibrometry data to microphone recordings of the cicada and man-made devices. These steps were used to demonstrate that the cicada sound system acoustic performance is far superior to current man-made acoustic systems. The micro CT scans were utilized to construct a finite element (FE) model of the cicada. In order to obtain material property values like density, and elastic and shear moduli for the FE model, tests were conducted on the tymbal due to the difficulty of measuring such a thin membrane. Thus, a novel dia-magnetic normal force procedure quantified the elastic modulus for the FE model. However, ABAQUS (FE software) could not compute a model at these frequencies with the material properties obtained. Regarding the sound propagation portion of this overall cicada research effort, there have been recent attempts at simulating how the sound would propagate within approximately a foot and a half or less from the cicada. The in-air propagation of acoustic signals generated by cicadas uses a numerical solver for viscous the Burgers' equation. The method employs a weighted essentially non-oscillatory (WENO) reconstruction to approximate the first and second derivatives of the semi-discrete operator. This choice is motivated by the non-smooth structure of the propagating waveform. This method showed very good agreement with the

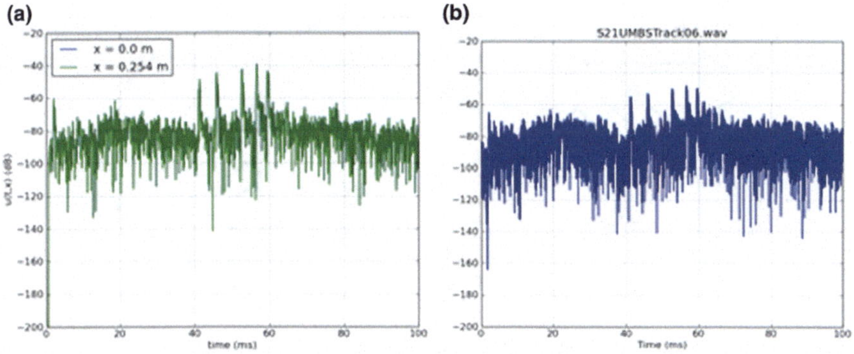

Fig. 22.6. **a** WENO result at 15 in from the cicada. **b** Recorded data at 15 in from the cicada

experimental cicada data and also indicated this model can be applied to further study the propagation of cicada mating calls.

22.5 Conversion of Cicada Mating Calls into Man-Made Projection

Subsequently, a preliminary mass-spring model was developed that captures the basic acoustic and structural dynamics exhibited in the insect. This requires understanding the general trends displayed in the cicada signals and translating those observable dynamics into known ordinary differential equations (ODEs) through Newtonian physics. Then, the ODEs are simulated and compared against actual experimental data and refined to account for existing frictional effects. Deriving and interpreting the appropriate type of damping and assigning the adequate damping coefficients are not trivial tasks. Therefore, generating this analytical mass-spring model requires considerable effort.

This research effort utilized an analytical model to establish an overall model to describe the bulk of the dynamics present in the cicada's sound generation to achieve an answer to bound the physical understanding of the cicada sound system to the first order. However, to provide further valuable insight into this incredibly unique sound system, the lumped mass spring system research is studied. The mass-spring system uses MATLAB with ODES to simulate the radiated sound loss of the cicada. For example, in Fig. 22.7, there is a general mass-spring sound production system. The R_{Rad} term is the sound loss from the mass-spring model shown.

The force is applied to the fixed support or arbitrary mechanical mass, and the system oscillates based on the viscous damping coefficient R_v and stiffness K of the model. If the mass-spring diagram is transferred into the transducer realm, this model is considerably similar to a Tonpilz transducer, which is illustrated in Fig. 22.8. The capacitance and voltage electrical equivalence for the force, spring and mass are translated into electrical equivalence such as voltage V, capacitance C_o and inductance. Figure 22.8 is not a complete depiction of the total electrical circuit; however, this

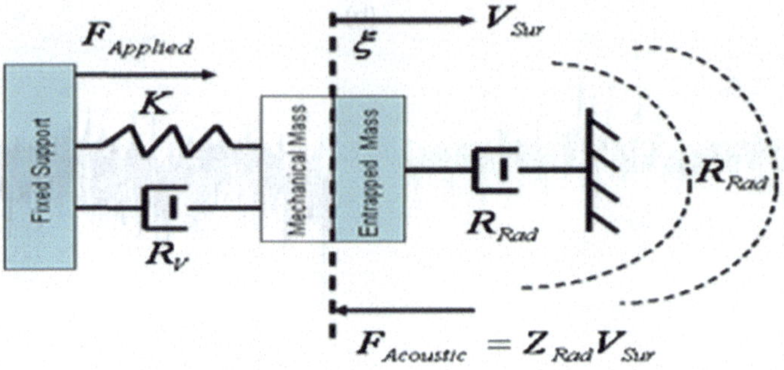

Fig. 22.7. Theoretical model of acoustic transmission

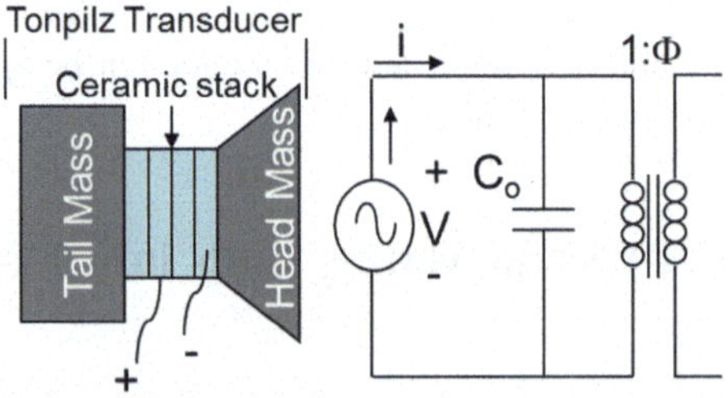

Fig. 22.8. Example of the Tonpliz transducer model

preliminary diagram indicates the transfer of motion to acoustics in the 1:∅ electrical transfer turns representation.

When a voltage is applied to the ceramic stack, the contraction and expansion of the assembly causes the head mass surface to oscillate back and forth. Thus, the loss mechanism of the system is in the sound that is produced (Fig. 22.9).

This mass-spring diagram accounts for the tymbal and tympanum as a two-body system to describe the physics that produces the velocity of the tymbal and tympanum at their faces. Figure 22.10 depicts the results of this mass-spring system being tuned to a similar frequency to create a beat frequency. The Newtonian ODEs are computed to produce the following plots in Figs. 22.10, 22.11 and 22.12, which describe the effects of altering the physical parameters of this analytical mass-spring diagram. Note that in Eqs. (22.17) and (22.18) the stiffness values are K_{tymb} and K_{tymp}. And, C_{tymb} and C_{tymp} are damping coefficients and M_{tymb} and M_{tymp} are the mass terms for the tymbal and tympanum, respectively. C_{airsac} is the damping term for air sack.

$$\ddot{M}_{tymb}\ddot{X}_{tymb} + C_{tymb}\dot{X}_{tymb} + K_{tymp}\left(X_{tymb} - X_{tymp}\right) + K_{tymb}X_{tymb} = F_{applied} \quad (22.17)$$

$$\ddot{M}_{tymp}\ddot{X}_{tymp} + C_{tymp}\left(\dot{X}_{tymp} - \dot{X}_{tymb}\right) + C_{airsac}\dot{X}_{tymp} + K_{tymp}\left(X_{tymp} - X_{tymb}\right) = 0 \quad (22.18)$$

In Fig. 22.10, the effect of the damping values for the tymbal is not significant for this damping range. Figure 22.11 shows the effect of heavy damping in this beat

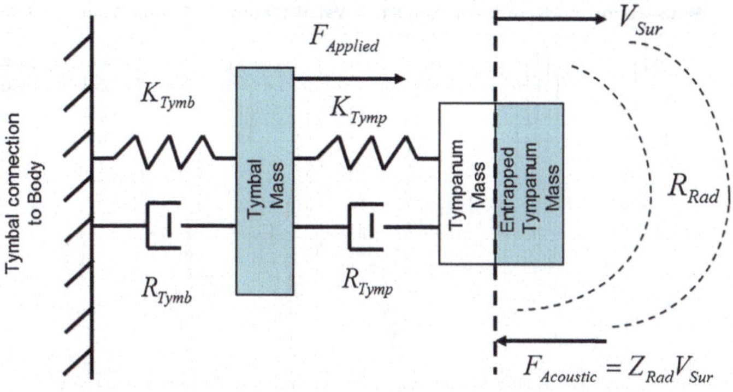

Fig. 22.9. Example of the mass-spring diagram for cicada

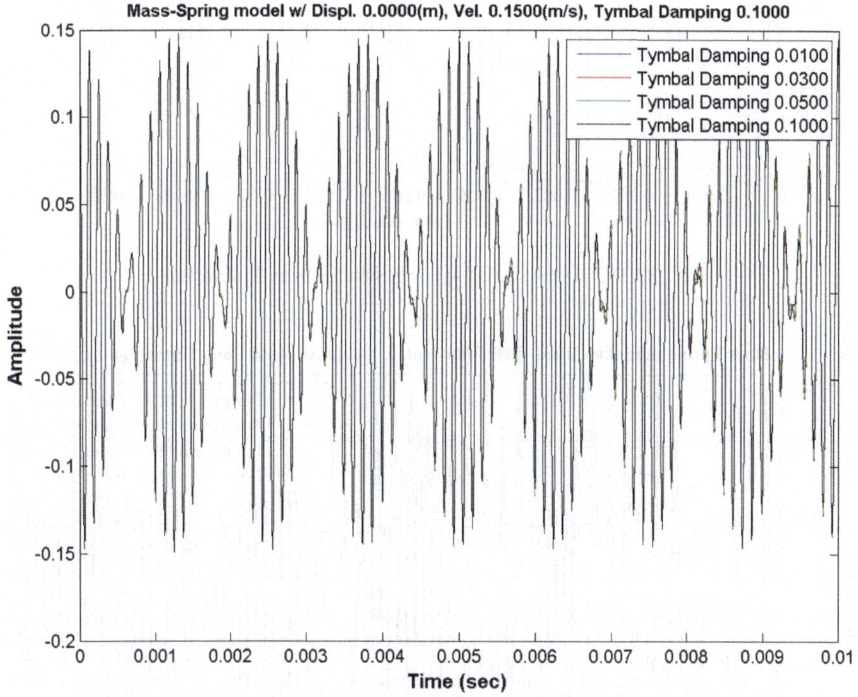

Fig. 22.10. Abdomen mass-spring system lightly damped

frequency system. These over-damped systems have a lag or phase shift between the pulses based upon the level of damping that starts at a tenth and proceeds to a critical damping coefficient of 1. Note at critical damping, the lag in the velocity of the tymbal is damped at 0.5. This value of 0.5 corresponds to the maximum angular frequency difference between the natural frequency of the tymbal and tympanum, which are tuned

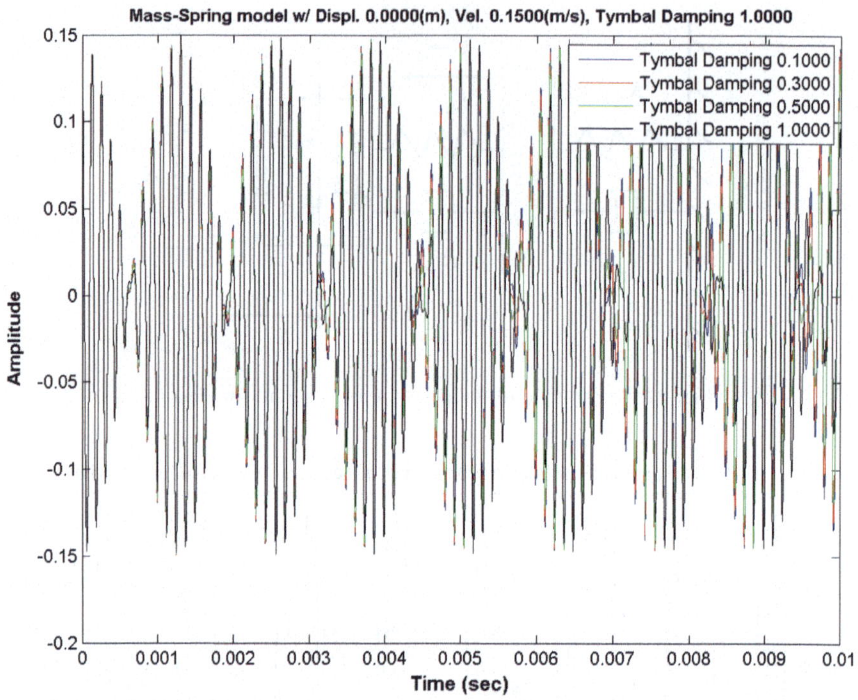

Fig. 22.11. Tymbal mass-spring system heavily damped

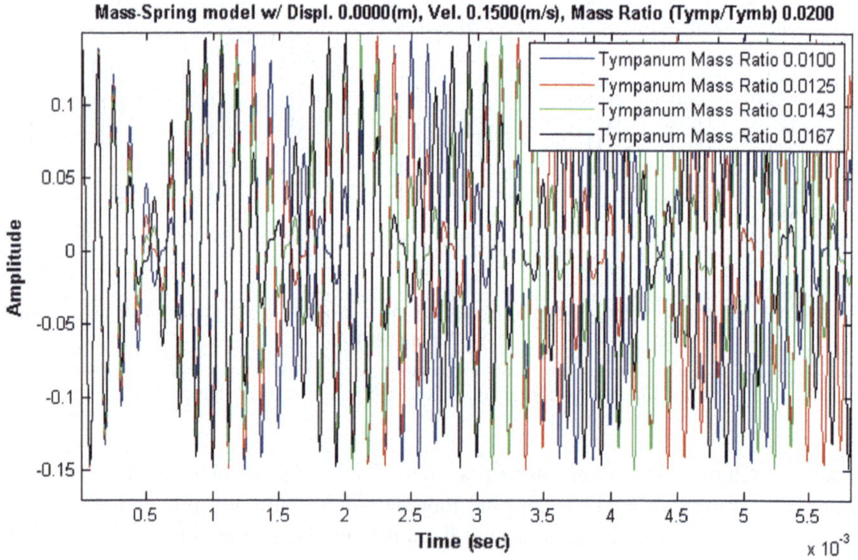

Fig. 22.12. Tympanum and Tymbal mass ratio plot

for a beat frequency. The beat frequency describes the cycles present in the pulses and the spacing of the pulses.

Another important parameter for this mass-spring system is the mass ratio between the tympanum and the tymbal. In Fig. 22.12, the mass ratios are indicated in the legend. The effect of the mass ratio is a critical parameter in altering the angular frequency of the beat frequency of the pulses.

Figure 22.12 indicates a change in the phase of the pulses as a result of the mass ratio, which has the tympanum initially at 100th of the tymbal and increases. This approach of using a ratio is required due to the fact that tympanum is a thin mucus-like membrane. The mass of the tympanum is very difficult to measure. This membrane breaks and deforms creating considerable uncertainty. The effect of the damping on the tymbal and mass ratio are shown; however, damping effects from the abdomen (air sack) and the combined effects of two parameters simultaneously require study as well. This information is not currently found in scientific literature. These material properties are necessary for successful and accurate transduction modeling. This basic science research proposes to investigate these important parameters in detail along with a comparison of these results with previously measured experimental data.

22.6 The Validation and Verification (V&V) of Cicada to Man-Made System Conversion with Nuttall-Volterra-Wiener (NVW) Model

In this new method, a procedure to characterize nonlinear systems with memory under time-invariant conditions is initially examined to validate and verify the cicada system as well as other systems. This time-invariant investigation uses the discretized standard Volterra form taken to third-order as shown:

$$\begin{aligned} y(n) &= h_0 + \sum_{k_1=0}^{\tilde{K}} h_1(k_1)x(n-k_1) + \sum_{k_1=0}^{\tilde{K}}\sum_{k_2=0}^{\tilde{K}} h_2(k_1,k_2)x(n-k_1)x(n-k_2) \\ &\quad + \sum_{k_1=0}^{\tilde{K}}\sum_{k_2=0}^{\tilde{K}}\sum_{k_3=0}^{\tilde{K}} h_3(k_1,k_2,k_3)x(n-k_1)x(n-k_2)x(n-k_3) \quad (1\text{A}) \\ &= y_0 + y_1(n) + y_2(n) + y_3(n), \end{aligned} \tag{22.19}$$

where $y(n)$ is the model output; $x(n-k_1)$ is the excitation input time-delayed by k_1 sample intervals; h_0 is the DC component of the zeroth-order kernel; $h_1(k_1)$ is the first-order kernel; $h_2(k_1,k_2)$ is the second-order kernel, and $h_3(k_1,k_2,k_3)$ is the third-order kernel. The $y_0, y_1(n), y_2(n), y_3(n)$ are the individual functionals of the zeroth-, first-, second-order, and third-order modeled outputs. The sampling frequency f_s is in units of Hz; and $\Delta = 1/f_s$ is the time sampling increment (in seconds). Also, $x_c(n\Delta) = x(n)$, $z_c(n\Delta) = z(n)$, where $x_c(t)$ and $z_c(t)$ are the continuous excitation and response of the nonlinear system. The memory length $L = \tilde{K}\Delta = \tilde{K}/f_s$ is in seconds for the model functionals. In general, large values for $\tilde{K}$ are required to realize adequate memory length for the model. However, large values of $\tilde{K}$ are computationally difficult at higher orders due to storage space and execution time requirements, which are consequences of the *Curse of Dimensionality* (CoD).

When confronting the CoD one must understand the relationship between memory length and the number of coefficients, the consideration of degrees of freedom and bandwidth. If a continuous low-pass real excitation $x_c(t)$ with an average voltage-density spectrum covering the band $(-W, W)$ Hz is used to excite a nonlinear system the duration of the excitation be T seconds. This waveform $x_c(t)$ can be sampled at time increment $1/2W$ seconds without loss of information, and the number of degrees of freedom is $DOF = T/(1/2W)$ in this excitation. Also, the number of (real) coefficients used in the first-order model kernel of the nonlinear system is K_1 and the desired memory length of this linear model is L_1 seconds. Since the frequency coverage of the model's transfer function (first-order frequency-domain kernel) can also be limited to $(-W, W)$ Hz, the model's impulse response (first-order time-domain kernel) can be sampled at time-delay increment $\Delta_\tau = 1/2W$ without loss of information. Namely, $L_1 = K_1(1/2W) \to K_1 = 2L_1W$. The DOF available in the excitation must exceed this number K_1 of unknown coefficients. Using a symmetric time-domain kernel, the total number of second-order coefficients is defined as $C_2 = K_2(K_2+1)/2 \sim K_2^2/2$. Thus, a second-order memory length L_2 in seconds (per dimension) would follow as $K_2 = 2L_2W$. Since the DOF must exceed C_2, number of coefficients, it follows $2TW > K_2^2/2 = 2(L_2W)^2$ and finally $T > L_2^2W$. Using the same Taylor series expansion approach for third-order coefficients, the observation time and memory length relationship is $T > 2L_3^3W^2/3$. Consequently, the observation interval T may be quite large depending on the received signal-to-noise ratio during the system excitation and characterization.

22.7 Modified Volterra First-Order Term

For example, alleviating the CoD, a modification of the first-order term in Eq. (22.19) reduces the number of kernel coefficients. Consider, the first-order continuous model output:

$$\mathrm{y}_{1\mathrm{c}}(\mathrm{t}) = \int d\tau h_{1c}(\tau)x_c(t-\tau) = \int df\ \exp(i2\pi ft)H_1(f)X(f). \tag{22.20}$$

Choosing an excitation voltage-density spectrum $X(f) = 0$ for frequencies $|f| > W$ the interest lies in characterizing the nonlinear system for frequencies $|f| < W$. Thus, there is no need to characterize a system beyond the frequency region. Without loss of generality this model first-order frequency domain kernel $H_1(f) = 0$ for $|f| > W$ is set. However, the frequency-domain kernel $H_{1a}(f)$ content can extend to higher frequencies. Note that the frequency kernel $H_{1a}(f)$ is only an estimate of the condition $|f| < W$ leading to the following expression for the first-order kernel:

$$h_{1c}(\tau) = \sum_{k=0}^{K} h_{1c}\left(\frac{k}{2W}\right) sinc(2W\tau - k). \tag{22.21}$$

Also, note that having achieved $L = K/2W$ and not $\tilde{K}/f_s$, this combats CoD. By taking advantage of the fact that $K < \tilde{K}$, equal values of memory length L are realized. This leads to the model output:

$$y_{1c}(t) = \frac{1}{2W}\sum_{k=0}^{K} h_{1c}\left(\frac{k}{2W}\right)x_c\left(t - \frac{k}{2W}\right) \equiv \sum_{k=0}^{K} h_1(k)x_c\left(t - \frac{k}{2W}\right). \tag{22.22}$$

The original convolution (22.22) has now been discretized in the time-delay variable τ, but has not been discretized in continuous time variable t. Thus, we can sample $y_{1c}(t)$ in (22.22) for any t values and fit to the measured data values $z_c(n\Delta) = z(n)$. Hence, the following expression for the first-order model solution is of the form:

$$\begin{aligned} y_1(n) : y_1(n) &\equiv y_{1c}(n\Delta) \\ &= \frac{1}{2W}\sum_{k=0}^{K} h_{1c}\left(\frac{k}{2W}\right)x_c\left(n\Delta - \frac{k}{2W}\right) \\ &\equiv \sum_{k=0}^{K} h_1(k)x\left(n - \frac{kf_s}{2W}\right). \end{aligned} \tag{22.23}$$

The last term $x()$ in (22.23) only contains integer values. Thus, f_s and/or W are chosen such that $f_s/2W$ is an integer to avoid interpolation of samples $x(n)$, which creates a computationally burdensome algorithm. Thus, the following generalized first-order convolution is obtained:

$$y_1(n) = \sum_{k=0}^{K} h_1(k)x(n - kI); \, Integer\, I = f_s/2W. \tag{22.24}$$

22.8 Explanation of Least Squares Computation on First Order

Least-squares fitting procedure is used to compute the minimum error between the actual measured response $z(n)$ and the modeled output $y(n)$. In the first-order model, the least squares approximation is

$$y_1(n) = \sum_{k=0}^{K} h_1(k)x(n - kI) \sim z(n) \text{for}\, n = 1 + KI : N \tag{22.25}$$

and begins incrementing at $n = 1 + KI$ instead of $n = 1$. By doing so, the modeled output avoids the requirement to interpolate the $y_1(n)$ solution. Therefore, the next step is to solve for the corresponding kernel values $h_1(k)$. This is done by solving the matrix equation approximation:

$$Ah_1 \sim \underline{z}, A'A\,h_1 = A'\underline{z} \tag{22.26}$$

For the best results, all columns of design matrix A should be uncorrelated with each other. Also, the condition number CN is monitored to assess the quality of A matrix. In an ideal sense, a condition number whose numerical value is one (i.e., $CN(A) = 1$) on an ensemble-average basis, would yield column-wise basis functions that are uncorrelated with each other. In a physical realization of a sample design matrix A, the objective is basis functions that yield as low a condition number as possible. Achieving low condition numbers at first order is not problematic. The challenge has always been to derive good basis functions for the higher-order components (i.e., $y_2(n)$ and $y_3(n)$ of Eq. (22.19)) whose CN values are reasonably low (i.e., of order 10).

22.9 The Effects of Noise on New Modified Volterra Technique

Frequently, the major limit on kernel-estimation accuracy by means of least squares is not the amount of data N, but rather the signal-to-noise ratio of the measured data $z(n)$; hence, the effects of the signal-to-noise ratio on kernel estimation are analyzed via simulation. As an example, this modified Volterra method computed simulated data $z(n)$. Cubic and quadratic cosine functions and a third-order passband nonlinearity with power levels were simulated. The simulated received $z(n)$ sequence time duration was 1 s, while the sampling frequency was 60 kHz. The center frequency of the transmission was 3 kHz, while the bandwidth W was 2 kHz. The number of model coefficients per dimension was chosen as $K_{11} = 50$, $K_{22} = 30$, $K_{20} = 29$, $K_{22} = 30$, $K_{31} = 19$, $K_{33} = 20$. The kernel indices K and subscripts indicate the system order and the center frequency relationship to that order and likewise with the kernels, $h's$, and simulated data $z's$. For example, the K_{33} kernel is the third-order component at three times the center frequency. Meanwhile, the total received data sequence $z(n)$ available for fitting purposes was the sum $z(n) = z_{11} + z_{20} + z_{22} + z_{31} + z_{33}$ with and without additive noise.

For this condensed manuscript, the kernels h_{11} and h_{22} are shown in Figs. 22.13 and 22.14, respectively. To illustrate and depict the effects of additive white noise on these kernel plots, simulated noise was injected to the level in which the noise free peak signal is submerged into the background. Hence, there are noise-free plots shown in Figs. 22.1a and 22.14a, and the additive noise plots are shown in Figs. 22.13b and 22.14b.

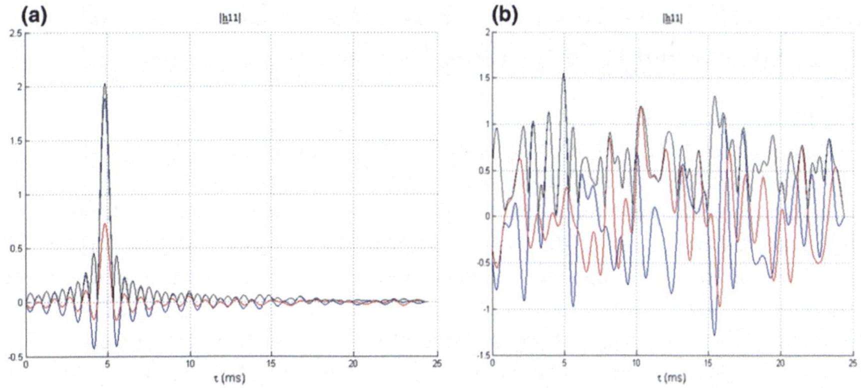

Fig. 22.13. **a** The first-order kernel with no noise **b** The first-order kernel with additive noise, where black lines represent magnitude of kernel, blue lines are real kernel component, and red is the imaginary component

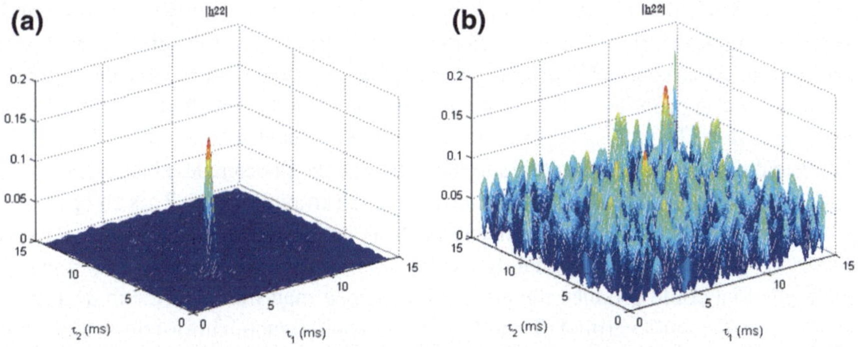

Fig. 22.14. **a** The second-order h_{22} kernel with no noise **b** The second-order h_{22} kernel with additive noise

In summary, inspired by investigations of the large sound production capabilities of an insect, the cicada, a new methodology for computing the higher-order terms in a modified Volterra expansion is described. This article highlights: (a) a substantial reduction in the CoD by sampling the kernel values at $1/2W$ vice $1/f_s$; and (b), the remarkable fact that the higher-order frequency components outside of the input excitation band are detectable for time-invariant nonlinear systems with memory. Thus, the computation of higher-order frequency intermodulation products at reduced cost and computational time is now possible.

22.10 Conclusions About the Conversion of Cicada Sound Generation to Man-Made Device

The structural acoustics generated by the anatomical members in the cicada have yet to be fully understood. The research described in this article explains the potential for the transducer design, wave propagation, and the verification process. However, in order to take advantage of the potential gains this bio-inspired transducer source offers for naval and commercial applications, the mathematical intricacies associated with partial differential equations for nonlinear frequency modulation and the structural acoustics accounting for high deflection amplitude to aperture size, and the complex and higher order modes from the unique shape of the tymbal must be further investigated.

After addressing the mathematical issues just highlighted, the physical system-level challenges related to creating a cicada-like transducer must also be addressed. The mathematical intricacies produce a number of mechanical concerns in the practical implementation of a man-made transduction device for applications, such as complex modes of the tymbal surface. A refined FE model will help develop the knowledge to understand how to generate a pressure wave with the associated amplitude, mode and beam pattern in air that is equivalent to the tymbal dynamics. Another unknown under current investigation is the effect of air on the tymbal surface. Does the air load the tymbal like water loads a submerged hydrophone? If so, the FE software must have the capability to maintain the appropriate frequency and surface velocity obtained with the tymbal experimental data. The physical parameters for the transducer such as material properties and physical dimensions are modified to meet the desired pressure levels as a function of frequency and propagation distance to compare to the empirical data. This comparison is met by adjusting the physical parameters such as material properties, dimensions and boundary conditions to fit the signal dynamics created by the cicada. There are a number of materials that could produce the appropriate scaled man-made device in accordance with the experimentally verified FE model, which would transmit information to meet the requirements for real-world applications. These materials span the gamut of shape memory alloys, 1–3 composites, and single crystal based elements. Essentially, an empirically validated FE model could simulate the tymbal membrane motion with sufficient accuracy to help design a man-made cicada transducer. This bio-inspired source applies the proper voltage in a prescribed time/space methodology with flexible state-of-the-art alloy materials for transducer elements to generate acoustic signals.

Finally, the NWV method is the V&V required to state confidently that the modification to the cicada sound production system to transformation its biological structures into a man-made device was not altered beyond the dynamics seen in nature. To date, the simulations are based upon acoustic theories that are supported by the empirical data generated during experiments to study the cicada anatomical sound generation system. However, anatomical biological structures operate in such a nonlinear fashion that it often requires considerable time to create the physical and mathematical models to accurately represent the physics demonstrated in nature. Ultimately, this research effort has also been encouraged by the development and discovery of a new (NWV) technique in which to characterize and quantify nonlinear (and linear) signal propagation dynamics in time-invariant systems with memory.

References

1. D. Young, Do cicadas radiate sound through their ear-drums? J. Exp. Biol. **151**, 41–56 (1990)
2. H.C. Bennet-Clark, Tymbal mechanics and the control of song frequency in the cicada Cyclochila australasiae. J. Exp. Biol. **200**, 1681–1694 (1997)
3. H.C. Bennet-Clark, A.G. Daws, Transduction of mechanical energy into sound energy in the cicada Cyclochila australasiae. J. Exp. Biol. **202**, 1803–1817 (1999)
4. K.D. Edoh, D.R. Hughes, R.A. Katz, Nonlinearity in cicada sound signals. J. Biol. Syst. **21**(1), 1350004 [13 pages] (2013)
5. A.D. Pierce, *Acoustics: An Introduction to Its Physical Principles and Applications* (Acoustical Society of America, Melville, 1989)
6. C. Shu, Essentially non-oscillatory and weighted essentially non-oscillatory schemes for hyperbolic conservation laws. ICASE Report 97-65, NASA/CR-97-206253, November 1997

Chapter 23
Fabrication of YBCO Josephson Junction Using Wet Etching

Teresa Emery-Adleman(✉) and Benjamin Taylor

Space and Naval Warfare Systems Center San Diego, Code 71740, 53560 Hull St, San Diego, CA 92152-5001, USA
{teresa.emery,benjamin.taylor4}@navy.mil

Abstract. Nano-bridge style Josephson junctions have the possibility of creating a dense high temperature superconducting circuits. The junctions are created using to two step etch. The first, wide area etch defines the overall shape of the Josephson junction and the second etch defines a small non-superconducting barrier by thinning the material to a thickness that cannot support the superconducting state. The wide area etching of the high temperature superconductor yttrium barium copper oxide (YBCO) is very sensitive to high temperature and water, which can destroy its superconducting properties. We have explored several different methods of wet etching of YBCO and determined their effects on the superconductivity of the material. The wet etches examined are weak acid etches of nitric acid, phosphoric acid, and a bromine alcohol solution. The bromine alcohol solution prove to be the ideal etch for our purposes.

23.1 Introduction

Josephson junctions are seen as the key elements for ultrafast computers using single flux quanta, and for sensitive and broadband electromagnetic field sensors in the form of superconducting quantum interference devices (SQUIDs) [1]. Josephson junctions are created from two superconducting regions separated by a thin barrier. The barrier can be an insulating material or simply a non-superconducting metal. Using an insulating material requires the barrier to be less than 30 angstroms thick. Using a non-superconducting metal allows the barrier to be up to a few microns thick, which is easier to fabricate. Some superconducting materials can lose their superconducting properties if there is a defect in their lattice or if the material becomes too thin to support superconductivity [2]. The three main designs using this idea to form a Josephson junction barrier are the ramp, ion damage, and nanobridge, as shown in Fig. 23.1. The ramp Josephson junction uses a "long" etched ramp to cause a lattice defect as the material knees over the edge. The junction works but it takes up a lot of space on

V. In et al. (Eds.): *Proceedings of the 5th International Conference on Applications in Nonlinear Dynamics*, Understanding Complex Systems, https://doi.org/10.1007/978-3-030-10892-2_23

the wafer and is not ideal for dense circuity. The ion damage Josephson junction uses a helium ion focused ion beam to damage the lattice of the superconducting material. This junction is compact, but currently very slow to fabricate. The third method is the nanobridge. The nanobridge Josephson junction thins an area of the superconductor until it no longer supports the superconducting state. This method can be written densely using electron beam lithography at a reasonable speed. This is the method we are pursuing to define Josephson junctions in the superconducting material yttrium barium copper oxide (YBCO).

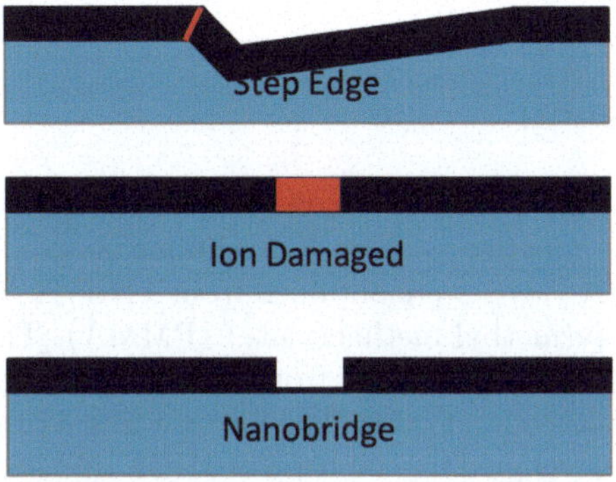

Fig. 23.1. Three different examples of damage Joesphson junctions

YBCO is a high temperature superconducting material operating at 87 K. Thin film deposition of YBCO is performed using several methods including radio frequency sputtering, chemical vapor deposition, and pulsed laser ablation. In each method, the material is sintered after deposit at around 900 C in an oxygen atmosphere to achieve a dense superconducting material [3]. With such a high temperature sinter, YBCO is usually the first layer deposited on the substrate. The layer of YBCO is then covered with a gold layer for protection from further processing.

Lithographic patterning of the YBCO into Josephson junction poses several challenges. YBCO is sensitive to humidity and heat, both will degrade its superconducting properties. This makes it unsuitable for use with most standard photoresists, which are usually baked at 105 C and are developed in a combination of tetramethylammonium hydroxide (TMAH) and deionized water. YBCO degradation can be addressed by using one of the few photoresists that develop with non-aqueous solvent and the bake temperature can be lowered by baking longer.

The next challenge is wide area etching of the YBCO to transfer the photoresist pattern. Dry etching usually uses accelerated plasma of chlorine or fluorine chemistry to induce chemical breakdown of the material to improve speed and selectivity of the etch. Unlike most materials YBCO does not chemically react with the halogen plasma. Instead, physical removal through Argon milling can be used to dry etch YBCO but the etch byproducts can be redeposited on the surface. This makes dry etching ineffective if the amount of material is large.

Although YBCO is too tough for dry etching, its chemical resistance to aqueous etchants is too low to easily perform a controlled etch. Since YBCO is rapidly dissolved in weak acids, etch rates are difficult to control. The speed of wet etch is important because the isotropic etch causes undercut, which is a big problem for the small features required for Josephson junctions. The various etchants such as HNO3, HCl, H3PO4, and ethylenediaminetetra-acetic acid (EDTA) are combined with water to etch YBCO. Although the acids are only 1% of the solution, the etch rates are in the 0.5–1 um/min. The etchants also are in water, which can damage the remaining YBCO. One alternative is a 1% bromine solution in ethyl alcohol [4].

23.2 Experimental

We have explored several different etchants to perform wide area wet etching for nanobridge Josephson junctions. 32 nm YBCO films on sapphire substrates with a 200 nm gold protective layer were obtained from THEVA. The wafers were covered with 950 kW poly(methyl-methacrylate) (PMMA). The pattern consisted of groupings lines that ranged in width from 500 nm to 2.5 microns, all 5 um long and placed between two contact pads. The pattern was written via e-beam (Vistec EBPG 5200) and developed in a mixture of MIBK and isopropanol. The wafer was loaded into a DC sputtering system and 30 nm of titanium was deposited forming a hard mask for the wet etching. Lift off was completed in an acetone bath and isopropanol wash to expose the area to be removed by the etch. The pattern was transferred to the gold layer using a wet etch of potassium iodine and iodine. At this point, the wafer was covered with a thick layer of photoresist and diced into several samples. The photoresist was removed in acetone and rinsed in isopropanol. Three different etchant baths were prepared, 1% phosphoric acid in deionized water (DI), 0.5% nitric acid in DI, and 1% bromine in ethyl alcohol. The samples were dipped in the etchant for 10 s intervals and then rinsed for 10 s in the corresponding solvent. The sample was checked under an optical microscope a after rinse to determine if the etch was complete, and the process was repeated if necessary After complete etching, samples were observed in a FEI scanning electron microscope to determine the quality of the etch (Fig. 23.2).

23.3 Results and Discussion

The first sample was etched for 10 s in a 0.5% solution of sulfuric acid in water. After 10 s the pattern had been mostly etched away. Lines under 2 microns were dissolved completely. The sulfuric etch is clearly too aggressive to etch small features consistently for this application.

The second sample was etched for a total of 20 s in a 1% solution of phosphoric acid in water. The pattern was over etched eliminating the smallest features. The 500 nm lines were etched away but all other features survived. As shown in

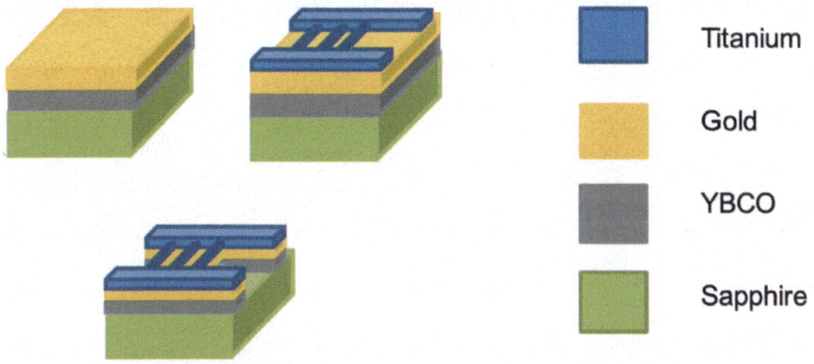

Fig. 23.2. Fabrication steps for the test samples. Received sample was a sapphire wafer with thin layers of YBCO and gold. Sample was pattern with photolithography techniques to have a titanium hard mask for argon ion milling

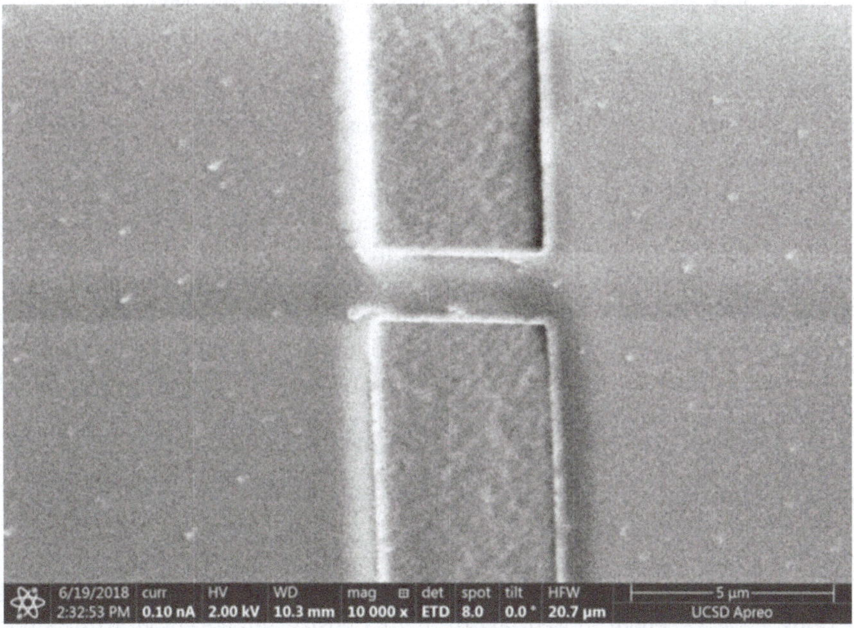

Fig. 23.3. Test sample etched using 1% phosphoric acid in DI water. Bright white edges show over etch

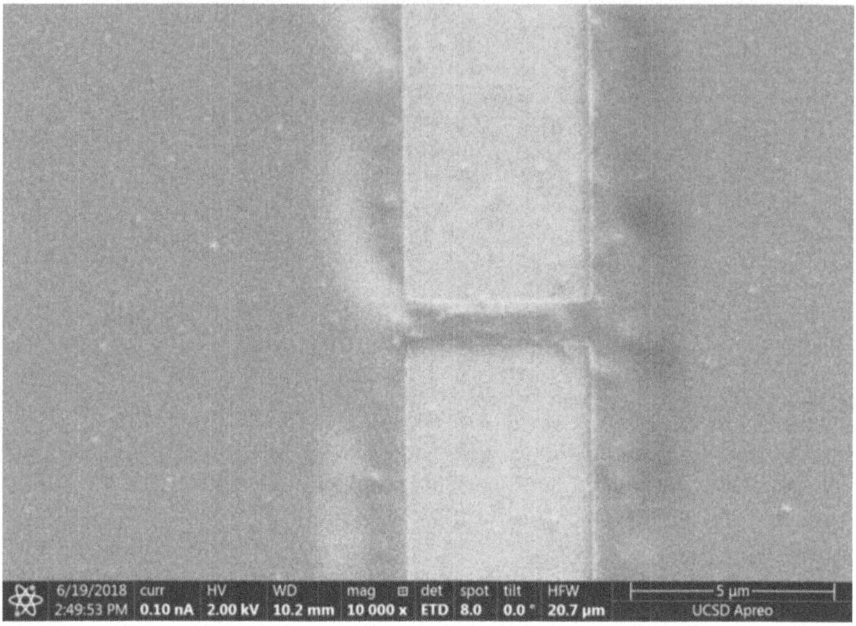

Fig. 23.4. Test sample etched using 1% bromine in ethanol

Fig. 23.3, the features are slightly over etched as seen by the bright edges indicating undercutting of the hard mask. The phosphoric etch is still too aggressive for our purposes.

The third sample was etched for 45 s in a 1% bromine solution in ethyl alcohol. This process was the slowest and most controlled etch of the group. This sample was not over etched as seen in Fig. 23.4. The bromine etch also etched the titanium layer on top of the gold. The titanium etching is not a problem for our processes since the titanium was used to mask the gold etch and will be removed before making the junction. The partial removal of the titanium layer causes a ripple in the SEM image that does not translate to the YBCO layer.

23.4 Conclusions

In this work we examined 3 different wet etching processes to perform wide area etching for formation of nanobridge Josephson junctions in YBCO. Nanobridge junctions have excellent potential for high density, high temperature superconducting circuits. Wet etching to define the overall junction area without degrading superconducting behavior requires a slow, controlled etch. From our experiments, the 1% bromine in ethyl alcohol proved to etch controllably and yield accurate pattern transfer. Additionally, the low temperature and non- aqueous chemistry of the etch should maintain the quality of the YBCO material.

This etching process is ideal to prepare the junction area for nanobridge definition using subsequent ion milling.

Acknowledgments. The author would like to thank the staff at CalIT2 Nano3 staff for help and use of their facilities.

References

1. G. Bednorz, K.A. Multer, Z. Fhvs. B **64**, 189 (1086)
2. Y. Yoshizaki, M. Tonouchi, T. Kobayashi, Jpn. Journal Appl. Phys. **26**(2), 9 (1987)
3. R.P. Vasquez, B.D. Hunt, M.C. Foote, Appl. Phys. Lett. **53**(26), (1988)
4. M.K. Wu, J.R. Ashburn, C.T. Torng, P.H . Hor, R.L. Meng, L. Gao, Z.J. Huang, Y.J. Wang, C.W. Chu, Phys. Rev. Lett. **58**, 908 (1987)

Chapter 24
Quasi-analytical Perturbation Analysis of the Generalized Nonlinear Schrödinger Equation

J. Bonetti[1,2], S. M. Hernandez[1], P. I. Fierens[2,3(✉)], E. Temprana[4], and D. F. Grosz[1,3]

[1] Instituto Balseiro (IB), Río Negro, Argentina
jbonetti@ib.edu.ar
[2] Consejo Nacional de Investigaciones Científicas y Técnicas (CONICET), Buenos Aires, Argentina
[3] Instituto Tecnológico de Buenos Aires (ITBA), Buenos Aires, Argentina
pfierens@itba.edu.ar
[4] Department of Electrical and Computer Engineering, University of California San Diego, San Diego, CA, USA

Abstract. The Generalized Nonlinear Schrödinger Equation (GNLSE) finds several applications, especially in describing pulse propagation in nonlinear fiber optics. A well-known and thoroughly studied phenomenon in nonlinear wave propagation is that of modulation instability (MI). MI is approached as a weak perturbation to a pump and the analysis is based on preserving those terms linear on the perturbation and disregarding higher-order terms. In this sense, the linear MI analysis is relevant to the understanding of the onset of many other nonlinear phenomena, but its application is limited to the evolution of the perturbation over short distances. In this work, we propose quasi-analytical approximations to the propagation of a perturbation consisting of additive white noise that go beyond the linear modulation instability analysis. Moreover, we show these approximations to be in excellent agreement with numerical simulations and experimental measurements.

24.1 Introduction

Pulse propagation in single-mode lossless nonlinear fibers is modeled by the Generalized Nonlinear Schrödinger Equation [1]

$$\frac{\partial A}{\partial z} - i\hat{\beta}A = i\hat{\gamma}A\int_{-\infty}^{\infty} R(T')\left|A(z,T-T')\right|^2 dT'. \tag{24.1}$$

V. In et al. (Eds.): *Proceedings of the 5th International Conference on Applications in Nonlinear Dynamics*, Understanding Complex Systems, https://doi.org/10.1007/978-3-030-10892-2_24

$A(z, T)$ is the pulse envelope, z is the direction of propagation and T is the time referred to a co-moving frame with group velocity $v_g = \beta_1^{-1}$ (i.e., $T = t - z\beta_1$). Linear dispersion is modeled by the operator $\hat{\beta}$, while $\hat{\gamma}$ is related to the third-order susceptibility:

$$\hat{\beta} = \sum_{k\geq 2} \frac{i^k \beta_k}{k!} \frac{\partial^k}{\partial T^k}, \quad \hat{\gamma} = \sum_{k\geq 0} \frac{i^k \gamma_k}{k!} \frac{\partial^k}{\partial T^k}. \tag{24.2}$$

Finally, $R(T)$ models instantaneous and molecular Raman responses.

Analytical solutions of Eq. (24.1) are known in a variety of simplified cases. For example, solitonic solutions can be found by means of the inverse-scattering method originally proposed by Zakharov and Shabat [2] (see also, e.g., [3]), but only under some simplifying assumptions such as neglecting higher-order dispersion ($\beta_k = 0$ for $k \geq 3$). An important family of periodic solutions, known as Akhmediev breathers [4], has attracted attention in relation to supercontinuum generation and rogue waves [5,6]. Although Akhmediev breathers were originally found for low-order dispersion cases, Eq. (24.1) has been found to be integrable in more complex cases (see, for example, [7–11] and references therein). However, the number of exactly integrable variations of the GNLSE is still very limited.

Although exact solutions of simplified versions of Eq. (24.1) provide important insight on many characteristics of the propagation of pulses in nonlinear fibers, they cannot give a precise description in general. For this reason, the GNLSE is usually studied by means of simulations based on efficient algorithms such as split-step Fourier (SSF) [1] or a fourth-order Runge–Kutta in the interaction picture (RK4IP) [12].

In this work, we propose analytical approximations to the solution of Eq. (24.1) that provide a precise description of pulse propagation for a particular case of great interest. Our analysis focuses on a continuous-wave (CW) laser pumping the fiber. This CW pump is always accompanied by technical and quantum noise. One possibility is to approach noise propagation as a perturbation of the CW state. First-order perturbation or linear stability analysis is related to the study of the modulation instability (MI) phenomenon [4,5,13–23,23–29] (see also Chapter 5 of Ref. [1] and references therein). Exact solutions of MI accounting for the complete GNLSE have also been developed [30,31]. The particular case of the propagation of additive noise has been dealt with in the literature (see, e.g., [32,33]).

The wave propagation analysis of a noisy CW pump in an MI setting has several limitations. The continuous-wave pump is assumed undepleted and, hence, results are valid for short propagation distances. Furthermore, as it is a first order perturbation analysis, it disregards the four-wave mixing 'cascading effect', in the sense that perturbations to the pump, in turn, act as pumps themselves as soon as they attain enough power. One alternative to incorporate such cascading effect is to solve the GNLSE through Picard's iterations. Resulting expressions are, nevertheless, not easily tractable and even evaluating them numerically may be an expensive computational effort as compared to pure numerical solutions obtained from the usual SSF or RK4IP algorithms. For this reason, we put forth several simplifications that allow a simpler analysis of higher-order perturbations. The validity of these simplifications is tested through numerical and experimental studies.

It must be mentioned that there are alternative approaches which are related to ideas presented in this work. In particular, many tools have been developed for the statistical analysis of optical wave turbulence (see, e.g., [34–38]).

The remaining of this paper is organized as follows. In Sect. 24.2 we develop a higher-order perturbation analysis of the GNLSE and motivate the simplifications that allow tractability. We validate our approach with experiments and simulations in Sect. 24.3. Finally, conclusions are presented in Sect. 24.4.

24.2 Higher-Order Perturbation

Let us again consider the generalized nonlinear Schrödinger equation. It is useful to normalize the propagation distance as $\zeta = \gamma_0 P_0 z$. We study the propagation of a small perturbation $a(\zeta, T)$ to the stationary solution of Eq. (24.1), i.e., we consider $A(\zeta, T) = \sqrt{P_0}\,[1 + a(\zeta, T)]\, e^{i\zeta}$. Fourier transformation (with respect to time T) leads to

$$\frac{\partial \tilde{\mathbf{a}}(\zeta, \Omega)}{\partial \zeta} = \mathbf{A}(\Omega)\tilde{\mathbf{a}}(\zeta, \Omega) + \tilde{\mathbf{N}}(\tilde{\mathbf{a}}(\zeta, \Omega)), \tag{24.3}$$

where $\tilde{\mathbf{a}}(\zeta, \Omega) = \left[\tilde{a}(\zeta, \Omega), \overline{\tilde{a}(\zeta, -\Omega)}\right]^T$, with $\tilde{a}(\zeta, \Omega)$ the Fourier transform of $a(\zeta, T)$. The linear and nonlinear terms in the right-hand side are defined by

$$\mathbf{A} = i \begin{bmatrix} B(\Omega) & C(\Omega) \\ -B(-\Omega) & -\overline{C(-\Omega)} \end{bmatrix}, \; \tilde{\mathbf{N}}(\tilde{\mathbf{a}}(\zeta, \Omega)) = \begin{bmatrix} \tilde{\gamma}(\Omega)\tilde{N}(\tilde{a}(\zeta, \Omega)) \\ \overline{\tilde{\gamma}(-\Omega)\tilde{N}(\tilde{a}(\zeta, \Omega))} \end{bmatrix}, \tag{24.4}$$

where $B(\Omega) = \tilde{\beta}(\Omega) + \tilde{\gamma}(\Omega)[1 + \tilde{R}(\Omega)] - 1$, $C(\Omega) = \tilde{\gamma}(\Omega)\tilde{R}(\Omega)$,

$$\tilde{\beta}(\Omega) = \frac{1}{\gamma_0 P_0} \sum_{m=2}^{M} \frac{(-1)^m}{m!} \beta_m \Omega^m, \qquad \tilde{\gamma}(\Omega) = \frac{1}{\gamma_0} \sum_{n=0}^{N} \frac{(-1)^n}{n!} \gamma_n \Omega^n, \tag{24.5}$$

$$\begin{aligned} \tilde{N}(\tilde{a}) = {} & \tilde{R}(\Omega) \left[\tilde{a}(\zeta, \Omega) * \overline{\tilde{a}}(\zeta, -\Omega)\right] + \\ & \tilde{a}(\zeta, \Omega) * \left[\tilde{R}(\Omega) \left(\tilde{a}(\zeta, \Omega) + \overline{\tilde{a}}(\zeta, -\Omega)\right)\right] + \\ & \tilde{a}(\zeta, \Omega) * \left[\tilde{R}(\Omega) \left[\tilde{a}(\zeta, \Omega) * \overline{\tilde{a}}(\zeta, -\Omega)\right]\right], \end{aligned} \tag{24.6}$$

and $\tilde{R}(\Omega)$ is the Fourier transform of $R(T)$. For the sake of simplicity, in this work we let $\tilde{R}(\Omega) = 1$, that is, we neglect stimulated Raman scattering in the analysis.

Let us focus on the case where $a(0, T)$ is white noise. In particular, we assume that the mean power spectral density $s = \langle |\tilde{a}(0, \Omega)|^2 \rangle$ is constant and that $\langle \tilde{a}(0, \Omega_1)\tilde{a}(0, \Omega_2) \rangle = 0$ and $\langle \tilde{a}(0, \Omega_1)\overline{\tilde{a}}(0, \Omega_2) \rangle = 0$ for $\Omega_1 \neq \Omega_2$. Using these hypotheses, it is simple to show [32,33] that the solution to Eq. (24.3) when the nonlinear term is neglected is given by

$$\langle|\tilde{a}_0(\zeta,\Omega)|^2\rangle = \left\{\cosh\left(2G_1(\Omega)\zeta\right) - \frac{\left(\frac{B(\Omega)+B(-\Omega)}{2}\right)^2 - G_1^2(\Omega) + \tilde{\gamma}^2(\Omega)}{\left(\frac{B(\Omega)+B(-\Omega)}{2}\right)^2 + G_1^2(\Omega) + \tilde{\gamma}^2(\Omega)}\right\}$$
$$\times \frac{\left(\frac{B(\Omega)+B(-\Omega)}{2}\right)^2 + G_1^2(\Omega) + \tilde{\gamma}^2(\Omega)}{2G_1^2(\Omega)} s, \tag{24.7}$$

where $G_1(\Omega)$ is the MI gain given by

$$G_1(\Omega) = \frac{\sqrt{4c(\Omega) - b^2(\Omega)}}{2}, \tag{24.8}$$

with $b(\Omega) = B(-\Omega) - B(\Omega)$ and $c(\Omega) = C(\Omega)C(-\Omega) - B(\Omega)B(-\Omega)$. Let us assume that there is gain, i.e., $G_1(\Omega) \in \mathbb{R}$, for some Ω. Then, we may approximate

$$\langle|\tilde{a}_0(\zeta,\Omega)|^2\rangle \approx s + \left(e^{2G_1(\Omega)\zeta} - 1\right)|A_1(\Omega)|^2 s. \tag{24.9}$$

where

$$|A_1(\Omega)|^2 = \frac{\left(\frac{B(\Omega)+B(-\Omega)}{2}\right)^2 + G_1^2(\Omega) + \tilde{\gamma}^2(\Omega)}{2G_1^2(\Omega)}. \tag{24.10}$$

Equations (24.9)–(24.10) suggest the perturbative *ansatz*

$$\tilde{a}(\zeta,\Omega) \approx \sqrt{s}e^{i\phi_0(\zeta,\Omega)} + \sum_{n=1}^{\infty}\left(e^{G_n(\Omega)\zeta} - 1\right)A_n(\Omega)\sqrt{s^n}e^{i\phi_n(\zeta,\Omega)}. \tag{24.11}$$

Substitution of Eq. (24.11) in Eq. (24.3), along with the formal computation of the mean power spectral density, allows the determination of A_n and G_n. Since the equations are quite involved, several simplifications must be made. One of the main simplifying assumptions is that $\langle\exp\{i(\phi_n(x,\mu) - \phi_m(y,\nu))\}\rangle = 0$ if either $n \neq m$, $x \neq y$ or $\mu \neq \nu$. After some tedious computations, it may be shown that, for $n \geq 2$

$$G_n(\Omega) \approx \max_{\mu}\left[G_1(\mu) + G_{n-1}(\Omega - \mu)\right], \tag{24.12}$$
$$|A_n(\Omega)| \approx \Delta_\Omega^{n-1} J(G_n(\Omega),\Omega), \tag{24.13}$$

where Δ_Ω is a positive constant and

$$J(g,\Omega) = \frac{\sqrt{\left|\overline{B}(-\Omega) - ig\right|^2 |\tilde{\gamma}(\Omega)|^2 + \left|\overline{C}(-\Omega)\right|^2 \left|\overline{\tilde{\gamma}}(-\Omega)\right|^2}}{\left|[B(\Omega) + ig]\left[\overline{B}(-\Omega) - ig\right] - C(\Omega)\overline{C}(-\Omega)\right|}. \tag{24.14}$$

Although we do not present the details of the calculations due to the lack of space, some intuition on Eq. (24.12) may be gained by referring to the nonlinear

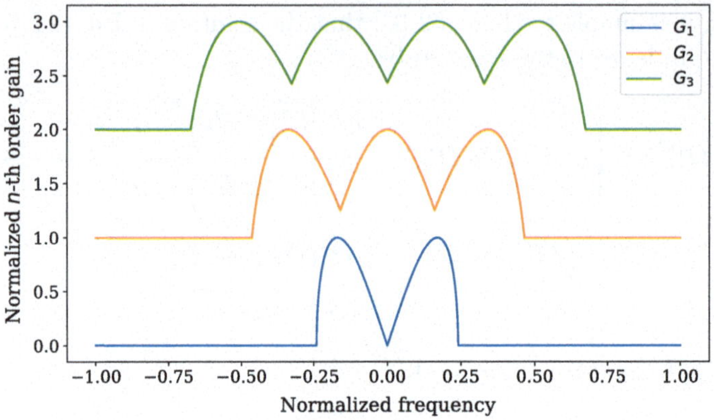

Fig. 24.1. Normalized gain for different perturbation orders. As the order increases, the gain captures the cascading effect of four-wave mixing

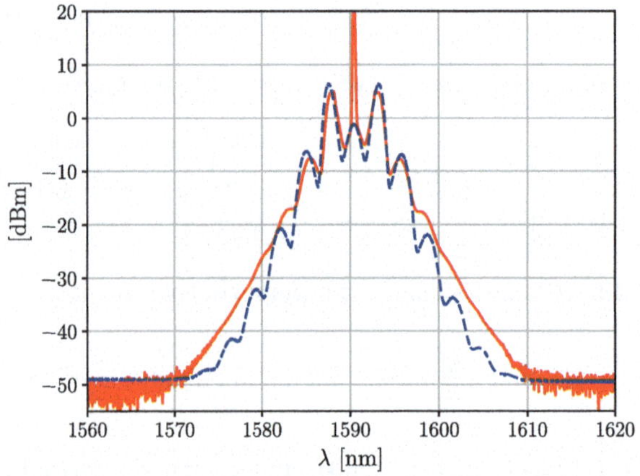

Fig. 24.2. Analytical approximation (blue dashed line) vs. experimental results (red solid line). A CW 30-dBm pump laser at 1590.4 nm was launched at the input end of the 770-m long dispersion-stabilized HNLF

operator in Eq. (24.6). The sum in Eq. (24.12) arises from the convolutions in the nonlinear operator. We are able to simplify the corresponding integrals by assuming that results are dominated by the largest gain and thus we take the maximum value. Figure 24.1 shows that, as the perturbation order n increases, G_n captures the cascading effect of four-wave mixing. Indeed, G_1 represents the well-known MI-gain due to the pump. G_{n+1} incorporates the gain due to the perturbations amplified by G_n acting as nth order 'pumps'.

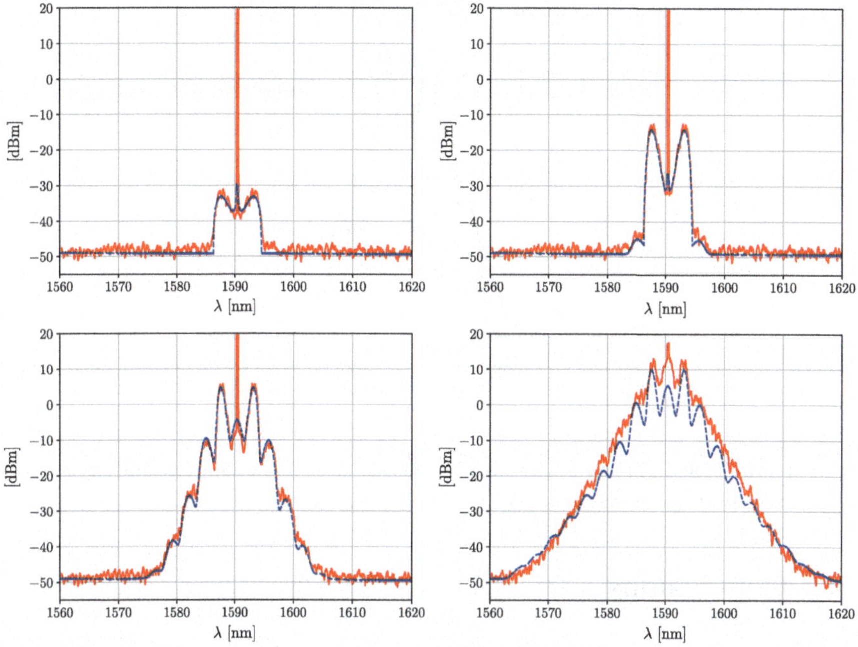

Fig. 24.3. Analytical approximation (blue dashed line) versus numerical results (red solid line) for different propagated distances: ∼0.25 km (top left), ∼0.50 km (top right), ∼0.75 km (bottom left) and ∼1 km (bottom right)

24.3 Experimental and Numerical Results

In order to test our approach we performed measurements of MI in a 770 m-long, dispersion-stabilized [39] Highly-Nonlinear Fiber (HNLF). A CW 30-dBm pump laser at 1590.4 nm was launched at the input end of the fiber. Figure 24.2 presents a comparison between the observed power (measured with 0.1-nm resolution) and the quasi-analytical approximation. The latter was obtained by using Eqs. (24.11)–(24.14) (adding up to $n = 8$) with $\gamma_0 = 8.7\ \mathrm{W}^{-1}\mathrm{Km}^{-1}$, $\gamma_k = 0$ for $k > 0$, $\beta_2 = -3.9198\ \mathrm{ps}^2/\mathrm{km}$, $\beta_3 = 0.1267\ \mathrm{ps}^3/\mathrm{km}$, $\beta_4 = 1.7594 \times 10^{-4}\ \mathrm{ps}^4/\mathrm{km}$ and $\beta_k = 0$ for $k > 4$. As it is readily observed, experimental, and analytical results are in excellent agreement.

In order to further explore the validity of the approximations, we performed computer simulations using the split-step Fourier algorithm. Figure 24.3 shows that the accuracy of the approximation decreases with the propagation distance, although reasonable good results are obtained even after 1 km. Figure 24.4 shows how approximations improve as the number of terms in Eq. (24.11) increases. Comparison to Fig. 24.1 helps to understand that the increasing detail is a consequence of the incorporation of the cascading four-wave mixing effect through higher-order perturbation gains G_n.

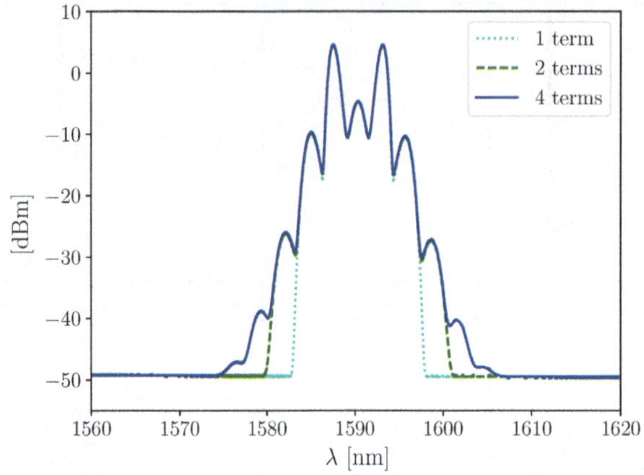

Fig. 24.4. Analytical approximation when increasing orders of approximation are used, at a propagation distance ∼0.75 km

24.4 Conclusions

A continuous-wave laser pump is always accompanied with technical and quantum noise. Thus, the propagation of a CW pump in a nonlinear optical fiber is a complex process. Its study is usually based on two different tools: numerical simulations and first-order linear stability (MI) analysis. While computer simulations are useful, they tend to hide the underlying basic physics. On the contrary, the modulation instability analysis gives some insights on the initial stages of propagation but fails at providing an accurate picture for longer propagated distances.

In this work, we put forth a perturbation analysis that offers both a precise description and meaningful physical insights. In particular, we showed our formulas to be accurate by comparing their predictions to actual experimental results. Furthermore, we validated our approximations with numerical simulations for propagated distances up to 1 km. The perturbation analysis also reveals the relevance of the cascading effect of four-wave mixing. In simple words, we might understand how produced MI gain spectra act as a new pumps further on.

The derivation of our approximation is complex and involves many simplifying assumptions. It is a matter of future work to look for a shorter path and less restrictive simplifications. It must be noted that, while those simplifications lead to extremely simple formulas, they may hide some interesting phenomena. For instance, it may be argued that the cascading effect of four-wave mixing is implicitly embedded in our choice of keeping only the largest gain in Eq. (24.12), but such an approximation might neglect relevant details appearing at longer distances (see Fig. 24.2). Finally, we believe our analysis to be of value when studying the early stages of supercontinuum generation and to contribute tools for the better understanding of rogue-wave formation.

Acknowledgments. We gratefully acknowledge S. Radic for hosting J. B.'s research stay at the Photonic Systems Group, UCSD, financial support from project PIP 2015, CONICET, Argentina, and from ONR Global through the Visiting Scientists Program.

References

1. G. Agrawal, *Nonlinear Fiber Optics*, 5th edn. Optics and Photonics (Academic, New York, 2012)
2. V.E. Zakharov, Sov. Phys. JETP **35**, 908 (1972)
3. M.A. Ablowitz, P.A. Clarkson, *Solitons, Nonlinear Evolution Equations and Inverse Scattering* (Cambridge University Press, Cambridge, 1991)
4. N. Akhmediev, V. Korneev, Theor. Math. Phys. **69**(2), 1089 (1986)
5. J.M. Dudley, G. Genty, F. Dias, B. Kibler, N. Akhmediev, Opt. Express **17**(24), 21497 (2009). https://doi.org/10.1364/OE.17.021497
6. N. Akhmediev, J.M. Soto-Crespo, A. Ankiewicz, Phys. Rev. A **80**, 043818 (2009). https://doi.org/10.1103/PhysRevA.80.043818
7. N. Akhmediev, A. Ankiewicz, M. Taki, Phys. Lett. A **373**(6), 675 (2009). https://doi.org/10.1016/j.physleta.2008.12.036
8. A. Ankiewicz, J.M. Soto-Crespo, M.A. Chowdhury, N. Akhmediev, J. Opt. Soc. Am. B **30**(1), 87 (2013). https://doi.org/10.1364/JOSAB.30.000087
9. A. Ankiewicz, N. Akhmediev, Phys. Lett. A **378**(4), 358 (2014). https://doi.org/10.1016/j.physleta.2013.11.031
10. A. Ankiewicz, Y. Wang, S. Wabnitz, N. Akhmediev, Phys. Rev. E **89**, 012907 (2014). https://doi.org/10.1103/PhysRevE.89.012907
11. A. Ankiewicz, D.J. Kedziora, A. Chowdury, U. Bandelow, N. Akhmediev, Phys. Rev. E **93**, 012206 (2016). https://doi.org/10.1103/PhysRevE.93.012206
12. J. Hult, J. Light. Technol. **25**(12), 3770 (2007). https://doi.org/10.1109/JLT.2007.909373
13. T.B. Benjamin, J.E. Feir, J. Fluid Mech. **27**, 417 (1967). https://doi.org/10.1017/S002211206700045X
14. A. Hasegawa, Phys. Rev. Lett. **24**, 1165 (1970). https://doi.org/10.1103/PhysRevLett.24.1165
15. V. Zakharov, A. Shabat, Sov. Phys. JETP **34**, 62 (1972)
16. A. Hasegawa, W. Brinkman, IEEE J. Quantum Electron. **16**(7), 694 (1980). https://doi.org/10.1109/JQE.1980.1070554
17. P.A.E.M. Janssen, Phys. Fluids **24**(1), 23 (1981). https://doi.org/10.1063/1.863242
18. D. Anderson, M. Lisak, Opt. Lett. **9**(10), 468 (1984). https://doi.org/10.1364/OL.9.000468
19. P.K. Shukla, J.J. Rasmussen, Opt. Lett. **11**(3), 171 (1986). https://doi.org/10.1364/OL.11.000171
20. K. Tai, A. Hasegawa, A. Tomita, Phys. Rev. Lett. **56**, 135 (1986). https://doi.org/10.1103/PhysRevLett.56.135
21. M.J. Potasek, Opt. Lett. **12**(11), 921 (1987). https://doi.org/10.1364/OL.12.000921
22. M. Erkintalo, K. Hammani, B. Kibler, C. Finot, N. Akhmediev, J.M. Dudley, G. Genty, Phys. Rev. Lett. **107**, 253901 (2011). https://doi.org/10.1103/PhysRevLett.107.253901
23. D. Solli, G. Herink, B. Jalali, C. Ropers, Nat. Photonics **6**(7), 463 (2012). https://doi.org/10.1038/nphoton.2012.126

24. D. Grosz, C. Mazzali, S. Celaschi, A. Paradisi, H. Fragnito, IEEE Photonics Technol. Lett. **11**(3), 379 (1999). https://doi.org/10.1109/68.748242
25. D. Grosz, J.C. Boggio, H. Fragnito, Opt. Commun. **171**(1–3), 53 (1999). https://doi.org/10.1016/S0030-4018(99)00494-0
26. K. Hammani, B. Wetzel, B. Kibler, J. Fatome, C. Finot, G. Millot, N. Akhmediev, J.M. Dudley, Opt. Lett. **36**(11), 2140 (2011). https://doi.org/10.1364/OL.36.002140
27. S.T. Sørensen, C. Larsen, U. Møller, P.M. Moselund, C.L. Thomsen, O. Bang, J. Opt. Soc. Am. B **29**(10), 2875 (2012). https://doi.org/10.1364/JOSAB.29.002875
28. J.M. Soto-Crespo, A. Ankiewicz, N. Devine, N. Akhmediev, J. Opt. Soc. Am. B **29**(8), 1930 (2012). https://doi.org/10.1364/JOSAB.29.001930
29. V.E. Zakharov, A.A. Gelash, Phys. Rev. Lett. **111**, 054101 (2013). https://doi.org/10.1103/PhysRevLett.111.054101
30. P. Béjot, B. Kibler, E. Hertz, B. Lavorel, O. Faucher, Phys. Rev. A **83**, 013830 (2011). https://doi.org/10.1103/PhysRevA.83.013830
31. S.M. Hernandez, P.I. Fierens, J. Bonetti, A.D. Sánchez, D.F. Grosz, IEEE Photonics J. **9**(5), 1 (2017). https://doi.org/10.1109/JPHOT.2017.2754984
32. P. Fierens, S. Hernandez, J. Bonetti, D. Grosz, in *Proceedings of the 4th International Conference on Applications in Nonlinear Dynamics (ICAND 2016)*, ed. by V. In, P. Longhini, A. Palacios (Springer, Berlin, 2016), pp. 265–276. https://doi.org/10.1007/978-3-319-52621-8_23
33. J. Bonetti, S.M. Hernandez, P.I. Fierens, D.F. Grosz, Phys. Rev. A **94**, 033826 (2016). https://doi.org/10.1103/PhysRevA.94.033826
34. V. Zakharov, F. Dias, A. Pushkarev, Phys. Rep. **398**(1), 1 (2004). https://doi.org/10.1016/j.physrep.2004.04.002
35. A. Picozzi, S. Pitois, G. Millot, Phys. Rev. Lett. **101**, 093901 (2008). https://doi.org/10.1103/PhysRevLett.101.093901
36. A. Picozzi, S. Rica, Opt. Commun. **285**(24), 5440 (2012). https://doi.org/10.1016/j.optcom.2012.07.081
37. A. Picozzi, J. Garnier, T. Hansson, P. Suret, S. Randoux, G. Millot, D. Christodoulides, Phys. Rep. **542**(1), 1 (2014). https://doi.org/10.1016/j.physrep.2014.03.002
38. J.M. Soto-Crespo, N. Devine, N. Akhmediev, Phys. Rev. Lett. **116**, 103901 (2016). https://doi.org/10.1103/PhysRevLett.116.103901
39. B.P.P. Kuo, J.M. Fini, L. Grüner-Nielsen, S. Radic, Opt. Express **20**(17), 18611 (2012). https://doi.org/10.1364/OE.20.018611

Chapter 25
Wave Turbulence: A Set of Stochastic Nonlinear Waves in Interaction

Eric Falcon(✉)

Université Paris Diderot, Université de Paris, CNRS, MSC, Paris, France
eric.falcon@univ-paris-diderot.fr

Abstract. Wave turbulence concerns the study of dynamical and statistical properties of a field of random nonlinear waves in interaction. Although it occurs in various situations (ocean surface waves, internal waves in geophysics, Alfvén waves in astrophysical plasmas, or nonlinear waves in optics), well-controlled laboratory experiments on wave turbulence are relatively scarce despite the experimental efforts of the last decade. At the ICAND2018 conference, I presented a short review on laboratory experiments on wave turbulence on the surface of a fluid. I notably discussed the role of strongly nonlinear waves to better describe the dynamics of ocean waves. Here, I report some results obtained by our group on wave turbulence, performed in different experimental systems.

25.1 Introduction

Wave turbulence is a domain rapidly expanding for several years. It focuses on the properties of a field of stochastic nonlinear waves undergoing resonant interactions. The latter transfer wave energy between spatial and temporal scales leading generally to a cascade of energy from a large (forcing) scale, up to a small (eventually dissipative) one. This phenomenon occurs in various situations ranging from spin waves in solids, nonlinear optics, internal or surface waves in oceanography up to plasma waves in astrophysics (for reviews, see [1–4]). The theory of weak wave turbulence, developed in the 1960s [5–7], leads to analytical predictions on the wave energy spectrum in an out-of-equilibrium stationary state, which have been applied in almost all domains of physics involving waves [2,3]. This theory assumes strong hypotheses such as weakly nonlinear and random waves, infinite size system, large number of waves, scale separation (no dissipation), constant energy flux, local interactions, etc. Moreover, the energy transfer between waves is assumed to be governed only by resonant wave interactions. In the past decade, an important experimental effort has been performed to test the domain of validity of weak turbulence theory on different wave systems (e.g. hydrodynamics, nonlinear optics, hydro-elastic or elastic waves) [8]. These well-controlled laboratory experiments have shown the limitations of the

V. In et al. (Eds.): *Proceedings of the 5th International Conference on Applications in Nonlinear Dynamics*, Understanding Complex Systems, https://doi.org/10.1007/978-3-030-10892-2_25

current theoretical framework, which in return, arouses a theoretical and numerical renewed interest.

Here, I present a brief overview of the experimental results obtained by our group in different experimental systems: hydrodynamics wave turbulence (in Sects. 25.2–25.4), hydroelastic wave turbulence (Sect. 25.5), and magnetic wave turbulence (Sect. 25.6).

25.2 Gravity-Capillary Wave Turbulence: Laboratory Experiments

We have experimentally studied and characterized gravity-capillary wave turbulence on the surface of a fluid to better understand the basic mechanisms of energy transfer between hydrodynamics waves.

We have observed in laboratory the regime of gravity-capillary wave turbulence [9], and have reported the first observation of intermittency in wave turbulence [10]. This small scale intermittency is shown to be enhanced by some coherent structures at large scale (wavebreakings, capillary bursts on steep gravity waves) [11,12], but its origin is still an open problem. Moreover, two major experimental challenges have been faced: the measurement of the injected power in the system [13], and a space and time resolved measurement of the wave field [14]. At the time, those quantities were not yet been measured directly for wave turbulence on the surface of a fluid. Two main results have then been obtained:

- The energy transfer mechanisms are not restricted to purely resonant wave interactions, as assumed by the theory, but involved other mechanisms related to the presence of strong nonlinear waves (sharp crested waves, bound waves, ...) [14],
- Large fluctuations of the power injected in the fluid are observed [13,15] that are not taken into account by weak turbulence theory. We showed that the probability distribution of these fluctuations is well described by a simple model, not restricted to wave turbulence since it describes also the energy flux distribution in other dissipative out-of-equilibrium systems [16,17].

We have then reported the first observation in laboratory of the direct gravity-capillary cascade when the fluid departs from the deep-water regime [18]. The study of the non-stationary regime of capillary wave turbulence, when the forcing is stopped, led to the first observation of decay wave turbulence [19]. Another optical method (different form Fourier Transform Profilometry used in [14]) called Diffusing Light Photography, combined with a high-speed camera, has been used to reconstruct the capillary wave field both in time and space. We have highlighted the role of strongly nonlinear capillary waves on the turbulent dynamics [20,21]. The study of 3-wave interactions between gravity-capillary waves allows us to validate experimentally, for the first time for noncollinear waves, the theory of 3-wave resonant interactions [22]. We have also obtain the first indirect measurement of the energy flux at each scale of the turbulent cascade from the dissipated energy spectrum [23]. The energy flux is then found to be non constant, dissipation occurring at each scale of the capillary cascade. A good agreement with weak turbulence theory is nevertheless found for the energy flux and the frequency scalings of the capillary wave spectrum. Indeed, no inconsistency appears since nonlinear wave interactions occur faster than viscous damping processes. The constant of the Kolmogorov–Zakharov spectrum was also inferred experimentally for the first time and compared with its theoretical value [23, 24]. We have also observed the occurrence of stochastic bursts in time transferring wave energy through the spatial scales within all the inertial range [20]. Numerical simulations of capillary wave turbulence were first performed from the kinetic equation or the Hamiltonian dynamics of weak turbulence [2, 25, 26]. We made the first direct numerical simulations of capillary wave turbulence from the two-phase Navier-Stokes equations [27]. These simulations confirm the validity of weak turbulence derivation when hypotheses are verified. Finally, we have studied for the first time wave turbulence on the interface between two immiscible fluids with free upper surface. We show that the coupling between free surface waves and interface waves modifies strongly the wave turbulence regime [28].

25.3 Gravity Wave Turbulence: Large Scale Experiments

Gravity wave turbulence is of primordial interest in oceanography but remains still not well understood. Although oceanography provides more and more data [29–32], the obtained wave spectra vary and depend on numerous and poorly constrained parameters (wind direction, oceanic current, fetch...). Laboratory experiments are much more relevant to accurately tune and control the system parameters [33].

Beyond the laboratory observation of the direct cascade of gravity wave turbulence (from the forcing scales to smaller scales), we showed that the frequency power-law wave spectrum is non-universal and depends on the wave steepness [9], as subsequently reported in other groups in different basin sizes (0.5–50 m) [24, 34–36]. Moreover, we experimentally showed that a spatially homogeneous forcing leads to a good agreement with theoretical predictions [37], contrary to previous observations with a localized forcing with wavemakers.

We have then performed experiments in a much larger basin size (50 m in length, 30 m in width, 5 m in depth) at Ecole Centrale Nantes, France, involving four French laboratories: Université Paris Diderot (MSC), Ecole Normale Supérieure (ENS, LPS), CEA Saclay (SPHINX), and Ecole Centrale Nantes (LHEEA). The stochastic wave field is experimentally found to strongly depend on the basin boundary conditions (absorbing, i.e with a beach, or reflecting, i.e with a wall) although their statistical and spectral properties are close [24]. Moreover, we have shown that the self-similar wave spectra, depending on the wave steepness, observed previously result from the modulation of coherent nonlinear structures (bound waves) [38]. This thus explains the departure from predictions of gravity wave turbulence, observed in oceanography and in numerous well-controlled experiments. In another series of experiments, we have also studied resonant interactions between nonlinear waves that are the fundamental mechanism that transfers energy in wave turbulence. By means of this experiment on 4-wave interactions between oblique gravity surface waves, we have validated experimentally, for the first time, the theory of 4-wave resonant interactions with no fitting parameter [39]. This strongly extends previous experimental results performed mainly for perpendicular or collinear wave trains [40–42]. For stronger nonlinearities, meaningful departures from this weakly nonlinear theory are observed [43].

Finally, an inverse cascade of wave action, from the forcing scales to larger scales, is expected theoretically for gravity wave turbulence [2,3]. It has been confirmed numerically [44]. We have reported the first laboratory observations of an inverse cascade of gravity wave turbulence [45], but on a limited inertial range due to the small container size used. Additional studies in the large-scale basin are currently in progress in Nantes.

25.4 Wave Turbulence in Low-Gravity Environments

Many laboratory experiments have been performed with surface waves on a horizontal layer of fluid. In this configuration, the dominant restoring force is gravity for large wavelength and capillarity for short wavelength. The transition between the two regimes occurs for the capillary length that depends on the acceleration of gravity, notably. Energy transfer mechanisms are different for gravity and capillary waves and this makes the cascade process of the energy more difficult to understand since the mechanisms change when one crosses the capillary length [9]. An advantage of experiments in reduced gravity is to increase the capillary length above the size of the container and thus to have capillary waves throughout the cascade. Another advantage is related to the geometry of the experiment. In low gravity, the fluid inside a spherical container wets the inner boundary and therefore takes the shape of a spherical fluid layer. Capillary waves thus propagate on its inner surface without meeting any lateral boundary in contrast to the configurations studied on Earth.

We have first studied purely capillary waves in a spherical container in low-gravity environment during CNES parabolic flight campaigns. We have observed capillary wave turbulence on a broad range of scales usually masked on Earth by the gravity wave regime [46]. When the forcing is periodic, various patterns (hexagons, lines) have been observed on the spherical fluid surface [46,47]. The main limitation of parabolic flights is related to the 20 s duration of each parabola that does not allow enough statistics. To reach much longer measurements, we have reported experiments conducted by ESA astronauts on the International Space Station (ISS). Using a new device, "FLUIDICS" (Fluid Dynamics in Space) developed by CNES and Airbus Defense and Space, they studied turbulence of capillary waves on the surface of a fluid in a spherical container. Power spectra of wave turbulence have been found to be in good agreement with weak turbulence theory [48]. Using higher frequency forcing will also allow us to test whether scales larger than the one of the forcing are in statistical equilibrium [49]. This work is currently pursued on ISS.

25.5 Hydroelastic Wave Turbulence

Hydroelastic waves, including gravity-bending waves, are found in various domains: on the surface of lakes or oceans covered by ice, or for very large floating structures in oceanography, flapping flags, or in biomedical applications such as heart valves. Hydroelasticity is defined by the coupling of the elastic medium with the hydrodynamics of the surrounding fluid.

We have reported results of laboratory experiments on nonlinear waves on the surface of a fluid covered by an elastic sheet (where both tension and bending are important). When a set of stochastic waves are in interaction, a regime of wave turbulence has been observed in this new experimental system [50]. The existence of 3-wave interactions, predicted theoretically in this system, has been also highlighted experimentally [51].

25.6 Magnetic Wave Turbulence

When wave amplitudes are high enough, weak turbulence theory predicts a nonlinear resonant process between waves that generates smaller wavelengths. For a ferrofluid (a liquid with a suspension of nanometric magnetic particles), the dispersion relation of surface waves was known to be tuned by applying a magnetic field. We have thus studied the dynamics of random waves propagating on the surface of a ferrofluid submitted to a magnetic field.

We have reported the first observation of a magnetic wave turbulence regime [52]. The existence domains of gravity and capillary wave turbulence are also documented as well as a triple point of coexistence of these three regimes. These results are understood using dimensional analysis since weak turbulence derivation has not been yet considered theoretically for the magnetic regime. Such an experimental system where the dispersion relation is tuned by the operator from a non-dispersive to a dispersive system is thus of primary interest to test the wave turbulence theory. The case of a magnetic field parallel to the fluid surface shows several differences with the normal case. The striking one is the meaningful broadening of the inertial domain of the magnetic wave turbulence regime [53].

25.7 Conclusion

The experiments presented here have raised the understanding of the regime of wave turbulence that occurs in various systems involving waves (e.g. hydrodynamics, hydroelastic or magnetic waves). It results that the weak turbulence theory gives a correct image of the underlying physical phenomena but its validity range in experiments appears limited. The progress realized in wave turbulence also sheds new light to certain similar problems in usual hydrodynamic turbulence (such as small scale intermittency or the statistical equilibrium of large scales). The future study of the interaction between wave turbulence and a flow (turbulent or not) paves the way to a better understanding of natural systems such as the coupling between the dynamics of ocean and that of the atmosphere, key ingredient for the climate modeling.

Acknowledgments. I thank all my co-authors quoted in the references of this article. This work was supported by the French National Research Agency via ANR DYSTURB project No. ANR-17-CE30-0004 (2017-2021), ANR TURBULON project No. ANR-12-BS04-0005 (2012-2016) and ANR TURBONDE project No. ANR-07-BLAN-0246 (2007-2011). The support of Novespace during Parabolic Flight Campaigns is acknowledged, as well as partial financial support by French National Space Agency (CNES).

References

1. E. Falcon, Laboratory experiments on wave turbulence. Discrete Contin. Dyn. Syst.-Ser. B **13**, 819 (2010)
2. V.E. Zakharov, V. L'vov, G. Falkovich, *Kolmogorov Spectra of Turbulence I: Wave Turbulence* (Springer, Berlin, 1992)
3. S. Nazarenko, *Wave Turbulence* (Springer, Berlin, 2011)
4. A.C. Newell, B. Rumpf, Wave turbulence. Annu. Rev. Fluid Mech. **43**, 59 (2011)
5. K. Hasselmann, On the non-linear energy transfer in a gravity-wave spectrum Part 1. General theory. J. Fluid. Mech. **12**, 481 (1962)
6. D.J. Benney, A.C. Newell, The propagation of non-linear wave envelopes. J. Math. Phys. **46**, 363 (1967)
7. V.E. Zakharov, N.N. Filonenko, Energy spectrum for stochastic oscillations of the surface of liquid. Sov. Phys. Dokl. **11**, 881–884 (1967)
8. V. Shrira, S. Nazarenko (eds.), *Advances in Wave Turbulence*, vol. 83 (World Scientific, Singapore, 2013)
9. E. Falcon, C. Laroche, S. Fauve, Observation of gravity-capillary wave turbulence. Phys. Rev. Lett. **98**, 094503 (2007)
10. E. Falcon, S. Fauve, C. Laroche, Observation of intermittency in wave turbulence. Phys. Rev. Lett. **98**, 154501 (2007)
11. E. Falcon, S.G. Roux, C. Laroche, On the origin of intermittency in wave turbulence. EPL (Eur. Lett.) **90**, 34005 (2010)
12. E. Falcon, S.G. Roux, B. Audit, Revealing intermittency in experimental data with steep power spectra. EPL (Eur. Lett.) **90**, 50007 (2010)
13. E. Falcon, S. Aumaître, C. Falcón, C. Laroche, S. Fauve, Fluctuations of energy flux in wave turbulence. Phys. Rev. Lett. **100**, 064503 (2008)
14. E. Herbert, N. Mordant, E. Falcon, Observation of the nonlinear dispersion relation and spatial statistics of wave turbulence on the surface of a fluid. Phys. Rev. Lett. **105**, 144502 (2010)
15. S. Aumaître, E. Falcon, S. Fauve, Fluctuations of the energy flux in wave turbulence, pp. 53–72, in [8]
16. C. Falcón, E. Falcon, Fluctuations of energy flux in a simple dissipative out-of-equilibrium system. Phys. Rev. E **79**, 041110 (2009)

17. A. García-Cid, P. Gutiérrez, C. Falcón, S. Aumaître, E. Falcon, Statistics of injected power on a bouncing ball subjected to a randomly vibrating piston. Phys. Rev. E **92**, 032915 (2015)
18. E. Falcon, C. Laroche, Observation of depth-induced properties in wave turbulence on the surface of a fluid. EPL (Eur. Lett.) **94**, 34003 (2011)
19. L. Deike, M. Berhanu, E. Falcon, Decay of capillary wave turbulence. Phys. Rev. E **85**, 066311 (2012)
20. M. Berhanu, E. Falcon, Space-time-resolved capillary wave turbulence. Phys. Rev. E **89**, 033003 (2013)
21. M. Berhanu, E. Falcon, L. Deike, Turbulence of capillary waves forced by steep gravity waves. J. Fluid Mech. **850**, 803 (2018)
22. F. Haudin, A. Cazaubiel, L. Deike, T. Jamin, E. Falcon, M. Berhanu, Experimental study of three-wave interactions among capillary-gravity surface waves. Phys. Rev. E **93**, 043110 (2016)
23. L. Deike, M. Berhanu, E. Falcon, Energy flux measurement from the dissipated energy in capillary wave turbulence. Phys. Rev. E **89**, 023003 (2014)
24. L. Deike, B. Miquel, P. Gutiérrez, T. Jamin, B. Semin, M. Berhanu, E. Falcon, F. Bonnefoy, Role of the basin boundary conditions in gravity wave turbulence. J. Fluid Mech. **781**, 196 (2015)
25. A.N. Pushkarev, V.E. Zakharov, Turbulence of capillary waves. Phys. Rev. Lett. **76**, 3320 (1996)
26. Y. Pan, D.K.P. Yue, Understanding discrete capillary-wave turbulence using a quasi-resonant kinetic equation. J. Fluid Mech. **816**, R1 (2017)
27. L. Deike, D. Fuster, M. Berhanu, E. Falcon, Direct numerical simulations of capillary wave turbulence. Phys. Rev. Lett. **112**, 234501 (2014)
28. B. Issenmann, C. Laroche, E. Falcon, Wave turbulence in a two-layer fluid: coupling between free surface and interface waves. EPL (Eur. Lett.) **116**, 64005 (2016)
29. M.A. Donelan, J. Hamilton, W.H. Hui, Directional spectra of wind-generated waves. Philos. Trans. R. Soc. Lond. A **315**, 509 (1985)
30. P.A. Hwang, D.W. Wang, E.J. Walsh, W.B. Krabill, R.N. Swift, Airborne measurements of the wavenumber spectra of ocean surface waves. Part I: spectral slope and dimensionless spectral coefficient? J. Phys. Ocean. **30**, 2753 (2000)
31. L. Romero, W.K. Melville, Airborne observations of fetch-limited waves in the Gulf of Tehuantepec. J. Phys. Ocean. **40**, 441 (2010)
32. F. Leckler, F. Ardhuin, C. Peureux, A. Benetazzo, F. Bergamasco, V. Dulov, Analysis and interpretation of frequency-wavenumber spectra of young wind waves. J. Phys. Ocean. **45**, 10 (2015)
33. S. Nazarenko, S. Lukaschuk, Wave turbulence on water surfaces. Annu. Rev. Condens. Matter Phys. **7**, 61 (2016)
34. P. Denissenko, S. Lukaschuk, S. Nazarenko, Gravity wave turbulence in a laboratory flume. Phys. Rev. Lett. **99**, 014501 (2007)
35. P. Cobelli, A. Przadka, P. Petitjeans, G. Lagubeau, V. Pagneux, A. Maurel, Different regimes for water wave turbulence. Phys. Rev. Lett. **107**, 214503 (2011)
36. Q. Aubourg, A. Campagne, C. Peureux, F. Ardhuin, J. Sommeria, S. Viboud, N. Mordant, Three-wave and four-wave interactions in gravity wave turbulence. Phys. Rev. Fluids **2**, 114802 (2017)
37. B. Issenmann, E. Falcon, Gravity wave turbulence revealed by horizontal vibrations of the container. Phys. Rev. E **87**, 011001(R) (2013)
38. G. Michel, B. Semin, A. Cazaubiel, F. Haudin, T. Humbert, S. Lepot, F. Bonnefoy, M. Berhanu, E. Falcon, Self-similar gravity wave spectra resulting from the modulation of bound waves. Phys. Rev. Fluids **3**, 054801 (2018)

39. F. Bonnefoy, F. Haudin, G. Michel, B. Semin, T. Humbert, S. Aumaître, M. Berhanu, E. Falcon, Observation of resonant interactions among surface gravity waves. J. Fluid Mech. (Rapids) **805**, R3 (2016)
40. M.S. Longuet-Higgins, N.D. Smith, An experiment on third-order resonant wave interactions. J. Fluid Mech. **25**, 417 (1966)
41. L.F. McGoldrick, O.M. Phillips, N.E. Huang, T.H. Hodgson, Measurements of third-order resonant wave interactions. J. Fluid Mech. **25**, 437 (1966)
42. H. Tomita, Theoretical and experimental investigations of interaction among deep-water gravity waves. Rep. Ship Res. Inst. **26**, 251 (1989)
43. F. Bonnefoy, F. Haudin, G. Michel, B. Semin, T. Humbert, S. Aumaître, M. Berhanu, E. Falcon, Experimental observation of four-wave resonant interactions in a wave basin. La Houille Blanche **5**, 56 (2017)
44. S.Y. Annenkov, V.I. Shrira, Direct numerical simulation of downshift and inverse cascade for water wave turbulence. Phys. Rev. Lett. **96**, 204501 (2006); A.O. Korotkevitch, Simultaneous numerical simulation of direct and inverse cascades in wave turbulence. Phys. Rev. Lett. **101**, 074501 (2008)
45. L. Deike, C. Laroche, E. Falcon, Experimental study of the inverse cascade in gravity wave turbulence. EPL (Eur. Lett.) **96**, 34004 (2011)
46. C. Falcón, E. Falcon, U. Bortolozzo, S. Fauve, Capillary wave turbulence on a spherical fluid surface in zero gravity. EPL (Eur. Lett.) **86**, 14002 (2009)
47. S. Fauve, E. Falcon, Gravity-capillary wave turbulence, in *Report to COSPAR (World Committee for Space Research), 37th Scientific Assembly, 13–20 July 2008, Montréal, Canada, CNES Ed.* (2008), pp. 90–91
48. M. Berhanu, E. Falcon, S. Fauve, Wave turbulence in microgravity, in *Report to COSPAR (World Committee for Space Research), 42th Scientific Assembly, 14–22 July 2018, Pasadena, USA, CNES Ed.* (2018), pp. 66–67
49. G. Michel, F. Pétrélis, S. Fauve, Observation of thermal equilibrium in capillary wave turbulence. Phys. Rev. Lett. **118**, 144502 (2017)
50. L. Deike, J.-C. Bacri, E. Falcon, Nonlinear waves on the surface of a fluid covered by an elastic sheet. J. Fluid Mech. **733**, 394 (2013)
51. L. Deike, M. Berhanu, E. Falcon, Observation of hydroelastic three-wave interactions. Phys. Rev. Fluids **2**, 064803 (2017)
52. F. Boyer, E. Falcon, Wave turbulence on the surface of a ferrofluid in a magnetic field. Phys. Rev. Lett. **101**, 244502 (2008)
53. S. Dorbolo, E. Falcon, Wave turbulence on the surface of a ferrofluid in a horizontal magnetic field. Phys. Rev. E **83**, 046303 (2011)

Chapter 26
Noise Benefits in Feedback Machine Learning: Bidirectional Backpropagation

Bart Kosko(✉)

Electrical and Computer Engineering Department, Signal and Image Processing Institute,
University of Southern California, Los Angeles, CA, USA
kosko@usc.edu

Abstract. The new bidirectional backpropagation algorithm converts an ordinary feedforward neural network into a simple feedback dynamical system. The algorithm minimizes a joint performance measure so that training in one direction does not overwrite training in the reverse direction. This involves little extra computation. The forward direction gives the usual classification or regression network. The new backward pass approximates the centroids of the input pattern classes in a neural classifier. The bidirectional algorithm can also approximate inverse point mappings in the rare cases where such mappings exist. Carefully injected noise can speed the convergence of the bidirectional backpropagation. This holds because backpropagation is a special case of the expectation-maximization algorithm for maximum likelihood and because such noise can always boost its convergence. The noise also tends to improve accuracy in classification and regression.

26.1 Bidirectional Neural Networks

Modern feedforward neural networks naturally define a feedback dynamical system if one uses the network in both the forward and backward directions. This leads to the new bidirectional backpropagation supervised learning algorithm [1,2]. The ordinary unidirectional backpropagation algorithm remains the most popular neural algorithm in modern machine learning [3–6]. Such unidirectional networks simply ignore the information that the network encodes in its backward direction. Figure 26.1 shows that a 3-layer neural network where the bidirectional backpropagation algorithm has learned the connection weights for the point-invertible 3-bit permutation mapping in Table 26.1. This learned network *exactly* represents the 3-bit permutation mapping and its inverse through the *same* set of connection weights. A basic theorem shows that a 3-layer threshold network can exactly represent any n-bit permutation and its inverse if it uses 2^n or exponentially many threshold neurons in its hidden layer [1].

V. In et al. (Eds.): *Proceedings of the 5th International Conference on Applications in Nonlinear Dynamics*, Understanding Complex Systems, https://doi.org/10.1007/978-3-030-10892-2_26

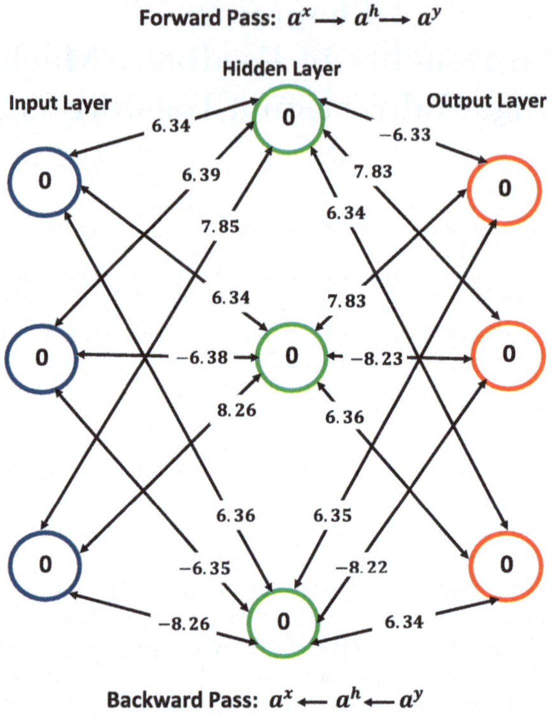

Fig. 26.1. Learned bidirectional representation of the 3-bit permutation in Table 26.1. The bidirectional backpropagation algorithm found this representation using the double-classification learning laws of [2]. All neurons were bipolar (emitting values 1 or −1) and had zero thresholds. The zero thresholding gave an exact representation of the 3-bit permutation in Table 26.1

Noise injection can in principle always help backpropagation and bidirectional-backpropagation training [7]. The probabilistic structure of all such neural networks allows the user to noise-boost their training. A noise-boost sufficient condition holds because the popular backpropagation neural learning algorithm turns out to be a special case of the generalized expectation-maximization (EM) algorithm [7]. The EM algorithm performs maximum likelihood for hidden variables or missing data by iteratively climbing the nearest hill of probability [8]. Carefully chosen noise can always boost the EM algorithm as it climbs a hill of probability [9,10]. The noise is not the blind-noise dither of stochastic resonance. It is just that noise or other perturbation that makes the current signal more probable. This follows from the gradient master equation in (26.1) that we present below for the neural network's probability density $p(\mathbf{y}|\mathbf{x}, \Theta_k)$ for a vector input x, a network output y, and the network parameters Θ_k at iteration k.

A modern neural network $N : R^n \rightarrow R^K$ is a feedforward mapping from the input vector space R^n to the output vector space R^K. The most common "deep"

Table 26.1. 3-bit bipolar permutation function f and inverse f^{-1} that the network in Fig. 26.1 encodes

Input x	Output t
$[++\,+]$	$[--\,+]$
$[++\,-]$	$[-+\,+]$
$[+-\,+]$	$[++\,+]$
$[+-\,-]$	$[+-\,+]$
$[-+\,+]$	$[-+\,-]$
$[-+\,-]$	$[--\,-]$
$[--\,+]$	$[+-\,-]$
$[--\,-]$	$[++\,-]$

neural networks are feedforward classifiers. They are deep if they contain at least two hidden layers of neurons. They map the input space to K output neurons that define a discrete K-dimensional probability vector. The network N classifies an input pattern vector x to pattern class j if and only if the jth output neuron has the largest activation and thus if it has the largest output probability. So the input image or other pattern vector x maps in one-shot fashion to an output density $N(x)$ in the simplex of K-dimensional probability vectors.

A natural way to turn the feedforward neural network $N : R^n \rightarrow R^K$ into a dynamical system is to pass the output $y = N(x)$ back through the network.

An earlier version of this bidirectional strategy was the bidirectional associative memory (BAM) in a two-layer neural network with a single connection matrix M [11,12]. The backward pass uses the matrix transpose M^T. Then the basic BAM theorem holds for standard threshold neurons or threshold-like neurons: Every matrix is globally bidirectionally stable [11]. All input perturbations quickly converge to a bidirectional fixed point (x_f, y_f). The equilibrium dynamics are more complicated when there are intervening hidden layers of neurons between the visible input and output layers. We here only mention that then BAM fixed points need not always occur. The 3-layer threshold network in Fig. 26.1 does produce the 8 input-output pairs of Table 26.1 as 8 bidirectional fixed points.

The bidirectional backpropagation (henceforth B-BP) algorithm usually operates in sequential synchronous mode. There are four main cases for learning depending on the type of neurons in the input layer and output layer: (1) classification-classification where both layers use soft-max (or threshold) neurons and encode targets with unit bit vectors, (2) regression-regression where both layers use identity neurons, (3) regression-classification where the input neurons are identity functions and the output neurons are softmax with 1-in-K unit-bit-vector encoding, and (4) classification-regression where the input neurons are softmax and the output neurons are identity functions. The networks in Figs. 26.1 and 26.2 used the classification-classification version of B-BP [2].

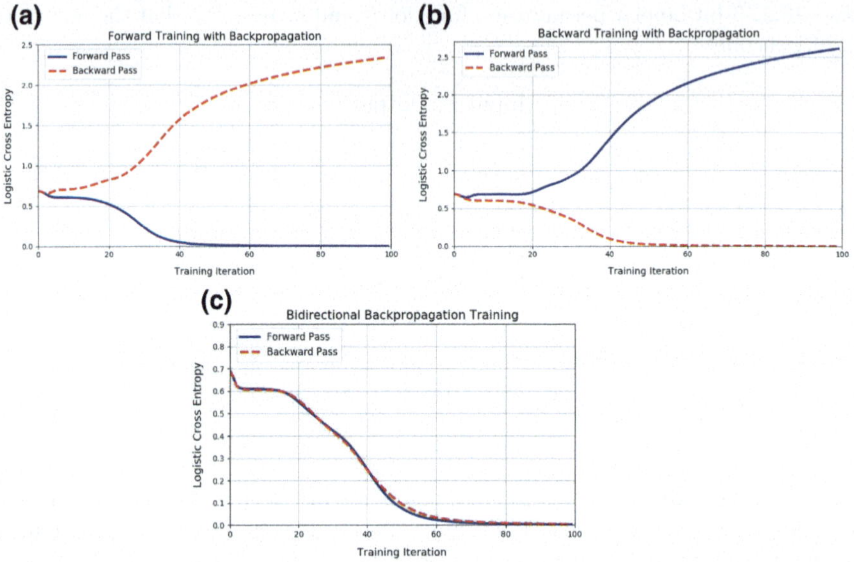

Fig. 26.2. Logistic-cross-entropy learning for double classification [2] using 100 hidden neurons with forward BP training, backward BP training, and bidirectional BP training. The trained network exactly represents the 5-bit permutation function in Table 26.2. **a** Forward BP tuned the network with respect to logistic cross entropy for the forward pass using E_f only. **b** Backward BP training tuned the network with respect to logistic cross entropy for the backward pass using E_b only. **c** Bidirectional BP training summed the logistic cross entropies for both the forward-pass error term E_f and the backward-pass error term E_b to update the network parameters. Only the bidirectional case (**c**) shows no overwriting of training in either direction

Figure 26.2 shows that unidirectional BP overwrites in the reverse direction using the usual error function for a given direction. The last plot in Fig. 26.2 shows that B-BP does not produce such overwriting in either direction.

The computational cost of B-BP is light because BP training in either direction has only $O(n)$ time complexity for n training samples. So BP scales well for problems. B-BP has the same linear complexity because $O(n) + O(n) = O(n)$.

The B-BP case (3) above describes the most common network set-up for a neural network. A user feeds an image or other pattern directly into the input identity neurons. Then the network maps that input vector to an output probability vector of K softmax neurons. The performance measure is cross entropy in the forward direction because the implied output probability is a one-sample multinomial or a single roll of a K-sided die. The implied performance measure in the backward direction is squared error. This holds because the input neurons are identity functions and because their implied probability is a conditional vector normal [7]. These probabilistic constraints imply that both classification and

Table 26.2. 5-bit bipolar permutation function from Fig. 26.2

Input x	Output t	Input x	Output t
[− − − − −]	[+ + − + +]	[+ − − − −]	[− + + + +]
[− − − − +]	[− − + − −]	[+ − − − +]	[− + − − −]
[− − − + −]	[− − − + −]	[+ − − + −]	[+ − − + −]
[− − − + +]	[+ + + − +]	[+ − − + +]	[− − + − +]
[− − + − −]	[+ + − + −]	[+ − + − −]	[− + − + +]
[− − + − +]	[+ − − + +]	[+ − + − +]	[+ + − − +]
[− − + + −]	[− + + − +]	[+ − + + −]	[+ + + + +]
[− − + + +]	[− − + + +]	[+ − + + +]	[− − + + −]
[− + − − −]	[+ − + + +]	[+ + − − −]	[+ + + − −]
[− + − − +]	[+ − − − +]	[+ + − − +]	[− + − + −]
[− + − + −]	[+ − + + −]	[+ + − + −]	[+ − − − −]
[− + − + +]	[− + + − −]	[+ + − + +]	[− − − + +]
[− + + − −]	[− + + + −]	[+ + + − −]	[− − − − −]
[− + + − +]	[+ + − − −]	[+ + + − +]	[− + − − +]
[− + + + −]	[+ − + − +]	[+ + + + −]	[+ + + + −]
[− + + + +]	[− − − − +]	[+ + + + +]	[+ − + − −]

regression have the *same* BP learning laws [2]. So we will not review them here. The next section summarizes the main probabilistic facts about BP and B-BP.

26.2 Backpropagation as Maximum Likelihood Estimation

We first show why backpropagation is a form of generalized Expectation-Maximization (EM) [7]. This new theorem gives insight into both algorithms and allows users to modify one by modifying the other. This key result states that the gradient of the network's log-likelihood $\log p(y|x, \Theta_k)$ equals the gradient of EM's surrogate likelihood function $Q(\Theta|\Theta_k)$:

$$\nabla_\Theta \log p(y|x, \Theta_k) = \nabla_\Theta Q(\Theta_k|\Theta_k) \tag{26.1}$$

at each iteration k for the network's total parameter vector Θ_k and input x.

The BP-EM gradient identity (26.1) follows if we expand the network likelihood $p(y|x, \Theta) = \frac{p(h,y|x,\Theta)}{p(h|y,x,\Theta)}$ for all hidden variables h in the network. Then EM takes expectations of the log-likelihood $\log p(y|x, \Theta)$ with respect to the hidden posterior $p(h|y, x, \Theta_k)$. This gives

$$\log p(y|x, \Theta_k) = Q(\Theta|\Theta_k) + H(\Theta|\Theta_k) \tag{26.2}$$

for EM's so-called surrogate likelihood $Q(\Theta|\Theta_k) = \mathbb{E}_{h|y,x,\Theta_k}[\log p(h, y|x, \Theta)]$ and for the entropy $H(\Theta|\Theta_k) = -\mathbb{E}_{h|y,x,\Theta_k}\{\log p(h|y, x, \Theta)\}$. A basic fact of EM is the "ascent property": Maximizing the surrogate likelihood Q can only increase the total log-likelihood $\log p(y|x, \Theta_k)$ [8]. The entropy inequality $H(\Theta_k|\Theta_k) \leq H(\Theta|\Theta_k)$ also holds for all Θ because of Jensen's Inequality and the concavity of the logarithm. So Shannon entropy minimizes cross entropy. Then taking the gradient gives $\nabla_\Theta H(\Theta|\Theta_k) = \mathbf{0}$ at $\Theta = \Theta_k$. Then taking gradients in (26.2) gives (26.1).

A neural classifier results if $p(y|x, \Theta)$ is a multinomial or categorical probability density with softmax output neurons and 1-in-K encoding. Then the log-likelihood $\log p(y|x, \Theta)$ equals negative cross entropy. So minimizing the cross entropy maximizes the log-likelihood. Then the gradient $\nabla \log p(y|x, \Theta)$ gives the usual BP learning law of backpropagation [3,4]. A neural regressor results if $p(y|x, \Theta)$ equals a K-dimensional Gaussian with identity output neurons. Then the log-likelihood $\log p(y|x, \Theta)$ equals the negative squared error of regression. Then taking the gradient also gives the same learning law [2].

We now show why a neural classifier uses a cross-entropy performance measure. The network's K output softmax neurons are independent because they have no intra-layer connections. Then the network likelihood $p_f(\mathbf{y}|\mathbf{x}, \Theta)$ factors into a product of K-many marginals [13]: $p_f(\mathbf{y}|\mathbf{x}, \Theta) = \prod_{k=1}^{K} p_f(y_k|\mathbf{x}, \Theta)$. Then taking logarithms gives

$$\log p_f(\mathbf{y}|\mathbf{x}, \Theta) = \log \prod_{k=1}^{K} p_f(y_k|\mathbf{x}, \Theta) \tag{26.3}$$

$$= \log \prod_{k=1}^{K} (a_k^y)^{y_k} \tag{26.4}$$

$$= \sum_{k=1}^{K} y_k \log a_k^y \tag{26.5}$$

$$= -E_f(\Theta) \tag{26.6}$$

because $\mathbf{y}$ is a 1-in-K-encoded unit bit vector. Then exponentiation gives $p_f(\mathbf{y}|\mathbf{x}, \Theta) = \exp\{-E_f(\Theta)\}$. So minimizing the forward cross entropy E_f is the same as maximizing the negative cross entropy $-E_f$. Minimizing E_f maximizes the forward network likelihood and vice versa.

The B-BP algorithm combines the above results into a compound or bidirectional network likelihood $p(y|x, \Theta)p(x|y, \Theta)$. Then taking logarithms gives the additive structure of the network's joint performance measure:

$$\log p(y|x, \Theta)p(x|y, \Theta) = \log p(y|x, \Theta) + \log p(x|y, \Theta). \tag{26.7}$$

Then the same neural network N can encode a forward classifier network through a multinomial likelihood $p(y|x, \Theta)$ and softmax output neurons while it also encodes a backward regression network through a Gaussian $p(x|y, \Theta)$ and identity input neurons.

B-BP does not depend on the existence of an inverse point-map. It works instead with the set-theoretic inverse as we now explain. Forward training of $N : X \to Y$ approximates some function $f : X \to Y$ from the input vector space X to the output space Y. But B-BP trains the set-theoretic pullback or inverse mapping $f^{-1} : 2^Y \to 2^X$ over the *same* connection weights and the same neural units of N as in the forward direction. A function f need not have a point inverse. Few functions do have point inverses because they are not bijective. But any function f does have a set-theoretic inverse $f^{-1} : 2^Y \to 2^X$ such that $f^{-1}(B) = \{x \in X : f(x) \in B\}$ for any $B \subset Y$. The backward pass $N^{-1}(y) \in X$ approximates the corresponding input vector x. The backward mapping of a neural classifier with K classes tends to approximate the centroids of the K classes [2].

26.2.1 Noise-Boosting Bidirectional Backpropagation via EM

We summarize last how carefully chosen noise can boost the EM algorithm and thereby boost the BP and B-BP algorithms. The Noisy EM Theorem shows that injecting noise or other perturbations can only speed up the EM algorithm on average at each iteration if the noise obeys the NEM positivity condition [9,10]. The noise need not be additive. It can be multiplicative or any other measurable function.

We state the basic result for additive noise for simplicity. The Noisy EM Theorem for additive noise states that a noise benefit holds at each iteration n if the following positivity condition holds:

$$\mathbb{E}_{\mathbf{x},\mathbf{h},\mathbf{N}|\Theta^*}\left[\ln\left(\frac{p(\mathbf{x}+\mathbf{N},\mathbf{h}|\Theta^n)}{p(\mathbf{x},\mathbf{h}|\Theta^n)}\right)\right] \geq 0\,. \tag{26.8}$$

Then the EM noise benefit

$$Q(\Theta^n|\Theta^*) \leq Q_N(\Theta^n|\Theta^*) \tag{26.9}$$

holds on average at iteration n:

$$\mathbb{E}_{\mathbf{x},\mathbf{N}|\Theta^n}\Big[Q(\Theta^n|\Theta^*) - Q_N(\Theta^n|\Theta^*)\Big] \leq \mathbb{E}_{\mathbf{x}|\Theta^n}\Big[Q(\Theta^*|\Theta^*) - Q(\Theta^n|\Theta^*)\Big]$$

where Θ^* denotes the maximum-likelihood vector of parameters. The NEM positivity condition (26.8) has a simple form for Gaussian mixture models [14] and for classification and regression networks [7].

The idea behind the NEM sufficient condition (26.8) is that some noise realizations n make a signal x more probable: $f(x+n|\Theta) \geq f(x|\Theta)$. Taking logarithms gives $\ln(\frac{f(x+n|\theta)}{f(x|\theta)}) \geq 0$. Then taking expectations gives a NEM-like positivity condition. The proof of the NEM Theorem uses Kullback–Liebler divergence to show that the noise-boosted likelihood is closer on average at each iteration to the optimal likelihood function than is the noiseless likelihood [10].

An important point is that the NEM positivity inequality (26.8) is not vacuous because the expectation conditions on the converged parameter vector Θ^*.

Vacuity would result in the usual case of averaging a log-likelihood ratio. Take the expectation of the log-likelihood ratio $\ln \frac{f(x|\Theta)}{g(x|\Theta)}$ with respect to the probability density function $g(x|\Theta)$ to give $E_g[\ln \frac{f(x|\Theta)}{g(x|\Theta)}]$. Then Jensen's inequality and the concavity of the logarithm give $E_g[\ln \frac{f(x|\Theta)}{g(x|\Theta)}] \leq \ln E_g[\frac{f(x|\Theta)}{g(x|\Theta)}] = \ln \int_X \frac{f(x|\Theta)}{g(x|\Theta)} g(x|\Theta)\ dx = \ln \int_X g(x|\Theta)\ dx = \ln 1 = 0$. So $E_g[\ln \frac{f(x|\theta)}{g(x|\theta)}] \leq 0$ and thus in this case strict positivity is impossible [15]. But the expectation in (26.8) does not in general lead to this cancellation of probability densities because the integrating density in (26.8) depends on the optimal maximum-likelihood parameter Θ^* rather than on just Θ^n. So density cancellation occurs only when the NEM algorithm has converged to a local likelihood maximum because then $\Theta^n = \Theta^*$.

The NEM Theorem simplifies for a classifier network with K softmax output neurons. Then the additive noise must lie above the defining NEM hyperplane where such noise adds directly to the training output targets in the cross-entropy (26.4) [7]. A similar NEM result holds for regression except that the noise-benefit region is a hypersphere [7, 16]. This same NEM noise-space geometry holds for the B-BP algorithm depending on whether the system design is that of a double classifier, double regressor, or a mixed regressor-classifier or classifier-regressor. NEM noise can also inject into the hidden neurons.

We refer the reader to [7, 16] for the detailed statements and illustrations of injecting NEM noise into classifiers and regressors. Extensive simulations with NEM-boosted B-BP have shown comparable improvement in speeding up training and in improving both classification and regression accuracy.

26.3 Conclusion

Bidirectional backpropagation allows a multi-layer neural network to exploit information in the backward direction as well as in the forward direction. Carefully injected noise can speed B-BP training on average because the backpropagation algorithm is a special case of the generalized expectation-maximization algorithm and because such noise can always speed the average convergence of the expectation-maximization algorithm as it iteratively climbs the nearest hill of probability or log-likelihood. The same likelihood method allows noise injection in recurrent networks [17, 18] for classification and regression [16] as well as noise injection in Markov-chain Monte Carlo estimation and simulated annealing [19].

References

1. O. Adigun, B. Kosko, Bidirectional representation and backpropagation learning, in *International Joint Conference on Advances in Big Data Analytics* (CSREA Press, 2016), pp. 3–9
2. O. Adigun, B. Kosko, Bidirectional Backpropagation. To appear in IEEE Trans. Syst. Man Cybern.: Syst. Man Cybern. (2018)

3. P.J. Werbos, Backpropagation through time: what it does and how to do it. Proc. IEEE **78**, 1550–1560 (1990)
4. D. Rumelhart, G. Hinton, R. Williams, Learning representations by back-propagating errors. Nature **323**, 533–536 (1986)
5. Y. LeCun, Y. Bengio, G. Hinton, Deep learning. Nature **521**, 436–444 (2015)
6. M. Jordan, T. Mitchell, Machine learning: trends, perspectives, and prospects. Science **349**, 255–260 (2015)
7. K. Audhkhasi, O. Osoba, B. Kosko, Noise-enhanced convolutional neural networks. Neural Netw. **78**, 15–23 (2016)
8. A.P. Dempster, N.M. Laird, D.B. Rubin, Maximum likelihood from incomplete data via the EM algorithm. J. R. Stat. Soc. Ser. B (Methodol.) **39**, 1–38 (1977)
9. O. Osoba, S. Mitaim, B. Kosko, The noisy expectation-maximization algorithm. Fluct. Noise Lett. **12**, 1350012 (2013)
10. O. Osoba, B. Kosko, The noisy expectation-maximization algorithm for multiplicative noise injection. Fluct. Noise Lett. **15**, 1650007 (2016)
11. B. Kosko, Bidirectional associative memories. IEEE Trans. Syst. Man Cybern. **18**, 49–60 (1988)
12. B. Kosko, *Neural Networks and Fuzzy Systems: A Dynamical Systems Approach to Machine Intelligence* (Prentice Hall, Englewood Cliffs, 1991)
13. C.M. Bishop *Pattern Recognition and Machine Learning* (Springer, Berlin, 2006)
14. K. Audhkhasi, O. Osoba, B. Kosko, Noisy hidden Markov models for speech recognition, in *Neural Networks* (2013), pp. 1–6
15. B. Kosko, K. Audhkhasi, O. Osoba, Noise can speed backpropagation learning and deep bidirectional pretraining **in review**
16. O. Adigun, B. Kosko, Using noise to speed up video classification with recurrent backpropagation (2017), pp. 108–115
17. S. Hochreiter, J. Schmidhuber, Long short-term memory. Neural Comput. **9**, 1735–1780 (1997)
18. C. Junyoung, G. Caglar, C. Kyunghyun, B. Yoshua, Gated feedback recurrent neural networks, in *Proceedings of the 32nd International Conference on Machine Learning (PMLR 37)* (2015), pp. 2067–2075
19. B. Franzke, B. Kosko, Using noise to speed up Markov chain Monte Carlo estimation. Procedia Comput. Sci. **53**, 113–120 (2015)

Chapter 27 Suppression of Stimulated Brillouin Scattering in Optical Fiber Using Boolean Chaos

Diana A. Arroyo-Almanza[1], Aaron M. Hagerstrom[2], Thomas E. Murphy[3], and Rajarshi Roy[4(✉)]

[1] Department of Industrial Engineering, Universidad Latina de Mexico, 20742 Celaya, Guanajuato, Mexico
diana_3a@hotmail.com

[2] Communications Technology Laboratory National Institute of Standards and Technology Boulder, Boulder, CO 80305, USA

[3] Department of Electrical and Computer Engineering, Institute for Research in Electronics and Applied Physics, University of Maryland, College Park, MD 20742, USA

[4] Institute for Physical Science and Technology, Department of Physics and Institute for Research in Electronics and Applied Physics, University of Maryland, College Park, MD 20742, USA
rroy@umd.edu

Abstract. Stimulated Brillouin scattering (SBS) limits the power that may be transmitted through an optical fiber because the pump is depleted as energy is transferred into the backward traveling Stokes wave. SBS occurs when the power in the pump wave exceeds a threshold power. The SBS threshold can be easily exceeded in practical contexts (e.g. 4 mW in a typical telecommunication fiber, 25 km in length). The SBS threshold can be increased by increasing the optical bandwidth of the pump wave. In this work, we propose and demonstrate a novel scheme for suppressing stimulated Brillouin scattering in optical fiber. We show that Boolean chaotic phase modulation, which is easily generated with a field-programmable gate array (FPGA), can raise the SBS threshold by >12 dBm.

27.1 Introduction

Stimulated Brillouin scattering (SBS) is a non-linear process in which a forward-traveling (Pump) light wave interacts with a backscattered (Stokes) light wave through an acoustic wave [1,2]. Stimulated Brillouin scattering (SBS) is the most important factor limiting the output power. In a passive fiber, SBS occurs when the product of intensity, fiber length, and Brillouin gain reaches a threshold

V. In et al. (Eds.): *Proceedings of the 5th International Conference on Applications in Nonlinear Dynamics*, Understanding Complex Systems, https://doi.org/10.1007/978-3-030-10892-2_27

value. The SBS threshold depends on the bandwidth of the light and on the material properties of the fiber, attenuation coefficient and its length. The SBS threshold can be increased by increasing the optical bandwidth of the pump wave [1,2]. The power-limiting effect of SBS is undesirable in many contexts, and several methods have been proposed to suppress SBS [3,4]. Many of these techniques focus on increasing the optical bandwidth of the input signal, which in turn increases the SBS threshold [5–8]. In general phase modulation helps to suppress SBS if the modulation signal has a wide electrical bandwidth. We use asynchronous Boolean networks to generate wideband electrical signals.

A Boolean network is a structure of nodes that can be in one of two Boolean states: "1" or "0" and their links are connected to nodes [9,10]. The dynamics of the network is determined by Boolean functions of the Boolean states; this includes the delays in signal propagation and the two methods commonly used are synchronous and autonomous Boolean networks. Synchronous Boolean networks evolve in discrete time steps. These can be experimentally realized using clocked logic circuits. Autonomous Boolean networks evolve in continuous time, experimentally realized with unclocked logic circuits. The processing delays in autonomous Boolean networks originate from processing times of the nodes and propagation delays along the links. Zhang and collaborators find that unlocked logic circuit with circuit elements that function on a timescale on the order of nanoseconds can generate periodic dynamics or deterministic chaotic dynamics depending on the delays in the circuit [11].

Boolean chaos offers several advantages for the suppression of SBS. We implement an experimental autonomous Boolean network. These Boolean networks are implemented with digital electronics, they run asynchronously without any external clock signal. This allows for the generation of higher bandwidth signals compared with clocked devices. Boolean networks can be realized in field-programmable gate arrays (FPGAs), which are easy to reconfigure, allowing for rapid and inexpensive development of experiments [12]. This report describes the applications of Boolean chaos to the suppression of SBS and the main results achieved.

27.2 Experiment and Results

Experiments were performed using the setup shown in Fig. 27.1. We measured the reflected and transmitted power through 25 km of optical fiber. The phase modulator (lithium niobate) has an electrical bandwidth of 10 GHz and an optical insertion loss <3.5 dB. We measured V_π for the phase modulator to be 6.8 V at 50 MHz. The electrical input to the phase modulator is a Boolean chaotic signal, which was generated by a field-programmable gate array (FPGA, Altera Cyclone III). This signal is amplified so that it has a peak-to-peak amplitude of 5.3 V. The modulated light was sent to an optical variable attenuator (VA) to control the input power. Then the laser was launched into a 25 km single-mode fiber by an optical circulator. One arm of the circulator gave us directly the reflected Stokes light (PR) measured by the power meter. The other arm of the circulator allows us to measure the transmitted power (PT) as shown.

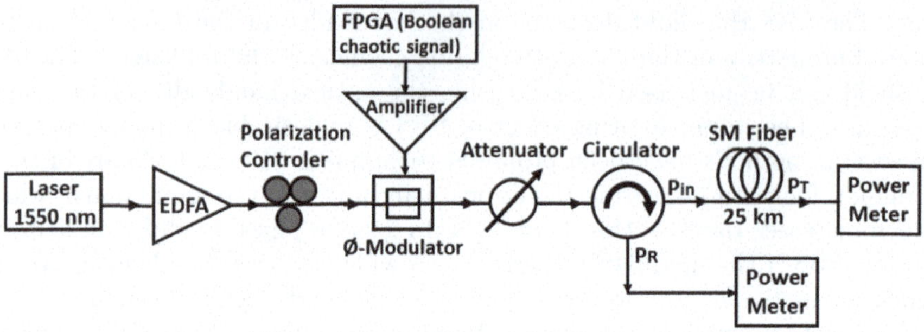

Fig. 27.1. Experimental setup to suppress SBS using Boolean chaos. Transmitted optical power (PT), and the power of the reflected Stokes wave (PR) are both measured. We use an FPGA to generate a Boolean chaotic signal asynchronously

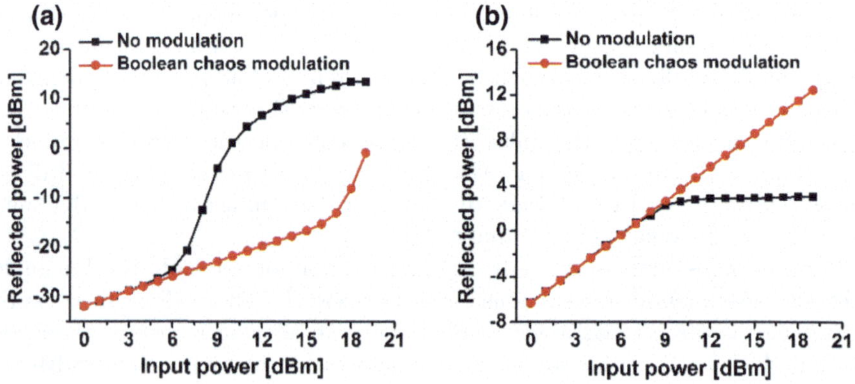

Fig. 27.2. **a** Measured reflected power and **b** transmitted power versus input optical power. Results for unmodulated (red dots) and unmodulated (black diamonds) pump are plotted

In Fig. 27.2a, b we show the measured transmitted and reflected power. In Fig. 27.2a, we can observe that for no phase modulation, the reflected Stokes wave has a low SBS threshold (6 dBm input power). When we phase modulate input light with the Boolean chaos signal, we find that the threshold SBS is increased by ~12 dBm compared with the case where there is no phase modulation.

In Fig. 27.2b, the transmitted power is plotted as a function of the input power. The results show that without phase modulation the output power from the 25 km of optical fiber quickly saturates due to SBS. When we phase modulate the input using the Boolean chaos we see a linear increase in a transmitted power.

27.3 Conclusions

In this work we show that the suppression of SBS depends on the bandwidth of the optical signal injected into the fiber and in the size and bandwidth of the Boolean chaotic phase modulation signal generated using an asynchronously operated FPGA. We show that the threshold for SBS can be raised significantly (by ~12 dB), reducing the reflected power signal and eliminating the saturation in the transmitted power. Boolean chaos signals help to achieve high bandwidth (~1 GHz) and since these signals are generated by asynchronous circuits, the spectrum has no artifacts from a periodic clock. An advantage to the use of FPGAs is that they can be reconfigured easily and can be implemented at a very low cost using digital logic components.

Acknowledgements. Diana A. Arroyo-Almanza gratefully acknowledges postdoctoral fellowship support from the Consejo Nacional de Ciencia y Tecnologia (CONACYT) Mexico. We would also like to thank to the Office of Naval Research for support.

References

1. G.P. Agrawal, *Nonlinear Fiber Optics* (Academic, San Diego, 2007)
2. X. Fu, S.C. Chan, Q. Liu, K.K.-Y. Wong, Appl. Opt. **50**, E92–E96 (2011)
3. M.W. Zmuda, Stimulated Brillouin Scattering (SBS) Suppression Techniques, Technical report, DTIC Document:1–22 (2007)
4. C. Zeringue, I. Dajani, S. Naderi, G.T. Moore, C. Robin, A theoretical study of transient stimulated Brillouin scattering in optical fiber seeded with phase modulated light Opt. Express **20**, 21196–21213 (2012)
5. C.E. Mungan, S.D. Rogers, N. Satyan, J.O. White, Time-dependent modeling of Brillouin scattering in optical fibers excited by a chirped diode laser. IEEE J. Quantum Electron. **48**, 1542–1546 (2012)
6. J.B. Coles, B.-P. Kuo, N. Alic, S. Moro, C.-S. Bres, J. Boggio, P. Andrekson, M. Karlsson, S. Radic, Bandwidth-efficient phase modulation techniques for stimulated Brillouin scattering suppression in fiber optic parametric amplifiers. Opt. Express **18**, 18138–18150 (2010)
7. J.O. White, A. Vasilyev, J.P. Cahill, N. Satyan, O. Okusaga, G. Rakulijic, C.E. Mungan, A. Yariv, Suppression of stimulated Brillouin scattering in optical fibers using a linearly chirped diode laser. Opt. Express **20**, 15872–15881 (2012)
8. J.O. White, D. Engin, M. Akbulut, G. Rakuljic, N. Satyan, A. Vasilyev, A. Yariv, Chirped laser seeding for SBS suppression in a 100-W pulsed erbium fiber amplifier. IEEE J. Quantum Electron. **51**, 6800110–6800120 (2015)
9. C.C. Walker, W.R. Ashby, On temporal characteristics of behavior in certain complex systems. Kybernetik **3**, 100–108 (1965)
10. S.A. Kauffman, Metabolic stability and epigenesis in randomly constructed genetic nets. J. Theor. Biol. **22**, 437–467 (1969)
11. R. Zhang, L.D. de Hugo, S. Cavalcante, Z. Gao, D.J. Gauthier, J.E.S. Socolar, M.M. Adams, D.P. Lathrop, Boolean chaos. Phys. Rev. E **80**, 045202 (2009)
12. D.P. Rosin, *Dynamics of Complex Autonomous Boolean Networks* (Springer International Publishing, 2015)

Chapter 28
The Influence of Entropy on the Classification Performance of a Non-linear Convolutional Neural Network

Iryna Dzieciuch(✉) and Daniel Gebhardt

Space and Naval Warfare Systems Center Pacific, Code 71750,
San Diego, CA 92152-6147, USA
iryna.dzieciuch@navy.mil

Abstract. The influence of noise levels on image classification with neural networks has been studied before. However, little is known about how different levels of entropy affect the performance of non-linear systems such as Convolutional Neural Networks (CNN), where the initial and final system states are predetermined and entropy represents a performance function. This study provides understanding on how a CNN system evolves from the original to the final state and explains the sensitive dependence on initial training conditions using the publicly available architecture and the MNIST dataset and also discusses the effects of entropy on side-scan sonar imagery. This paper describes a method of testing the effects of varying degrees of entropy on the performance of a non-linear neural network system. This approach allows the comparison of performance of the "black box" system under four states: (1) original non-altered dataset with minimal interclass variance, when a CNN trained on an original dataset is tested on images with added levels of entropy, (2) when a CNN trained on a dataset with varying levels of entropy and (3) tested to recognize the original labeled class and (4) to recognize the labeled class with varying levels of entropy. The advantage of this approach is that we can trace the performance of a single architecture CNN under varying levels of entropy, we can demonstrate the ability of the system to use noise to learn more abstract and complex features of the input space, and we can discuss the results in the light of a theoretical physical system.

Keywords: Feature selection convolution neural network (CNN) · Entropy · Accuracy · Information theory · Black box · Physical system

28.1 Introduction

Entropy has many interpretations such as "measurement of order" or "measurement of information". The definition of entropy used in information theory

V. In et al. (Eds.): *Proceedings of the 5th International Conference on Applications in Nonlinear Dynamics*, Understanding Complex Systems, https://doi.org/10.1007/978-3-030-10892-2_28

is directly analogous to the definition used in statistical thermodynamics [1]. Entropy is also a way to describe the number of states of a system. A non-linear system, such as Convolutional Neural Network (CNN) may have many or a few entropy levels. In this experiment we look at the entropy as the amount of uncertainty about a digit image texture associated with a given probability distribution (from 0 to 1) or as a measure of 'disorder' in the image [2]. We hypothesize, that as the level of disorder rises, the entropy rises and the pattern recognition of the digit image should become less likely with a non-linear neural network. For demonstration we can write down entropy as:

$$H(s_m) = -\sum_{n=1}^{256} p_n(s_m) log_2(p_n(s_m)), \quad m = 1, \dots, M \tag{28.1}$$

where, $H(s_m)$ is the entropy of the random variable s_m. Here $p_n(s_m)$ is the probability that outcome s_m happens and m are all the possible outcomes. The probability density pn is calculated using the gray level histogram with levels from 1 to 256.

In image processing, entropy is used to change the view of feature maps textures from each non-linear convolutional layer, a certain texture might have a certain entropy as certain patterns repeat themselves in approximately certain ways. In the context of this paper, low entropy $H(s_m)$ means low disorder, *low variance* within the component m [2]. A component with low entropy is more homogenous than a component with high entropy. Another way of looking at image entropy is to view it as the measure of *information content* ΔI. [2] A vector I with relatively 'low' entropy is a vector with relatively low information content for pixel values [0 1 0 1 1 1 0]. A vector I with relatively 'high' entropy is a vector for pixel values with relatively high information content. It might be [0 242 124 222 149 13] [2]. In this study we examine case with varying information content for a database of digit images size 28 by 28 pixels.

Convolutional neural network use non-linear set filters to process images to derive textures or feature maps that best describe the image class. These textures are then processed with a soft-max function that represents the categorical distribution over K different possible outcomes of the class. We wanted to test the performance of the CNN under different levels of entropy in the input and output space, to see if increasing entropy or information content of the dataset would affect the performance measure of CNN.

28.2 Dataset and CNN Architecture

The MNIST database of handwritten digits is publicly available and popular dataset that is used for testing performance of different pattern recognition algorithms. It contains training set with 60,000 examples of handwritten digits, and a test set of 10,000 examples. The digits have been size-normalized and centered in a fixed-size image Fig. 28.1 (Table 28.1).

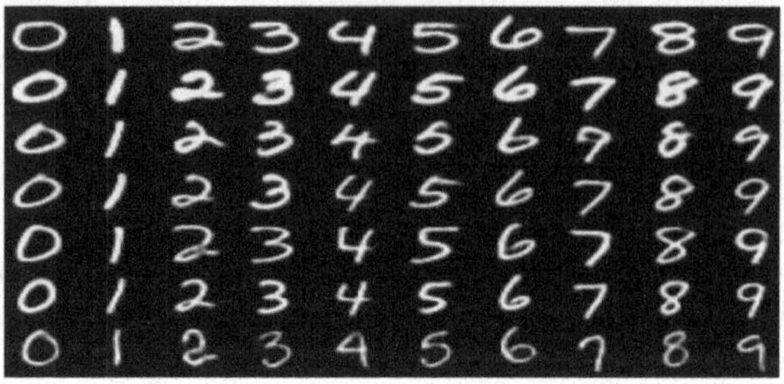

Fig. 28.1. Example of original MNIST dataset. Image shows a subset of the database with 60,000 handwritten digits

Table 28.1. MNIST dataset, each class represent digits from 0 to 9. Each digit class has an original dataset low variance component m, which will be artificially increased by randomly mixing order of the pixels within each class until each class becomes homogeneous

Images	*m*
	0
	0.1
	0.2
	0.3
	0.4
	0.5
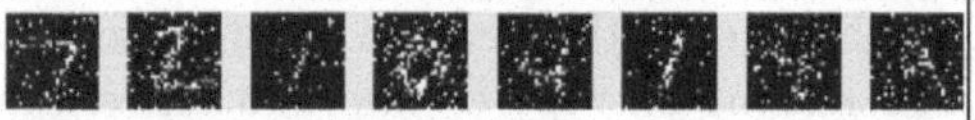	0.6
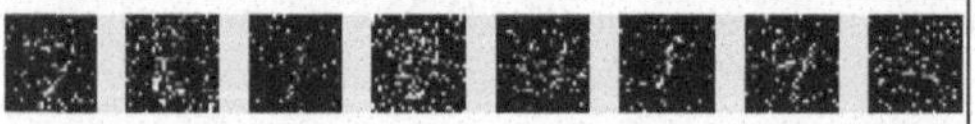	0.7
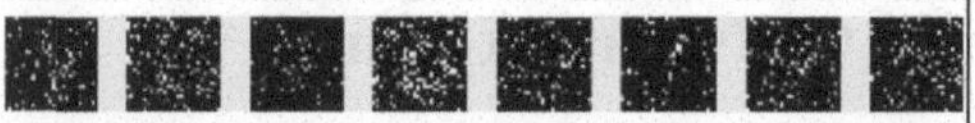	0.8
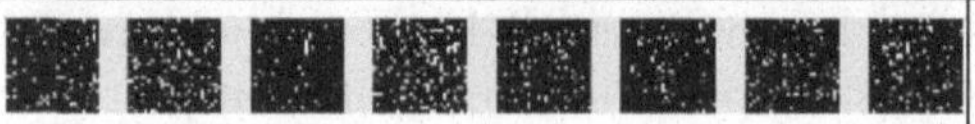	0.9
	1

28.3 CNN Architecture

The CNN configuration was specified as follows and visualized in Fig. 28.2. The input was a 1-dimensional series of 784 values that range between [0, 9]. This corresponds to a normalized 2-dimensional input image of 28×28 pixels. The first convolutional layer uses a set of 20 filters of size 5x5 pixels. Each filter corresponds to a set of weights (5×5 in this case) that are convolved across the image pixels in both the X and Y dimensions, producing an output value for every stride.

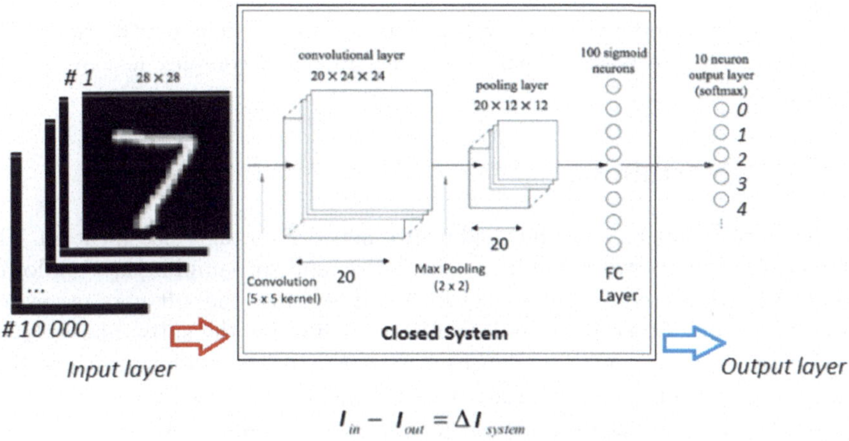

Fig. 28.2. CNN architecture showing 1 layer network of the MNIST image input. Efficiency of the non-linear system to correctly predict image class is represented by ΔI_{system}. Where I_{in} is information context used for training, and I_{out} is information context used for testing

The next layer is a pooling layer, non-overlapping, contiguous 2×2 pooling region, configured to max-pool the activations within a 5×5 pixel window. It effectively subsamples the output, reducing each dimension's resolution by half, choosing the maximum activation value within the window.

It partitions the input image into a set of non-overlapping rectangles and, for each such sub-region, outputs the maximum. The intuition is that the exact location of a feature is less important than its rough location relative to other features [2]. The pooling layer serves to progressively reduce the spatial size of the representation, to reduce the number of parameters and amount of computation in the network, and hence to also control overfitting. It is common to periodically insert a pooling layer between successive convolutional layers in a CNN architecture. The pooling operation provides another form of translation invariance [3].

The final layer is fully-connected (FC) to the pooling layer's output, producing N outputs, where N is the number of classes in the dataset. In this example, the number of classes is 9 (digits 0–9). For a given image example at the CNN input layer, the output will most strongly activate the output neuron in this layer corresponding to the object type it perceives as the most likely classification.

Basic CNN configuration and the MNIST dataset dimensions for this task were chosen for simplicity. For the initial experiment we decided to concentrate on training a CNN network under 4 states: (1) original non-altered dataset with "natural" interclass variance of a dataset; (2) when a CNN trained on an original dataset is tested on images with added levels of entropy; (3) when a CNN trained on a dataset with varying levels of entropy and tested to recognize the original labeled class and (4) to recognize the labeled class with varying levels of entropy. The advantage of this approach is that we can trace the performance of a single architecture CNN under varying levels of entropy, we can demonstrate the ability of the system to use noise to learn more abstract and complex features of the input space.

28.4 CNN Training

We have implemented a convolutional neural network for digit classification. The architecture of the network will be a convolution and subsampling layer followed by a densely connected output layer which will feed into the soft-max regression and cross entropy objective. We used mean pooling for the subsampling layer. We used the back-propagation algorithm to calculate the gradient with respect to the parameters of the model. Finally we trained the parameters of the network with stochastic gradient descent and momentum. Training and test were used with the same parameters except for the change in entropy level [2].

28.5 The Experiment

During the experimentation we would like to evaluate performance of CNN under different entropy levels (0–1) for the same non-linear network with different training and testing configurations.

During first experiment we train with entropy and test with entropy (TRWE/TWE). We have selected to train and test CNN under varying 0–1 entropy levels. Similarly, CNN can be viewed as a closed system where entropy levels $I_{\text{in}} = I_{\text{out}}$.

During second experiment, we have trained the CNN without any introduction of entropy but have tested its performance on images with varying levels of entropy (TRWOE/TWE). Similarly, the CNN can be viewed as a closed system, where $I_{\text{in}} < I_{\text{out}}$.

Finally, we have trained CNN under varying levels of entropy (0-1) and test its performance on unaltered images. (TRWE/TWOE). Similarly, CNN can be viewed as a closed system, where $I_{\text{in}} > I_{\text{out}}$.

Results of the experiment are shown in graph 1. With long enough training, accuracy peaks near 98%. A relatively smaller batch size of 256 images stays constant and runs for 3 epochs yielding high accuracy. Not surprisingly, for all

3 experiments the accuracy of correct digit declines with level of entropy, but at different rates: the CNN with $I_{in} < I_{out}$ gives the worst result. The CNN with $I_{in} = I_{out}$ performs at the same level as all other tests until the entropy level reaches 0.3, and then starts to diverge with up to 20% at entropy level 0.8 with $I_{in} < I_{out}$ and up to 50% with $I_{in} > I_{out}$.

The performance of the CNN with $I_{in} > I_{out}$ showed the highest level of resilience towards inflicted noise when compared to the other two performance curves, reaching 60 and 40% better performance at entropy level 0.8 when compared with other ΔI (Fig. 28.3).

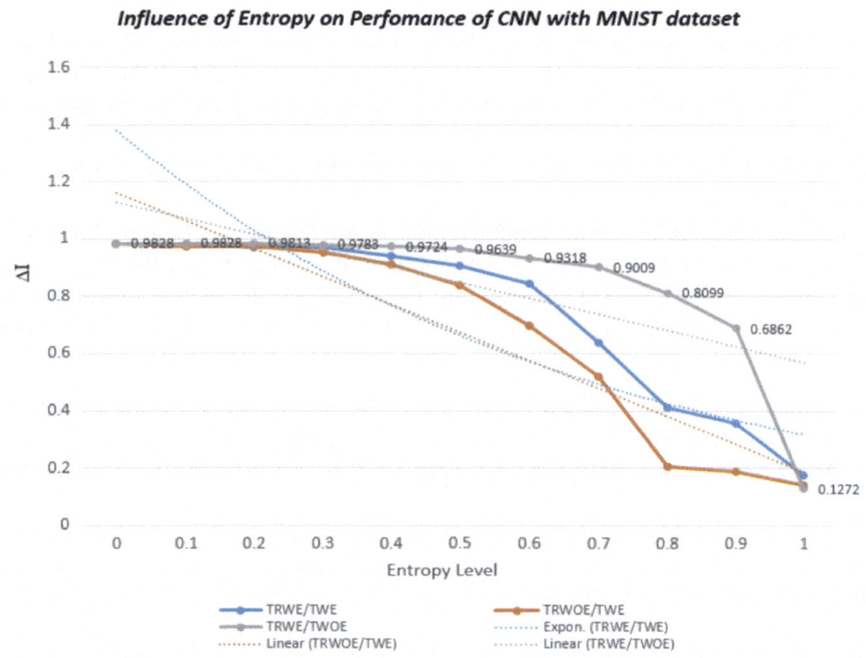

Fig. 28.3. Performance curves ΔI for a single-layer CNN: given all CNNs remain the same, only changing in levels of entropy I_{in} and I_{out} for non-linear convolutional neural network

28.6 Conclusions

This paper describes a method of testing the effects of varying degrees of entropy on the performance of a non-linear neural network system. This approach allows the comparison of performance of the "black box" system under four states, original data set ($I_{in} = I_{out}$), when a CNN trained on an original dataset is tested on images with added levels of entropy ($I_{in} < I_{out}$), when a CNN trained on a dataset with varying levels of entropy and tested to recognize the original labeled class ($I_{\text{in}} > I_{\text{out}}$), when CNN trained on a dataset with varying levels of entropy and, recognize the labeled class with varying levels of entropy ($I_{\text{in}} = I_{\text{out}}$).

A CNN learns a set of shared-parameter non-linear filters to produce feature maps that provide a fully-connected output layer with salient features it uses to determine the class label of the input. From the following graph we can conclude that:

1. The overall performance of the non-linear convolutional neural network ΔI decreases with introduction of entropy to I_{in} and /or I_{out} or to simply put high informational context will negatively alter the probability of class prediction and low informational context will positively alter class prediction up to the point (overfitting problem).
2. By artificially increasing informational context I_{in} into a CNN non-linear network, we can substantially increase the performance ΔI of the non-linear CNN. By altering interclass variability of the training set we can expand available data points, allowing us to *reduce overfitting error*. Appropriate levels of induced noise used during training phase will increase the performance of the CNN non-linear system.
3. Generally speaking, the accumulation of data available for training and testing of neural networks across all ML applications will increase. By introducing upper and lower limits of entropy into the CNN, we can have a better understanding of the behavior for different CNN architectures under varying levels of noise with a more definite method then current "random tuning".

A measure of the unavailability of a CNN non-linear system's energy to do work is similar to the physical system. In nature, for systems to become disordered and for less energy to be available for use as work because of that (Fig. 28.4).

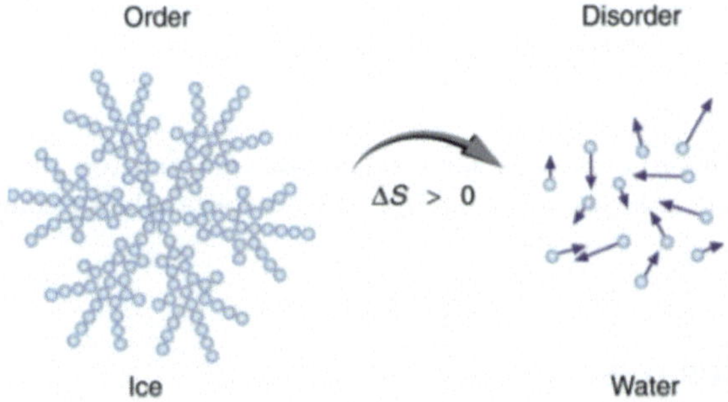

Fig. 28.4. Performance curves ΔI for a single-layer CNN: given all CNNs remain the same, only changing in levels of entropy I_{in} and I_{out} for non-linear convolutional neural network

When ice melts, it becomes more disordered and less structured. The systematic arrangement of molecules in a crystal structure is replaced by a more random and less orderly movement of molecules without fixed locations or orientations. Its entropy increases because heat transfer occurs into it. Energy becomes available. This is a gradual increase in entropy accompanying an increase in disorder [4].

If we use images of snowflakes in the CNN system we would to classify images of:

1. snowflakes among snowflakes (no entropy) ($I_{\text{in}} = I_{\text{out}}$),
2. different stage–melted snowflakes among different stage melted snowflakes (same entropy), ($I_{\text{in}} = I_{\text{out}}$),
3. melted snowflake ($I_{\text{in}} < I_{\text{out}}$) among snowflakes
4. and snowflakes among melted snowflakes ($I_{\text{in}} > I_{\text{out}}$).

If we draw an analogy, that after looking at the representation of 10,000 different types of snowflakes at melted different states, *we are more likely to find a structure of a snowflake among melted ones* then a structure of melted snowflake among structures of melted snowflakes, or even less so melted snowflakes among snowflakes. May it be possible, that there is be innate energy conservation phenomenon for some areas of a snowflake structure that allows to conserve energy, or non-crystalline lattices which carry higher energy potential then others?

Similarly, the mass within the boundary is the information system remains constant and only information energy transfer may take place between the system and its surrounding. A thermodynamic quantity representing the unavailability of a system's thermal energy for conversion into mechanical work, often interpreted as the degree of disorder or randomness in the system. Similarly, high levels of energy in the system represent the unavailability of the system information energy for conversion into pattern recognition work. Information entering non-linear system and information leaving non-linear system define the efficiency of non-linear system.

References

1. Entropy. Wikipedia, Wikimedia Foundation (2018). http://en.wikipedia.org/wiki/Entropy. 11 June 2018
2. D.J.C. MacKay, *Information Theory, Inference and Learning Algorithms* (Cambridge University Press, Cambridge, 2017)
3. I. Dzieciuch et al., *Non-Linear Convolutional Neural Network for Automatic Detection of Mine-Like Objects in Sonar Imagery* (Springer, Dordrecht, 2016). https://doi.org/10.1007/978-3-319-52621-8_27. 28 Aug 2016
4. College Physics Textbook Equity Edition Volume 2 of 3: Chapters 13–24. Google Books. http://books.google.com/books?id=rbBKBgAAQBAJ

Chapter 29
Enhanced Anti-stokes Raman Gain in Nonlinear Waveguides

A. D. Sanchez[1,4], S. M. Hernandez[2], J. Bonetti[2,3], D. F. Grosz[3], and P. I. Fierens[4(✉)]

[1] Instituto Balseiro (IB), Bariloche, Argentina
alfredo.sanchez@ib.edu.ar
[2] IB, Bariloche, Argentina
[3] IB and Consejo Nacional de Investigaciones Científicas y Técnicas (CONICET), Buenos Aires, Argentina
[4] Instituto Tecnológico de Buenos Aires (ITBA) and CONICET, Buenos Aires, Argentina
pfierens@itba.edu.ar

Abstract. We show that, under certain conditions, modulation instability in nonlinear waveguides gives rise to the usual double-sideband spectral structure, but with a Raman gain profile. This process is enabled by the energy transfer from a strong laser pump to both Stokes and anti-Stokes sidebands in a pseudo-parametric fashion. We believe this striking behavior to be of particular value in the area of Raman-based sensors which rely on sensitive measurements of the anti-Stokes component.

29.1 Introduction

Pulse propagation in a lossless nonlinear waveguide is well described by the generalized nonlinear Schrödinger equation (GNLSE) [1]

$$\frac{\partial A(z,T)}{\partial z} - i\hat{\beta}A(z,T) = i\hat{\gamma}A(z,T)\int_{-\infty}^{\infty} R(T')\left|A(z,T-T')\right|^2 dT', \tag{29.1}$$

where $A(z,T)$ is the slowly-varying envelope, z is the spatial coordinate, and T is the time coordinate in a comoving frame at the group velocity. $\hat{\beta}$ and $\hat{\gamma}$ are operators related to the dispersion and nonlinearity, respectively, and are defined by

$$\hat{\beta} = \sum_{m\geq 2} \frac{i^m}{m!}\beta_m \frac{\partial^m}{\partial T^m}, \qquad \hat{\gamma} = \sum_{n\geq 0} \frac{i^n}{n!}\gamma_n \frac{\partial^n}{\partial T^n}. \tag{29.2}$$

β_m are the coefficients of the Taylor expansion of the propagation constant $\beta(\omega)$ around a central frequency ω_0. Similarly, γ_n are the coefficients of the Taylor

V. In et al. (Eds.): *Proceedings of the 5th International Conference on Applications in Nonlinear Dynamics*, Understanding Complex Systems, https://doi.org/10.1007/978-3-030-10892-2_29

expansion of the nonlinear parameter. It is usually sufficient to consider the expansion up to the first term. Under this setting, it can be shown that the total number of photons is conserved if $\gamma_1 = \gamma_0/\omega_0$ [2], which is the usual approximation.

The function $R(T)$ models the Raman response of the medium. Stimulated Raman scattering is a non-parametric process that involves the excitation of molecular vibration modes of the waveguide and it does not conserve the energy of the wave. However, it does conserve the number of photons. Qualitatively speaking, the energy exchange experienced by a strong continuous-wave laser involves the annihilation of a pump photon and the simultaneous creation of another photon in a low-frequency (also known as Stokes) band. Similarly, a photon in a high-frequency (anti-Stokes) band is annihilated and another photon is created at the pump frequency. As a result, a *gain is observed only in the Stokes band*, enabling the application of stimulated Raman scattering in optical amplification [3].

First order linear perturbation analysis of the GNLSE reveals that, under certain conditions (*viz.*, anomalous dispersion), continuous-wave (CW) solutions are unstable. This phenomenon, known as modulation instability (MI) [4–11], is a parametric process where two photons from a CW pump are transferred to both low- and high-frequency bands, one photon each. As a result, *MI gain is observed in both sides of the pump.* It has been shown [12,13] that, when γ_1 is included, there is a power cutoff above which the MI gain vanishes.

In a recent work [14], we proved that there is still gain beyond the MI power cutoff when Raman scattering is taken into account. Moreover, we showed that the gain mimics the shape of the Raman response in the Stokes band. Here we extend these observations to the anti-Stokes band. Indeed, in the next section we show that there is *MI gain in both sides of the pump with a Raman spectral shape.* Further, we show this to be a pseudo-parametric process, that is, *a truly MI-like process where the anti-Stokes gain is not the result of one mediated by spectral generation in the Stokes band followed by four-wave-mixing generation in the anti-Stokes band.*

29.2 Raman and Modulation Instability

A few simulations may help to understand the behavior of stimulated Raman scattering. Figure 29.1 shows simulation results of an average over 50 noise realizations of a CW pump with additive white Gaussian noise. The signal was propagated a distance L_{R}, defined as the inverse of the peak Raman gain, in a normal dispersion regime. In particular, $\beta_2 = 50\,\mathrm{ps}^2/\mathrm{km}$, $\beta_m = 0$ for $m > 2$, $\gamma_0 = 100\,1/\mathrm{W/km}$, $\gamma_1/\gamma_0 = \omega_0^{-1}$. The pump power was set to $P_0 = 50\,\mathrm{W}$, its frequency to $\omega_0/2\pi = 376.73\,\mathrm{THz}$, and the signal-to-noise power ratio was 50 dB. For the Raman response [1], we used $R(T) = (1 - f_R)\,\delta(T) + f_R h_R(T)$, where f_R weights the contributions of the instantaneous (electronic) and delayed Raman response of the medium. We used the damped-oscillator approximation $h_R(t) \propto e^{-t/\tau_2} \sin(t/\tau_1)\,\Theta(t)$, where $\Theta(t)$ is the unit step function. We fixed $f_R = 0.031$, $\tau_1 = 15.5\,\mathrm{fs}$ and $\tau_2 = 230.5\,\mathrm{fs}$.

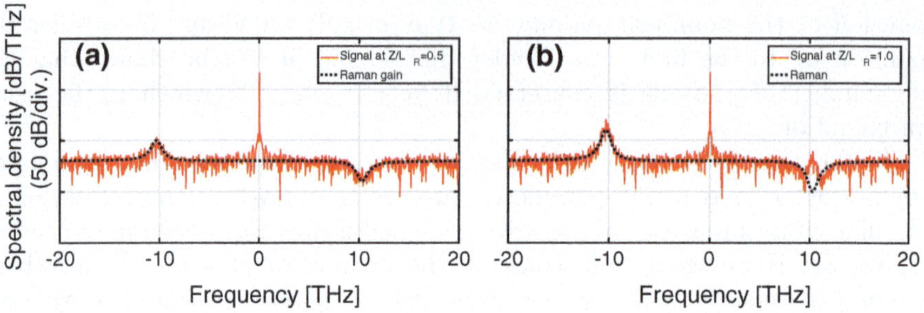

Fig. 29.1. Simulation results of an average over 50 noise realizations of a CW pump with additive white Gaussian noise: **a** at a propagated a distance $L_{\mathrm{R}}/2$; **b** at L_{R}. L_{R} is defined as the inverse of the peak Raman gain. The shape of the theoretical Raman gain (black dashed line) is also presented for comparison

Figure 29.1a shows the spectral density at $L_{\mathrm{R}}/2$, as a function of frequency deviations with respect to ω_0. We observe that noise in the Stokes band (negative frequencies) grows following the Raman gain as expected. However, in the anti-Stokes band noise decreases as photons are annihilated and new photons are created at the pump frequency. In Fig. 29.1b, after the signal propagates the remaining distance, it can be observed the growth of the anti-Stokes band through a third-order parametric process known as four-wave mixing (FWM). FWM involves the interaction between two pump photons with a Stokes and an anti-Stokes photon. In this sense, modulation instability (in the absence of Raman) is usually regarded as a four-wave mixing process.

A complete analysis of modulation instability includes the complex interplay between high-order dispersion, nonlinearity, and Raman scattering (see, e.g., [15,16]). For the sake of simplicity, let us consider the case where $\beta_m = 0$ for $m > 2$ and $\gamma_n = 0$ for $n > 1$. It can be shown that the MI gain is given by [17]

$$g_{\mathrm{MI}}(\Omega) = 2\max\{-\mathrm{Im}\{K_1(\Omega)\}, -\mathrm{Im}\{K_2(\Omega)\}, 0\}, \tag{29.3}$$

$$K_{1,2}(\Omega) = \frac{p|\beta_2|}{\tau}\Omega(1 + \tilde{R}(\Omega)) \pm |\beta_2\Omega|\sqrt{\frac{\Omega^2}{4} - \frac{p\tilde{R}(\Omega)}{\tau^2} + \frac{p^2\tilde{R}^2(\Omega)}{\tau^2}}, \tag{29.4}$$

where Ω is the deviation from the pump frequency ω_0 and $\tilde{R}(\Omega)$ is the Fourier transform of the Raman response $R(T)$. For convenience, γ_1 and the pump power P_0 have been normalized as $\tau = \gamma_1/\gamma_0$ and $p = P_0/P_{\mathrm{c}}$, with

$$P_{\mathrm{c}} = \frac{|\beta_2|\gamma_0}{\gamma_1^2}. \tag{29.5}$$

In the absence of Raman scattering, $\tilde{R}(\Omega) = 1$. In this case, it is easy to verify that there is no gain when $p > 1$, that is, when the pump power P_0 is beyond the power cutoff P_{c}. However, in the presence of Raman scattering ($\tilde{R}(\Omega) \neq 1$), there exists MI gain even for $p > 1$.

In order to understand the nature of the processes involved, it is convenient to study the number of photons, a quantity conserved when $\tau = \omega_0^{-1}$, as it was already explained. Let us define the quantity

$$\Psi(\Omega) = \frac{|A(z,\Omega)|^2}{\hbar\,(\Omega + \omega_0)}, \tag{29.6}$$

which is proportional to the number of photons at frequency Ω. Figure 29.2a shows simulation results for the propagation of a pump and two seeds located at the Stokes and anti-Stokes frequencies ($\mp$10.7 THz) under the same setting as that of Fig. 29.1 (both seeds have the same number of photons at $z = 0$.) It is observed that initially the number of photons at the anti-Stokes frequency decreases as a consequence of Raman scattering, and then begins to increase (at a distance $z \sim 0.4\,L_{\mathrm{R}}$) due to FWM. Figure 29.2b–c show the evolution of the same quantity in a purely parametric process such as MI in the absence of Raman. The normalized pump power is $p = 0.8$, the fiber dispersion is anomalous, $\beta_2 = -50\,\mathrm{ps}^2/\mathrm{km}$, and γ_0 and ω_0 are as in Fig. 29.1. The propagated distance is the characteristic MI length, defined as $L_{\mathrm{MI}} = \max(g_{\mathrm{MI}}^{-1})$. In Fig. 29.2b, $\gamma_1 = 0$ and, given that the number of photons is not conserved, seeds grow unevenly. On the contrary, in Fig. 29.2c, $\gamma_1 = \gamma_0/\omega_0$ and both seeds grow evenly.

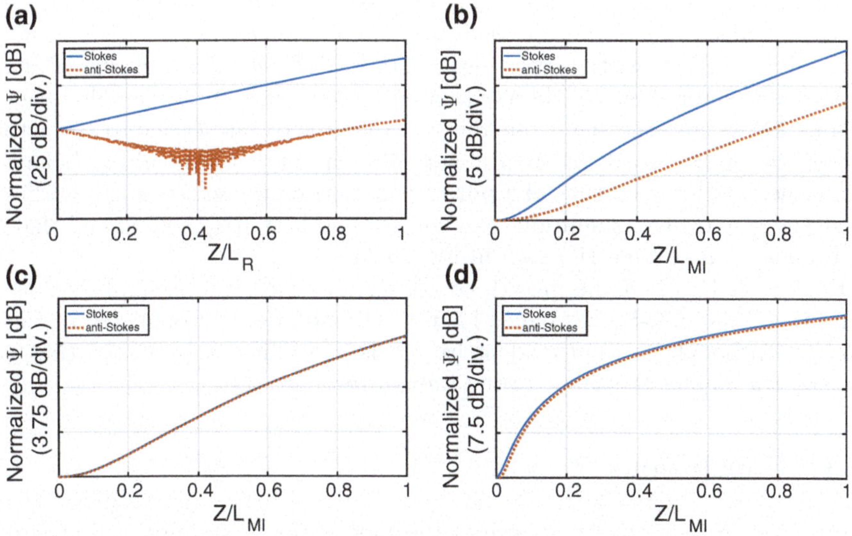

Fig. 29.2. Number of photons vs. normalized propagated distance for a pump and two seeds located at the Stokes and anti-Stokes frequencies ($\mp$10.7 THz): **a** results for the normal dispersion regime; **b** anomalous dispersion regime with $p = 0.8$, $\gamma_1 = 0$ and no Raman scattering; **c** anomalous dispersion regime with $p = 0.8$, $\gamma_1 = \gamma_0/\omega_0$ and no Raman scattering; **d** anomalous dispersion regime with $p = 1.1$, $\gamma_1 = \gamma_0/\omega_0$ and Raman scattering

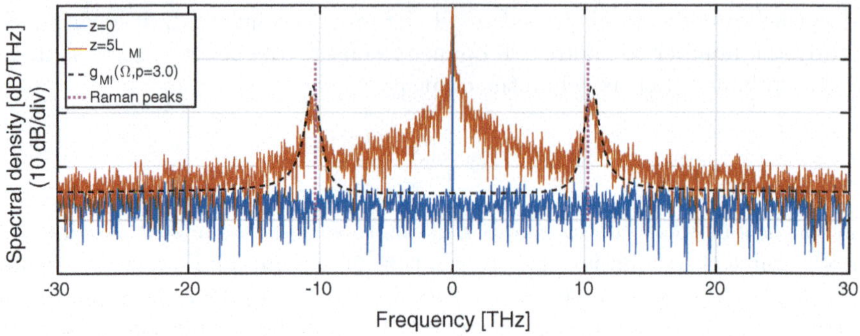

Fig. 29.3. Noise growth on the Stokes and anti-Stokes bands beyond the power cutoff ($p = 3.0$). Raman peaks at the Stokes and anti-Stokes at ∓ 10.7 THz are shown (dotted line). The MI-gain spectrum (black dashed curve) is also shown for comparison

If the effect of Raman scattering is included, MI gain cannot be the result of a purely parametric process. Recall Fig. 29.2a where the pump is shown to contribute photons only to the Stokes band, as it is the case with conventional Raman (non-parametric) amplification, and eventually the anti-Stokes band is amplified by means of a FWM interaction between the pump and the Stokes sideband. However, in the anomalous dispersion region of the waveguide, *we can have Raman amplification at both low- and high-frequencies simultaneously.* Indeed, Fig. 29.2d shows the evolution of both seeds for $p = 1.1$ and when Raman scattering is factored in. We observe that both seeds grow almost simultaneously (*cf.* Fig. 29.2a), and the slight difference in the growth rate is due to the actual gain of the Stokes band due to Raman. We may view the resulting behavior as intermediate between that of a purely parametric process, such as Fig. 29.2c, where the gain evolves simultaneously for low and high frequencies, and that of the Raman (non-parametric) gain in Fig. 29.2a.

Finally, in Fig. 29.3 the growth of noise shows clearly the amplification of both Stokes and anti-Stokes bands for $p = 3.0$ and after a propagated distance of $5\,L_{\mathrm{MI}}$. Although it is not evident from this figure, it can be shown that the gain spectra mimics the shape of the Raman response [14].

29.3 Conclusions

In this work we showed that beyond the modulation instability power cutoff nonlinear waveguides exhibit a gain with a Raman-like spectral shape. Inclusion of the higher-order nonlinear term γ_1 allows for the growth of both Stokes and anti-Stokes bands to be even and simultaneous, conserving the number of photons, as if in the presence of a pseudo-parametric process. As such, the nonlinear waveguide exhibits Raman gain in the anti-Stokes band, a striking feature that could find applications in the sensitivity enhancement of a wide variety of Raman sensors that rely on the monitoring of the anti-Stokes spectral component.

Acknowledgements. We gratefully acknowledge financial support from ONR Global through the Visiting Scientists Program.

References

1. G. Agrawal, *Nonlinear Fiber Optics*, 5th edn. Optics and Photonics (Academic Press, London, 2012)
2. K. Blow, D. Wood, IEEE J. Quantum Electron. **25**(12), 2665 (1989). https://doi.org/10.1109/3.40655
3. M. Ikeda, Opt. Commun. **39**(3), 148 (1981). https://doi.org/10.1016/0030-4018(81)90044-4
4. A. Hasegawa, W. Brinkman, IEEE J. Quantum Electron. **16**(7), 694 (1980). https://doi.org/10.1109/JQE.1980.1070554
5. K. Tai, A. Hasegawa, A. Tomita, Phys. Rev. Lett. **56**, 135 (1986). https://doi.org/10.1103/PhysRevLett.56.135
6. A. Demircan, U. Bandelow, Opt. Commun. **244**(1), 181 (2005)
7. J.M. Dudley, G. Genty, F. Dias, B. Kibler, N. Akhmediev, Opt. Express **17**(24), 21497 (2009). https://doi.org/10.1364/OE.17.021497
8. D. Solli, C. Ropers, P. Koonath, B. Jalali, Nature **450**(7172), 1054 (2007)
9. K. Hammani, C. Finot, B. Kibler, G. Millot, IEEE Photonics J. **1**(3), 205 (2009). https://doi.org/10.1109/JPHOT.2009.2032150
10. N. Akhmediev, J.M. Soto-Crespo, A. Ankiewicz, Phys. Rev. A **80**, 043818 (2009). https://doi.org/10.1103/PhysRevA.80.043818
11. S.T. Sørensen, C. Larsen, U. Møller, P.M. Moselund, C.L. Thomsen, O. Bang, J. Opt. Soc. Am. B **29**(10), 2875 (2012). https://doi.org/10.1364/JOSAB.29.002875
12. P.K. Shukla, J.J. Rasmussen, Opt. Lett. **11**(3), 171 (1986). https://doi.org/10.1364/OL.11.000171
13. C.D. Angelis, G. Nalesso, M. Santagiustina, J. Opt. Soc. Am. B **13**(5), 848 (1996). https://doi.org/10.1364/JOSAB.13.000848
14. A.D. Sánchez, S.M. Hernandez, J. Bonetti, P.I. Fierens, D.F. Grosz, J. Opt. Soc. Am. B **35**(1), 95 (2018). https://doi.org/10.1364/JOSAB.35.000095
15. P. Béjot, B. Kibler, E. Hertz, B. Lavorel, O. Faucher, Phys. Rev. A **83**, 013830 (2011). https://doi.org/10.1103/PhysRevA.83.013830
16. J. Bonetti, S.M. Hernandez, P.I. Fierens, D.F. Grosz, Phys. Rev. A **94**, 033826 (2016). https://doi.org/10.1103/PhysRevA.94.033826
17. S.M. Hernandez, P.I. Fierens, J. Bonetti, A.D. Snchez, D.F. Grosz, IEEE Photonics J. **9**(5), 1 (2017). https://doi.org/10.1109/JPHOT.2017.2754984

Chapter 30
Intrinsic Localized P-Mode in Forced Nonlinear Oscillator Array

Edmon Perkins[1(✉)] and Timothy Fitzgerald[2]

[1] Auburn University, 354 War Eagle Way, Auburn, AL 36849, USA
edmon@auburn.edu

[2] Gonzaga University, 502 E. Boone Ave., Spokane, WA 99258, USA
fitzgeraldt@gonzaga.edu

Abstract. Intrinsic localized modes (ILMs) are energy localizations that may occur in arrays of discrete, nonlinear oscillators. When present in physical systems, these energy localizations may cause undesirable dynamics or damaging effects. If properly understood, ILMs may be used to increase the sensing capacity of inertial sensors, store information, or move energy through an array. Depending on the system parameters, ILMs may have a variety of profiles (e.g., the symmetric ST-mode or the antisymmetric P-mode). Using the method of restricted normal modes, a displacement profile is calculated for the P-mode. After performing numerical simulations using the P-mode profile as initial conditions, the P-mode is found to be persistent when forced at 3 times the linear natural frequency. Although persistent, this P-mode ILM is found to have chaotic properties. This ILM may have been previously overlooked because of its positive Lyapunov exponent, meaning that there might be larger ranges of parameters capable of supporting these energy localizations.

30.1 Introduction

Intrinsic localized modes (also called discrete breathers in the physics literature) are localized vibratory modes involving a small number of oscillators in an array, and they may occur in spatially extended, perfectly periodic, discrete systems [1]. These energy localizations have been observed in a range of physical systems [2], including antiferromagnets [3], Josephson junctions [4–6], photonic lattices [7,8], and even biopolymer chains [9]. They have been realized in both microscale [10] and macroscale oscillator arrays [11,12].

These localizations may occur in different mode shapes, such as the Seivers-Takeno (ST-) mode [13] and the Page (P-) mode [14]. The ST-mode may be considered to be a forced nonlinear vibratory mode [15]. In addition, some stability studies have been performed on these modes [16–18].

In this paper, the P-mode ILM will be studied. The rest of the paper is organized as follows. First, the mode shape will be determined through the

V. In et al. (Eds.): *Proceedings of the 5th International Conference on Applications in Nonlinear Dynamics*, Understanding Complex Systems, https://doi.org/10.1007/978-3-030-10892-2_30

method of restricted normal modes for the unforced system in Sect. 30.2. Next, the effects of adding forcing will be studied in Sect. 30.3. A discussion of the results will be presented in Sect. 30.4.

30.2 Restricted Normal Mode Analysis

The oscillator array under consideration has both nonlinear nominal stiffness and nonlinear coupling. The equation of motion for the ith oscillator may be written as

$$\begin{aligned} &\ddot{x}_i + c\dot{x}_i + \alpha_1 x_i + \beta_1 x_i^3 + \alpha_2(x_i - x_{i+1}) \\ &+ \alpha_2(x_i - x_{i-1}) + \beta_2(x_i - x_{i+1})^3 + \beta_2(x_i - x_{i-1})^3 = Fcos(\Omega t) \end{aligned} \tag{30.1}$$

In Eq. 30.1, c is the damping, α_1 is the linear onsite stiffness, α_2 is the linear intersite stiffness, β_1 is the nonlinear onsite stiffness, β_2 is the nonlinear intersite stiffness, F is the forcing amplitude, and Ω is the forcing frequency.

To perform a restricted normal mode analysis of the P-mode, several initial assumptions must be made. Previously, the authors performed this analysis on the ST-mode, which is symmetric [19]. In this case, the center of the ILM was located at the 0th oscillator, and the oscillators to the left and right of the center were identical due to the symmetry condition. Proceeding in a similar fashion for the P-mode ILM, the center of the ILM is located between oscillators +1 and −1, and these oscillators are of equal and opposite amplitude. Two further assumptions are that oscillators +2 and −2 are also of equal and opposite amplitude, and oscillators +3 and −3 are equal to zero. With these assumptions, the solution of the restricted normal mode analysis is ensured to be the anti-symmetric P-mode ILM.

Now, setting the forcing and damping terms equal to zero, Eq. 30.1 for oscillators 1 and 2 becomes

$$\begin{cases} \ddot{x}_1 + \alpha_1 x_1 + \beta_1 x_1^3 + \alpha_2(x_1 - x_2) + \alpha_2(x_1 - x_{-1}) \\ \quad + \beta_2(x_1 - x_2)^3 + \beta_2(x_1 - x_{-1})^3 = 0 \\ \ddot{x}_2 + \alpha_1 x_2 + \beta_1 x_2^3 + \alpha_2(x_2 - x_3) + \alpha_2(x_2 - x_1) \\ \quad + \beta_2(x_2 - x_3)^3 + \beta_2(x_2 - x_1)^3 = 0 \end{cases} \tag{30.2}$$

And after enforcing the assumptions stated earlier and rearranging, Eq. 30.2 becomes

$$\begin{cases} \ddot{x}_1 + \alpha_1 x_1 + (\beta_1 + 8\beta_2)x_1^3 + \alpha_2(3x_1 - x_2) + \beta_2(x_1 - x_2)^3 = 0 \\ \ddot{x}_2 + \alpha_1 x_2 + (\beta_1 + \beta_2)x_2^3 + \alpha_2(2x_2 - x_1) + \beta_2(x_2 - x_1)^3 = 0 \end{cases} \tag{30.3}$$

Then, assuming that the two central oscillators and the oscillators directly adjacent to them all respond with the same frequency, the assumed solution is

$$\begin{aligned} x_1(t) &= A\cos(\omega t) \\ x_2(t) &= B\cos(\omega t) \end{aligned} \tag{30.4}$$

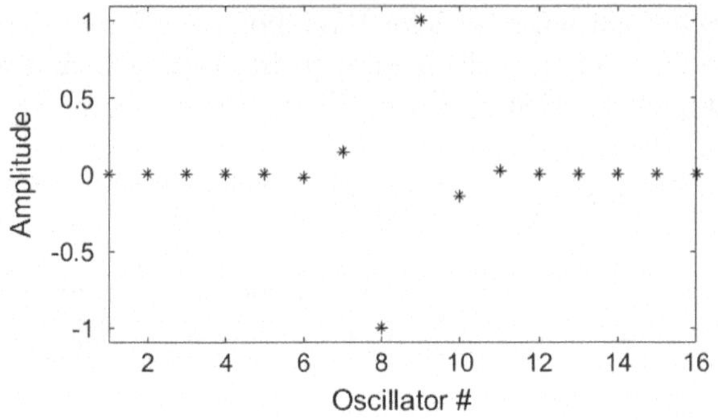

Fig. 30.1. By utilizing the restricted normal mode approach, an ILM profile was found for an array of sixteen oscillators. This profile was used as the initial conditions in Sect. 30.3

After substituting Eq. 30.4 into Eq. 30.3, ω^2 may be eliminated. To enforce that A and B are a half period out-of-phase, they are represented as

$$\begin{aligned} A &= R\cos(\theta) \\ B &= R\sin(\theta) \end{aligned} \tag{30.5}$$

where R^2 is the total energy of the system and $p = \frac{B}{A} = tan(\theta)$. Substituting Eqs. 30.4 and 30.5 into Eq. 30.3, the following polynomial may be found:

$$p^4 + \left(\frac{\alpha_2 - R^2\beta_1 + R^2\beta_2}{-\alpha_2 - R^2\beta_2}\right)p^3 + (0)p^2 + \left(\frac{\alpha_2 + R^2\beta_1 + 6R^2\beta_2}{-\alpha_2 - R^2\beta_2}\right)p + \left(\frac{\alpha_2 + R^2\beta_2}{-\alpha_2 - R^2\beta_2}\right) = 0 \tag{30.6}$$

In order to obtain a profile for the P-mode ILM, R and θ are found such that $A = 1$ by solving $Rcos(arctan(p)) = A = 1$, where p is a root of Eq. 30.6. Choosing $\alpha_1 = 1$, $\alpha_2 = 0$, $\beta_1 = 1$, and $\beta_2 = 1$, the P-mode ILM profile presented in Fig. 30.1 was calculated for an array of sixteen oscillators.

30.3 Effects of Forcing

Using the ILM profile obtained from the restricted normal mode approach (Fig. 30.1) as initial conditions, this system was numerically integrated in MATLAB. For the unforced case, the ILM is stable. However, the central oscillators respond with a different frequency than the non-central oscillators, due to the large nonlinearity of the system. The response and its Fast Fourier Transform are presented in Fig. 30.2.

By setting $F = 0.15$ and $c = 0.001$ in the simulations and using the same initial conditions as in Fig. 30.1, the ILM is still present. In this case, the frequency of maximal power (as determined from the FFT) for each oscillator is now approximately the same, as expected with a forced system. The simulation results are presented in Fig. 30.3.

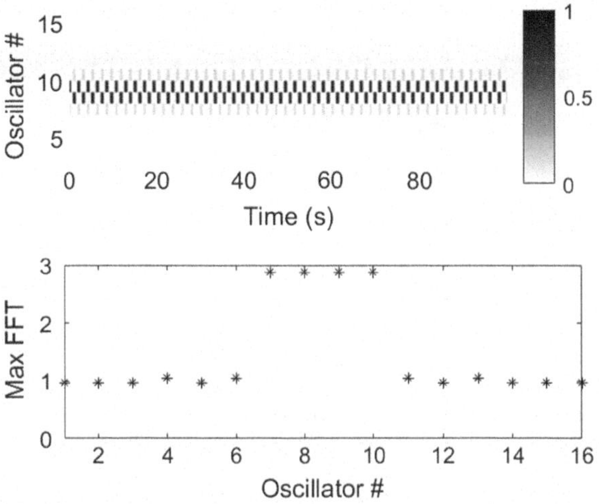

Fig. 30.2. Without forcing or damping, the ILM is persistent. However, with the large amount of nonlinearity, the oscillators participating in the ILM oscillate with a frequency much higher than the other oscillators

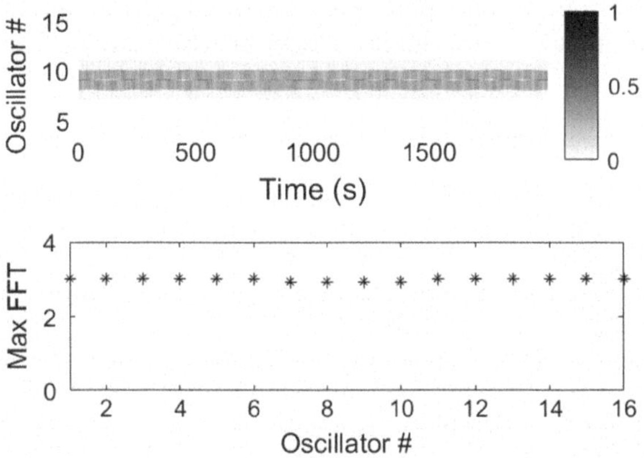

Fig. 30.3. With forcing and damping, the ILM is still obtained in the simulations. With forcing, the maximum frequency for each oscillator is approximately the same

Although the ILM in Fig. 30.3 is quite persistent, the difference in the central oscillators' amplitude has chaotic attributes. This difference by adding the amplitudes of x_9 and x_8, and the results are shown in Fig. 30.4. In this figure, it appears that the difference in these peak values is random, and moreover, the FFT of this difference has broadband characteristics. Further, it was found that the largest Lyapunov exponent for the time series in Fig. 30.4 was $\lambda_1 = 0.15 > 0$, as calculated in the method described in [20]. For these reasons, it appears that although this forced P-mode ILM is persistent, it is also chaotic.

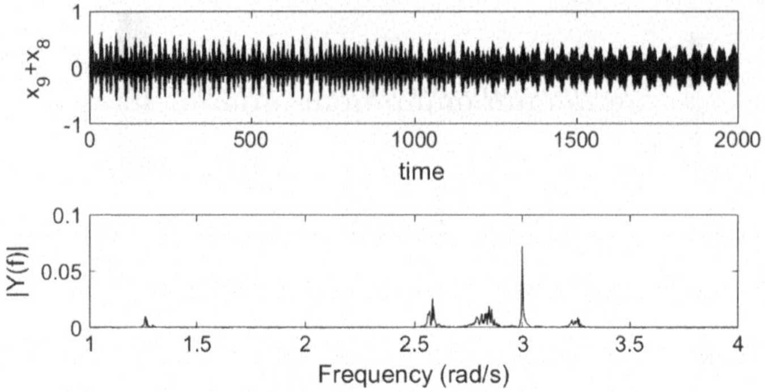

Fig. 30.4. With forcing and damping, the ILM is still persistent. With forcing, the maximum frequency for each oscillator is approximately the same

30.4 Conclusions

Although typically undesirable because of their damaging effects, ILMs could lead to technological innovation in the realms of sensing, computation, and energy transportation. While some work pertaining to stability has been performed for ILMs, this paper exhibits a case in which an ILM is persistent yet chaotic.

References

1. P.W. Anderson, Absence of diffusion in certain random lattices. Phys. Rev. **109**(5), 1492 (1958)
2. S. Flach, A.V. Gorbach, Discrete breathers–advances in theory and applications. Phys. Rep. **467**(1–3), 1–116 (2008)
3. M. Sato, A. Sievers, Direct observation of the discrete character of intrinsic localized modes in an antiferromagnet. Nature **432**(7016), 486 (2004)
4. A. Ustinov, Solitons in Josephson junctions. Phys. D: Nonlinear Phenom. **123**(1–4), 315–329 (1998)
5. A. Ustinov, Imaging of discrete breathers. Chaos: Interdiscip. J. Nonlinear Sci. **13**(2), 716–724 (2003)
6. P. Binder, D. Abraimov, A. Ustinov, S. Flach, Y. Zolotaryuk, Observation of breathers in Josephson ladders. Phys. Rev. Lett. **84**(4), 745 (2000)
7. J.W. Fleischer, M. Segev, N.K. Efremidis, D.N. Christodoulides, Observation of two-dimensional discrete solitons in optically induced nonlinear photonic lattices. Nature **422**(6928), 147 (2003)
8. S.F. Mingaleev, Y.S. Kivshar, R.A. Sammut, Long-range interaction and nonlinear localized modes in photonic crystal waveguides. Phys. Rev. E **62**(4), 5777 (2000)
9. S.F. Mingaleev, Y.B. Gaididei, P.L. Christiansen, Y.S. Kivshar, Nonlinearity-induced conformational instability and dynamics of biopolymers. EPL (Europhys. Lett.) **59**(3), 403 (2002)
10. M. Sato, B. Hubbard, L.Q. English, A. Sievers, B. Ilic, D. Czaplewski, H. Craighead, Study of intrinsic localized vibrational modes in micromechanical oscillator arrays. Chaos: Interdiscip. J. Nonlinear Sci. **13**(2), 702–715 (2003)

11. E. Perkins, M. Kimura, T. Hikihara, B. Balachandran, Effects of noise on symmetric intrinsic localized modes. Nonlinear Dyn. **85**(1), 333–341 (2016)
12. M. Kimura, T. Hikihara, Coupled cantilever array with tunable on-site nonlinearity and observation of localized oscillations. Phys. Lett. A **373**(14), 1257–1260 (2009)
13. A. Sievers, S. Takeno, Intrinsic localized modes in anharmonic crystals. Phys. Rev. Lett. **61**(8), 970 (1988)
14. J. Page, Asymptotic solutions for localized vibrational modes in strongly anharmonic periodic systems. Phys. Rev. B **41**(11), 7835 (1990)
15. A. Dick, B. Balachandran, C. Mote, Intrinsic localized modes in microresonator arrays and their relationship to nonlinear vibration modes. Nonlinear Dyn. **54**(1–2), 13–29 (2008)
16. S. Bickham, S. Kiselev, A. Sievers, Stationary and moving intrinsic localized modes in one-dimensional monatomic lattices with cubic and quartic anharmonicity. Phys. Rev. B **47**(21), 14206 (1993)
17. K. Sandusky, J. Page, Interrelation between the stability of extended normal modes and the existence of intrinsic localized modes in nonlinear lattices with realistic potentials. Phys. Rev. B **50**(2), 866 (1994)
18. M. Kimura, T. Hikihara, Stability change of intrinsic localized mode in finite nonlinear coupled oscillators. Phys. Lett. A **372**(25), 4592–4595 (2008)
19. B. Balachandran, E. Perkins, T. Fitzgerald, Response localization in micro-scale oscillator arrays: influence of cubic coupling nonlinearities. Int. J. Dyn. Control **3**(2), 183–188 (2015)
20. M.T. Rosenstein, J.J. Collins, C.J. De Luca, A practical method for calculating largest lyapunov exponents from small data sets. Phys. D: Nonlinear Phenom. **65**(1–2), 117–134 (1993)

Chapter 31
Bifurcation Analysis of Spin-Torque Nano Oscillators Parallel Array Configuration

Brian Sturgis-Jensen[1(✉)], Antonio Palacios[1], Patrick Longhini[2], and Visarath In[2]

[1] San Diego State University, San Diego, CA 92182, USA
bsturgisjensen@sdsu.edu
[2] Space and Naval Warfare Systems Center Pacific, Code 71740, 53560 Hull Street, San Diego, CA 92152-5001, USA

Abstract. The ability for a Spin Torque Nano Oscillator (STNO) to perform as a nano-scaled microwave voltage oscillator continues to be the focus of exten- sive research. Due to their small size (on the order of 10 nm), low power consumption, and ultrawide frequency range STNOs demonstrate significant potential for applications in microwave generation. To date, the ability for a STNO to produce microwave signals is achievable, however, the low power output produced by a single STNO currently renders them inoperable for applications. In response, various groups have proposed the synchronization of a network of STNOs such that the coherent signal produces a strong enough microwave signal at the nanoscale. Achieving synchronization, however, has proven to be a challenging task and raises complex problems related to the field of Nonlinear Dynamical Systems. In this work we analyze the problem of synchronization for networks of STNOs connected in parallel. Bifurcation diagrams for small networks of STNOs are computed which depicts bistability between in-phase and out-of- phase limit cycle oscillations for much of the phase space. In order to extend the analysis for large networks of STNOs, we exploit the S_N symmetry ex- hibited by the system all-to-all coupled STNOs. We develop implicit analytic expressions for Hopf bifurcations which yield synchronized limit cycle oscil- lations, allowing for the computation of the Hopf loci for an arbitrarily large network of oscillators. Through stability analysis we determine the parameter space for which the Hopf bifurcation is supercritical and exhibits a stable center-manifold. This analysis is completed for large arrays and used to nu- merically demonstrate synchronization in up to $N = 1000$ STNOs. These results should help guide future experiments and, eventually, lead to the design and fabrication of a nanoscale microwave signal generator.

V. In et al. (Eds.): *Proceedings of the 5th International Conference on Applications in Nonlinear Dynamics*, Understanding Complex Systems, https://doi.org/10.1007/978-3-030-10892-2_31

31.1 Introduction

An elementary Spin Torque Nano Oscillator (STNO) consists of two ferromagnetic layers separated by a nonferromagnetic spacer, see Fig. 31.1 (Left). In one ferromagnetic layer, the magnetization vectors are held fixed whereas the second ferromagnetic layer remains free in order to the Giant Magnetoresistive (GMR) effect. Under the influence of a biased current and applied magnetic field the free layer may exhibit steady state precessional motion. In turn, such dynamics will generate an oscillating resistance which, by Ohm's Law, the voltage across the resistor must also oscillate as well, thus yielding a microwave voltage oscillator.

Perspective advantages of STNOs are their small size (on the order of 100 nm), broad tunable frequency range, small output linewidth, and low power consumption [12]. These benefits establish STNOs as desirable commercial products for the many fields which utilize microwave voltage oscillators, e.g., wireless devices, radar, air traffic control, weather forecasting, and navigation systems. However, the power output measured in experiments are still an order of magnitude short of what is required with on-chip GHz applications [6]. A promising solution, as proposed by various group [4,5,10,11,13], is to synchronize a network of STNOs so that the coherent signal generated from the network will yield a greater power output.

Initial insights into achieving synchronization came in 2005 from two adjoining papers in *Nature Letters* [5,8], which showed that two STNOs tend to phase-lock into a single resonance when they are in close proximity. This work was followed by Grollier [4] who computationally analyzed the dynamics of a 1 D series array of $N = 10$ coupled STNOs that were magnetically uncoupled but electrically connected in series. The results showed that the microwave power output increases as N^2, where N is number of oscillators in the array. Most recently, in 2017, work in Turtle et al. [14,15] established an analytical and computational approach for achieving synchronization which is valid for networks of arbitrary size. The work that follows utilizes the techniques developed in Turtle et al. [14,15] to extend the analysis to the case of a parallel arrayed network of STNOs. The motivation is for both completeness purposes and to help guide the design and fabrication process in current ongoing experiments. The coupling term for the parallel array is distinct from the series array and produces results which are unique from previous works.

31.2 Modeling

For a single STNO, see Fig. 31.1 (Left), an electric current, I, is applied to the fixed magnetic layer whose magnetization is represented by $\mathbf{e}_p$. As the electrons pass through the fixed layer, their spins become aligned with the direction of the local ferromagnetic moment, thus creating a *spin-polarized current.* In turn, the polarized current exerts a spin-transfer torque on the free magnetic layer, $\mathbf{m}$, which may lead to steady state precession. The free layer magnetization vector $\mathbf{m} = [m_1, m_2, m_3]$ for a single STNO is governed by the Landau–Lifshitz–Gilbert–Slonczewski equation

$$\frac{d\mathbf{m}}{dt} = -\mathbf{m} \times \mathbf{H}_{\mathrm{eff}} + \lambda \mathbf{m} \times \frac{d\mathbf{m}}{dt} + I\mu \mathbf{m} \times (\mathbf{m} \times \mathbf{e}_p) . \tag{31.1}$$

Here, γ is the gyromagnetic ratio, λ is magnitude of the Gilbert damping term, μ contains material parameters, and $\mathbf{H}_{\mathrm{eff}}$ is the effective magnetic field. The term $\mathbf{H}_{\mathrm{eff}}$ consists of anisotropy, demagnetization field, and applied field. The anisotropy is defined as $\mathbf{H}_{\mathrm{an}} = \kappa (\mathbf{m} \cdot \mathbf{e}_{\mathrm{an}}) \mathbf{e}_{\mathrm{an}}$, where κ is the strength of the anistropy, which we set to be $\kappa = 45$ Oe [9], and $\mathbf{e}_{\mathrm{an}}$ is the preferred direction of magnetization, which for this work we set to be $\mathbf{e}_{\mathrm{an}} = [0, 0, 1]$ [14]. $\mathbf{H}_{\mathrm{d}}$ is a demagnetization field and we set $\mathbf{H}_{\mathrm{d}} = -4\pi \mathrm{S}_0 (N_x \mathrm{m}_x \mathrm{e}_x - N_y \mathrm{m}_y \mathrm{e}_y - N_z \mathrm{m}_z \mathrm{e}_z)$, where $\mathrm{S}_0 = 8400/4\pi$ is the constant magnitude of the average magnetization vector $\mathrm{S}(t)$ such that $\mathbf{m} = \mathrm{S}/\mathrm{S}_0$, N_x, N_y, and N_z are dimensionless constants satisfying $N_x + N_y + N_z = 1$ [14]. Additionally, e_x, e_y, and e_z are the orthonormal unit vectors. Lastly, $\mathbf{H}_{\mathrm{app}}$ is an applied magnetic field defined by $\mathbf{H}_{\mathrm{app}} = h_a [0, \sin\theta_H, \cos\theta_H]^T$, which is assumed to lie on the yz-plane where θ_H is the angle from the z-axis. Furthermore, h_a has units of oersted, and the direction of the fixed layer is chosen to point in the z-direction, i.e. $\mathbf{e}_p = [0, 0, 1]$.

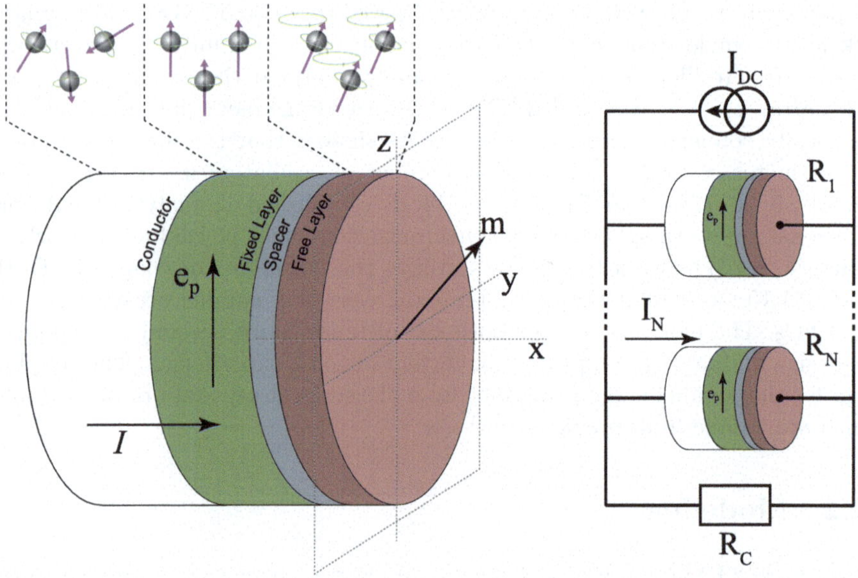

Fig. 31.1. (Right) Spin-valve with fixed layer in direction $\mathbf{e}_\mathrm{p}$ and free layer $\mathbf{m}$. (Left) Parallel arrayed STNOs with input current I_{DC} and output resistance R_c [15]

We now consider a network of STNOs coupled in a parallel array configuration, see Fig. 31.1 (Left). Here, the input current I is now replaced by I_j.

Assuming each STNO to be identical we apply Kirchhoff's laws to calculate the current passing through the j_{th} STNO as

$$I_j = \frac{\dfrac{1}{\dfrac{1}{R_c} + \sum\limits_{k=1,k\neq j}^{N} \dfrac{1}{R_k}}}{\dfrac{1}{\dfrac{1}{R_c} + \sum\limits_{k=1,k\neq j}^{N} \dfrac{1}{R_k}} + R_j} I_{DC}, \tag{31.2}$$

where $R_k = R_{0k} - \Delta R_k (\mathbf{m} \cdot \mathbf{e}_p)$ is the resistance of the k_{th} STNO, and R_0 is the mean resistance with ΔR the maximum variance in resistance. Substituting Eq. (31.2) into Eq. (31.1) yields a system of equations which models the dynamics for N coupled oscillators. Furthermore, in order to simplify future analysis we convert to complex stereographic coordinates using the change of variables $z_j = (m_{j1} + im_{j2})/(1 + m_{j3})$, which produces

$$\dot{z}_j = (1 + i\lambda)\Bigg[ih_{az}z_j + \frac{h_{ay}}{2}\left(1 + z_j^2\right) + i\kappa\frac{1 - |z_j|^2}{1 + |z_j|^2}z_j + \mu\tilde{I}_j I_{DC}$$
$$- \frac{i}{1 + |z_j|^2}\left(\frac{N_x - N_y}{2}\left(z_j^3 - \bar{z}_j\right) + \left(1 - \frac{3N_x + 3N_y}{2}\right)\left(z_j - z_j|z_j|^2\right)\right)\Bigg], \tag{31.3}$$

with

$$\tilde{I}_j = \frac{\dfrac{1}{\dfrac{1}{R_c} + \sum\limits_{k=1,k\neq j}^{N} \dfrac{1}{R_0 - \Delta R\frac{1-|z_k|^2}{1+|z_k|^2}}}}{\dfrac{1}{\dfrac{1}{R_c} + \sum\limits_{k=1,k\neq j}^{N} \dfrac{1}{R_0 - \Delta R\frac{1-|z_k|^2}{1+|z_k|^2}}} + \left(R_0 - \Delta R\dfrac{1 - |z_j|^2}{1 + |z_j|^2}\right)} z_j.$$

31.3 Bifurcation Analysis

31.3.1 Computational Bifurcation Diagram

This sections begins with the computational bifurcation diagram for a system of $N = 2$ STNOs. The expectancy is that certain dynamics may generalize to systems of larger N. The magnetic field is applied with $\theta_h = \pi/4$, which corresponds to the angle of the applied magnetic field, $\boldsymbol{H}_a$, from the z-axis in the direction of the y-axis. Settings for the demagnetization factors are defined to be

$N_x = 1, N_y = N_z = 0$ such that the free layer resembles the yz-plane. In practice, it is found that these settings provide an ideal configuration for fabrication. Using the complex stereographic representation, Eq. (31.3), the input current I_{DC} is varied to compute a one-parameter bifurcation diagram, see Fig. 31.2.

Referring first to large negative values of the input current the magnetization direction settles, as expected, to a stable equilibrium state marked as a solid red line. Following this stable branch into positive I_{DC} values, it loses stability through the onset of back-to-back Hopf bifurcations labeled HB_1 and HB_2, occurring at $I_{\mathrm{DC}} = 95.4$ and $I_{\mathrm{DC}} = 107$ respectively. As a result, the corresponding solution branches yield limit cycle solution trajectories. Green solid circles indicated stable synchronized oscillations, whereas the blue open circles signify unstable out-of-phase oscillations. It is emphasized that the initial parameter space containing stable synchronized limit cycle oscillations (green filled circles) exhibits no other stable solutions branches. In turn, the synchronized oscillations occurring from the Hopf bifurcation HB_1 do not appear to compete with any other stable solutions. Additionally, it is noted that HB_1 occurs at a small value of the current I_{DC}. Thus, in applications a relatively small current strength may be needed to produce oscillations implying that the system could operate at lower power. Consequently, this region exhibits multiple promising advantages for achieving synchronization in practice.

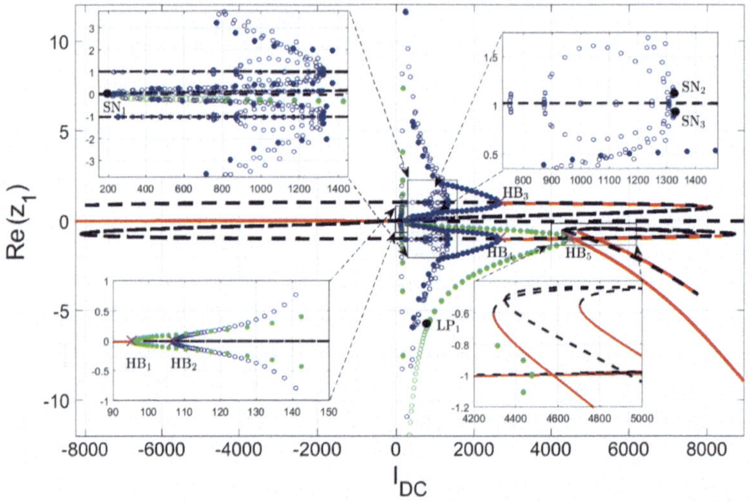

Fig. 31.2. One parameter bifurcation diagram in I_{DC} (μA) for a parallel array of $N = 2$ STNOs ($\theta_h = \pi/4$). Blue circles indicate out-of-phase oscillations while green circles indicate synchronized limit cycle oscillations. Filled-in (empty) circles indicate stable (unstable) oscillations

It is briefly mentioned that results from numerical bifurcation diagrams of the series array configuration with $N = 2$ STNOs no such dynamics are exhibited.

That is the parameter spaces which contain stable synchronized oscillations also display competing stable solution branches yielding out-of-phase oscillations. This further illustrate the uniqueness of the results for the system of coupled STNOs in a parallel array configuration. Lastly, it is noted that continuing in the positive I_{DC} direction, the bifurcation diagram displays an overlay of multiple solutions branches indicating that the ending dynamics are dependent on the choice of initial conditions. Hence, in applications, synchronization becomes increasingly difficult to achieve within this parameter space.

31.3.2 Conditions for Hopf Bifurcations

This section provides an overview the analysis for determining the existence and stability of synchronized oscillations for an arbitrarily large array of N oscillators. Once again, the motivation is that a system of STNOs oscillating in complete synchrony will generate a larger power output, thus meeting the necessary power increase for applications. Now as a result of Kirchhoff's Law, and the assumption of identical STNOs, Eq. (31.3) exhibits all-to-all coupling such that any permutation of the oscillators in the array leaves the coupling term invariant [14]. Consequently, the network of parallel arrayed STNOs has symmetry group S_N, that is the group of all permutations of N objects. Defining $\mathbf{z} = (z_1, \bar{z}_1, z_2, \bar{z}_2, \ldots, z_N, \bar{z}_N) \in \mathbb{C}^N$ allows Eq. (31.3) to be written as $\dot{z}_j = f_j(\mathbf{z})$. Furthermore, by the assumption that all STNO's are identical, it follows that $f_1 = f_2 = \ldots = f_N$, which yields the system of equations for N oscillators in the vector form as

$$\dot{\mathbf{z}} = \mathbf{f}(\mathbf{z}). \tag{31.4}$$

Next, let $\mathbf{z}_0 = (z_0, \bar{z}_0, z_0, \bar{z}_0, \ldots, z_0, \bar{z}_0)$ be an equilibrium solution of Eq. 31.4 with isotropy subgroup $\mathbf{S}_N$. Then the linearization at $\mathbf{z}_0$ is given by

$$\mathrm{L} := \begin{bmatrix} \mathrm{A} & \mathrm{B} & \ldots & \mathrm{B} \\ \mathrm{B} & \ddots & \ddots & \vdots \\ \vdots & \ddots & \ddots & \mathrm{B} \\ \mathrm{B} & \ldots & \mathrm{B} & \mathrm{A} \end{bmatrix},$$

where $\mathbf{A} = (df_{j_j})_{\mathbf{z}=\mathbf{z}_0}$ and $\mathbf{B} = (df_{j_k})_{\mathbf{z}=\mathbf{z}_0}$ are 2×2 Jacobian matrices of f_j with $j \neq k$. Next, using symmetry techniques, we block-diagonalize $\mathbf{L}$ to a form which maintains the symmetry invariant subspaces of $\mathbf{S}_N$. Let $\mathbf{P}$ be the change-of-coordinates matrix, then applying $\mathbf{P}$ to L results in a block diagonalization of the linear part of Eq. 31.4 as

$$\tilde{\mathrm{L}} := \mathbf{P}^{-1}\mathbf{L}\mathbf{P} = \mathrm{diag}\left\{\mathbf{A} + (N-1)\mathbf{B}, \mathbf{A} - \mathbf{B}, \ldots, \mathbf{A} - \mathbf{B}\right\}. \tag{31.5}$$

The block diagonal structure of $\tilde{\mathrm{L}}$ implies that eigenvalues of the blocks $\mathrm{A} + (N-1)\,\mathrm{B}$ and $\mathrm{A} - \mathrm{B}$ are also eigenvalues of $\tilde{\mathrm{L}}$. For $\mathrm{A} + (N-1)\,\mathrm{B}$, the corresponding eigenspace is $v_0 = [v, \ldots, v]^T$ and the symmetry group S_N acts trivially on v_0. Hopf bifurcations associated with this block correspond to a symmetry-preserving Hopf bifurcation which yields limit cycle solutions exhibiting complete synchrony. Here, each STNO oscillates with the same wave form, amplitude, and phase. For A − B, the eigenvalues have, generically, multiplicity $N-1$ and the emerging patterns of oscillations arise via symmetry-breaking Hopf bifurcations [3, 14].

Combining the conditions for an equilibrium solution with those that generate purely imaginary eigenvalues for the blocks $\mathrm{A} + (N-1)\,\mathrm{B}$ and using polar coordinates $z_0 = r\,(\cos\theta + i\sin\theta)$, yields the following set of Hopf conditions as a function of $(r, \cos\theta, I_{\mathrm{DC}}, \theta_H)$:

$$\begin{aligned}\mathrm{Re}\,(f_j) &= 0\\ \mathrm{Im}\,(f_j) &= 0\\ \mathrm{Tr}\,(\mathbf{A} + (N-1)\,\mathbf{B}) &= 0\end{aligned} \qquad (31.6)$$

To determine the analytical expressions for the Hopf boundary curves, we solve Eqs. (31.6) implicitly for the state variables (r, θ) as functions of the parameters $(I_{\mathrm{DC}}, \theta_H)$. Furthermore, a change is made by setting the configuration of demagnetization field to $N_x = N_y = 0.5, N_z = 0$ in order to achieve a solvable form of Eqs. (31.6). Next, through a series of substitutions we are able to reduce this system of three equations with four unknowns to a single expression with two variables (r, θ_H). Analytic expression for the Hopf loci are then solved in MAPLE using the function *implicitplot*(). Once the boundary curves are computed, a series of back substitutions are carried out to compute the point values $(I_{\mathrm{dc}}, \theta_h)$ and it is verified that $\det(\mathbf{A} - \mathbf{B}) > 0$ and $\det(\mathbf{A} + (N-1)\mathbf{B}) > 0$. Lastly, using the continuation software AUTO [1, 2], the movement of the Hopf loci as a function of the continuation parameter N_ϵ, with $N_x = 0.5 + N_\epsilon$ and $N_y = 0.5 - N_\epsilon$. In this way, at $N_\epsilon = 0.5$ we arrive at the physically relevant configuration of easy-plane anisotropy. The Hopf loci curves for $N_\epsilon = 0.5$ are depicted in Fig. 31.3 (Top) for up to $N = 1000$ STNOs.

Having computed the boundary curves containing the Hopf loci for an arbitrarily large network of STNOs, the focus now becomes determining the Hopf criticality and the stability of the synchronization manifold. The Hopf criticality is categorized as supercritical or subcritical which leads to stable or unstable synchronized oscillations, respectively. To determine the Hopf criticality we invoke the Lyapunov constant formula [7]. Specifically, if the Lyapunov constant is negative, the Hopf bifurcation is supercritical, whereas a positive Lyapunov constant leads to a subcritical Hopf bifurcation. Next, the stability properties of the synchronization manifold is determined by the eigenvalues transverse to the manifold which are given by the $N-1$ copies of the eigenvalues of the block A−B. It follows that the synchronized oscillations are asymptotically stable/unstable whenever the above mentioned eigenvalues are negative/positive. The calculations of the Lyapunov constant and the transverse eigenvalues are technical and lengthy and may be found in Ref. [14, 15]. The results of the Hopf criticality and asymptotic stability of the synchronization manifold are depicted in Fig. 31.3 (Bottom-Left) and Fig. 31.3 (Bottom-Right), respectively.

The stability analysis of Fig. 31.3 is now used to demonstrate numerical validation for synchronized oscillations of large systems of STNOs. Numerical simulations suggests the common equilibrium state for large arrays has a large basin of attraction for large negative values of I_{dc}. Therefore, the simulations start at a large negative values of I_{dc} in order to achieve rapid convergence. Next, guided by the results of Fig. 31.3, the strength of I_{dc} is increased until the parameter space is in a region that exhibits both a supercritical Hopf bifurcation and stable asymptotic behavior of the synchronization manifold . Using this strategy, synchronization is demonstrated for systems of $N = 1000$ STNOs, see Fig. 31.4. The network of oscillators exhibit a high level of synchrony indicating that this method could prove to be a useful path for achieving synchronization in experiment work.

31.4 Conclusion

The synchronization of a network of coupled STNOs is a viable solution for achieving the required power output for applications. Here we present the bifurcation analysis for a system of STNOs connected in a parallel array configuration. The computational bifurcation diagram for $N = 2$ STNOs depicts a favorable parameter space for achieving synchronization in experimental works. Using equivariant bifurcation theory we calculated the existence and stability

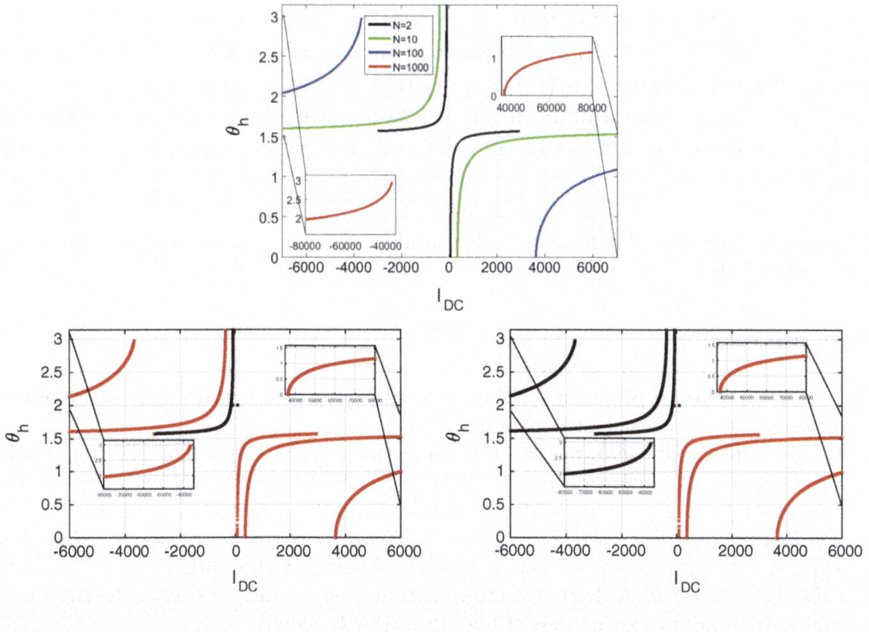

Fig. 31.3. (Top) Hopf loci corresponding to $\mathrm{Tr}(\boldsymbol{A} + (N-1)\boldsymbol{B}) = 0$, at $N_\epsilon = 0.5$. (Left) Criticality of Hopf: red - supercritical, black - subcritical. (Right) Transverse Lyapunov exponents: red - attractive, black - repulsive

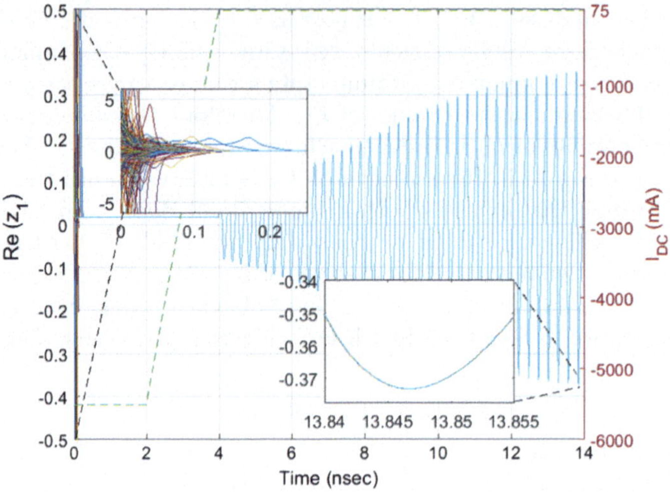

Fig. 31.4. Locking into synchronization with $N = 1000$ STNOs. Top inset: Zoom-in on the set of random initial conditions for the STNOs and evolution for small time values showing rapid convergence to a synchronized equilibrium. Bottom inset: Zoom-in on the bottom part of the oscillation showing a high level of synchronization between all the STNOs

of synchronized oscillations for an arbitrarily large array of N oscillators. These results were validated by numerically demonstrating synchronization for a system of $N = 1000$ STNOs. In future work we intend to develop a rigorous method for computing the basins of attraction in order to analyze the phase space which leads to synchronization. Furthermore, we desire to study the stability of the system by introducing variations in the material parameters and stochastic effects such as noise.

Acknowledgements. We recognize the support from the Office of Naval Research Grant N00014-16-1-2134.

References

1. E. Doedel, Auto: a program for the automatic bifurcation analysis of autonomous systems. Congr. Numer. **30**, 265–284 (1981)
2. E.J. Doedel, A.R. Champneys, T.F. Fairgrieve, Y.A. Kuznetsov, B. Sandstede, X. Wang et al., *Continuation and Bifurcation Software for Ordinary Differential Equations (with Homcont)*, (AUTO97, Concordia University, Canada, 1997)
3. M. Golubitsky, I. Stewart, D.G. Schaeffer, *Singularities and Groups in Bifurcation Theory*, vol. 2 (Springer Science & Business Media, Berlin, 2012)
4. J. Grollier, V. Cros, A. Fert, Synchronization of spin-transfer oscillators driven by stimulated microwave currents. Phys. Rev. B **73**, (2006)
5. S. Kaka, M.R. Pufall, W.H. Rippard, T.J. Silva, S.E. Russek, J.A. Katine, Mutual phase-locking of microwave spin torque nano-oscillators. Nature **437**, 389–392 (2005)

6. J. Katine, E.E. Fullerton, Device implications of spin-transfer torques. J. Magn. Magn. Mater. **320**, 1217–1226 (2008)
7. Y.A. Kuznetsov, *Elements of applied bifurcation theory*, vol. 112 (Springer Science & Business Media, Berlin, 2013)
8. F. Mancoff, N. Rizzo, B. Engel, S. Tehrani, Phase-locking in double-point-contact spin-transfer devices. Nature **437**, 393 (2005)
9. S. Murugesh, M. Lakshmanan, Spin-transfer torque induced reversal in magnetic domains. Chaos Solitons Fractals **41**, 2773–2781 (2009)
10. J. Persson, Y. Zhou, J. Akerman, Phase-locked spin torque oscillators: impact of device variability and time delay. J. Appl. Phys. **101**, 09A503 (2007)
11. W.H. Rippard, M.R. Pufall, S. Kaka, T.J. Silva, S.E. Russek, J.A. Katine, Injection locking and phase control of spin transfer nano-oscillators. Phys. Rev. Lett. **95**, 067203 (2005)
12. W.H. Rippard, M.R. Pufall, S.E. Russek, Comparison of frequency, linewidth, and output power in measurements of spin-transfer nanocontact oscillators. Phys. Rev. B **74**, 224409 (2006)
13. C. Serpico, R. Bonin, G. Bertotti, M. d'Aquino, I. Mayergoyz, Theory of injection locking for large magnetization motion in spin-transfer nano-oscillators. IEEE Trans. Magn. **45**, 3441–3444 (2009)
14. J. Turtle, P.-L. Buono, A. Palacios, C. Dabrowski, V. In, P. Longhini, Synchronization of spin torque nano-oscillators. Phys. Rev. B **95**, 144412 (2017)
15. J.A. Turtle, Synchronization in coupled spin-torque nano oscillators: nonlinear dynamics analysis. Ph.D. thesis, Diego State University, San, 2016

Chapter 32
Adventures in Stochastics

Derek Abbott(✉)

School of Electrical & Electronic Engineering, University of Adelaide, 5005 Adelaide, SA, Australia
derek.abbott@adelaide.edu.au

Abstract. This chapter describes my personal journey in the area of stochastic phenomena and how it has been impacted by Mike Shlesinger, to honor his 70th birthday for this *Festschrift.* I discuss my early explorations with Brownian ratchets and how this gave birth to the first paper on Parrondo's paradox. I then describe how this led to the next part of my journey in the areas of quantum game theory, suprathreshold stochastic resonance, and stochastic mixtures. Finally, I wrap up with discussion of our latest Bayesian analysis showing that too many confirmatory observations can paradoxically result in reduced confidence in an outcome.

32.1 In the Beginning

I started my career at the GEC Hirst Research Labs, London, UK, in the late seventies where I encountered luminaries such as Cyril Hilsum, who played an important role in getting gallium arsenide off the ground [1], and Mike Pepper who made a key step in the discovery of the quantum Hall effect [2]. One of my early tasks was measuring semiconductor device noise, and I successfully developed the first fully computer automated $1/f$ noise measurement set up there—possibly amongst the earliest in the world. I was mainly immersed in the literature of van der Ziel and the Dutch noise 'mafia' of the time. This was my first contact with the arcane world of stochastics and its enigmatic motley crew. The intellectual environment at Hirst was outstanding, though Hirst was not so generous with conference travel and I only got to attend one. My first noise conference was in the early 80s at a tiny local IEE meeting in London. There I met Lode Vandamme and he was the very first international noise person I came into contact with. About that time a Hirst colleague, by the name of Canute Moglestue, scored a better trip to a different noise conference, and returned with the news that Karel van Vliet was now to be called Carolyne.

Due to this lack of travel, I unfortunately missed many of the early conferences centered round stochastic resonance (SR) and nonlinear dynamics. I eventually left Hirst, and began research at the University of Adelaide where I had greater freedom. Unaware of some of the key conferences of the time I

V. In et al. (Eds.): *Proceedings of the 5th International Conference on Applications in Nonlinear Dynamics*, Understanding Complex Systems, https://doi.org/10.1007/978-3-030-10892-2_32

alas missed all the excitement of Saratov [3]. However, I attended the Unsolved Problems of Noise (UPoN) meeting in Szeged, Hungary [4], in 1996, and it was there that I first met a then very hirsute Mike Shlesinger for the first time. I had no idea who this guy with the tongue-twister name was, and I would later affectionately refer to him simply as 'Shles.' At the Szeged meeting I clearly remember Shles doing a banquet speech on the history of Brownian motion—his talk had me totally hooked. I did not even have any idea what 'ONR' stood for back then, but I instantly recognized Shles' great breadth and this led to many enjoyable discussions.

32.2 Feynman's Ratchet

The story behind how my *Nature* paper on Parrondo's paradox [5] came about is quite an amusing one and the space here only allows an abbreviated summary. The story begins around 1979 when I first read Chapter 46 of *Feynman's Lectures on Physics* on the ratchet and pawl. Essentially the chapter says that at thermal equilibrium, the probability of clockwise rotation balances the probability of counterclockwise rotation. Feynman heuristically stated, without proof, that this probability is $e^{-\epsilon/kT}$—this innocent looking Boltzmann factor fascinated me and I set about proving it from first principles for the ratchet system. Ten years later I was still obsessively going around in circles, unable to formally prove it.

So I started consulting those physicists, who I regarded a lot cleverer than myself, to see if anyone could actually do this calculation. I got an off-hand comment from David Mermin telling me that "Feynman is always right" and that I must have made a simple mistake. In fact that was the typical response I got from most physicists. Then I contacted Aephraim M. Steinberg who got my instant admiration as he actually tried the calculation himself. Alas, he gave up after a week. He sent me a tantalising email stating, "if the ratchet was a quantum system I would know exactly how to solve this, but because it's classical it's harder." Then Cosma R. Shalizi, a future student of Jim Crutchfield, offered to have a go at the challenge saying he would have it solved in "two days." I never heard back from him.

So needless to say when I was at the 1996 meeting in Szeged I asked around in vain to see if anyone could help with this problem. Then when I departed Szeged on the train to Budapest, Peter Jung happened to sit next to me. I told him the problem and he suggested I contact the ratchet guru Peter Hänggi. At the time, I had no idea who this guy with the double-g name and an umlaut was, but when I got home I emailed 'The Great Hänggi' and asked him if he had ever tried the calculation. His email reply simply stated "yes." In frustration I emailed him back and said "well, did you get it out?" After a pregnant silence of a few days that seemed to last eternity he replied, "No, but if you are really interested in solving this problem talk to Juan Parrondo."

I had no idea who this Parrondo guy was, but I flew to Madrid in 1997 and sat him down for coffee. I showed him the problem and he replied it was too hard to solve right away, but that we could work on it over time. So we did and published a paper on it two years later [6], ironically building on a level crossing statistics method from one of Hänggi's old papers. But it wasn't straight forward and one essentially ends up with a set of unconstrained equations that look intractable at first sight. One of my former PhD advisors, Bruce R. Davis, gets the credit for suggesting a cute trick that finally slew the dragon: we had to add a superfluous term to one the equations and then let it tend to zero at the end of the calculation! Who would have thought a sneaky math trick well-known to quantum field theorists would be needed for a 'simple' classical problem? Whilst we formally demonstrated detailed balance for the ratchet in equilibrium, using level crossing statistics, we still to this day do not have a way to demonstrate it using the Boltzmann factor that Feynman suggested.

32.3 The Genesis of Parrondo's Paradox

I often get asked why I didn't contact Feynman himself, given that he was alive until 1988. This simple answer is that in 1988 I was still at the stage where I thought I had made a naive error. Anyway, let us continue the story in 1997 in the Madrid coffee shop with Parrondo—we couldn't solve the ratchet detailed balance problem back then, so Parrondo started talking to me about the latest developments in Brownian ratchets and showed me his games of chance that illustrated the ratchet mechanism. He claimed it was possible to randomly switch between two different losing games and win. I have to admit I was skeptical and could not believe that it was possible to mix two losing games, and yet win. After all, everyone knows two wrongs don't make a right, so I thought at the time. Indeed, linear superposition would indicate that two wrongs should always make wrong—however, this is no longer true if we chose a nonlinear parameter space to work with.

Nevertheless, at the time I was sufficiently intrigued that I promised Parrondo that I would try to confirm the effect on return to Australia. I told him that if I could write a convincing paper demonstrating that it works, it would definitely get into *Nature* as it is so remarkable. I asked him, "Suppose I get it into *Nature*, would you prefer to be a co-author or instead shall we put your name in the title of the paper?" After a short pause, he replied, "put it in the title." The rest is history, and by 1999 the *Nature* paper was born.

Shles comes into the story here, as he provided a strong letter of support to ONR Asia to fund the next UPoN conference in 1999, Adelaide, Australia. With Shles' help, we won the funding and UPoN 99 was launched, where both the paper on Parrondo's paradox [7] and also our solution to the Feynman detailed balance problem were first published [6]. This then became the mathematical support for getting the Parrondo's paradox paper into *Nature*.

The paper has now been widely cited and the ideas have been extended to exciting areas from the control of chaos [8] through to population genetics [9]. My interest in quantum game theory [10] evolved by considering the question of Parrondo's paradox in the quantum domain [11]. Quantum versions of the game are impacting on the theory of quantum walks [12]. The idea that you can get finite channel capacity by combining two quantum channels of zero channel capacity is another area of related interest [13].

In the early days, one of my concerns was to constantly probe around to ensure the idea really was original and really surprising to people. Due to Shles' great breadth he was one of the many people I bounced this off. He loved the idea from the beginning and saw possible connections to the so-called 'chaos game' [14]. In the early noughties we met in Bethesda, Washington DC, for lunch and discussed a vast range of topics from Parrondo's games to protein folding.

Take the example of stochastic resonance (SR), where some authors have pointed out a connection between Debye's work on the dielectric properties of polar molecules and SR. This of course does not imply that Debye knew SR, or that he even had an SNR curve, only that his work can now be generalised in hindsight to connect with SR. Similarly, with Parrondo's paradox one can see in hindsight the ubiquity of the effect and quaint examples in the old literature [15]. However, prior to Parrondo, nowhere do we see a game-theoretic framework where the rate of losing reverses when we randomly mix losing games.

On the other hand, as the playwright Elias Canetti once said, "It doesn't matter how new an ideas is, what matters it how new it becomes." Parrondo's games seem to be just getting newer and newer. There are known connections between Parrondo's paradox and volatility pumping [16] on the stock market. Essentially they are both ratcheting mechanisms where randomness can be rectified by an asymmetry [15]. A new Parrondian effect called the *Allison mixture* [17] where a random mixture of two number sequences, possessing zero autocovariance, results in a paradoxical increase in autocovariance [18].

Random mixing can result in a reduction in randomness. A necessary ingredient for this to work is an asymmetry in the transition probabilities that describe the random switching between the two sequences. There are deep connections between the mixing of these numbers, irreversible thermodynamic processes, and information theory. In conclusion, Parrondo's games, Brownian ratchets, Allison mixtures, and volatility pumping are all examples where noise conspires with an asymmetry to produce directed motion in some variable. Physicists have traditionally sought symmetry in nature. A new paradigm is to now search for asymmetries and observe how they interact with noise or random behavior.

32.4 Suprathreshold Stochastic Resonance

It may be noted that Parrondo's games are also a form of stochastic resonance (SR) [19], and that my group's major studies in SR [20] developed out of this interest. In the early days of SR I was a skeptic, because the signal has to be subthreshold, and should it rise above the threshold the signal degrades. In practical applications, one has little control over where an arbitrary signal is going to be and this was my reasoning at the time.

However, I saw the light in 1999 when I was blown off my seat during a seminal talk [21] by Nigel G. Stocks at the *Stochaos* meeting in Ambleside, UK.

This meeting as organized by Peter E. V. McClintock and was pivotal for me. Nigel Stocks talked about suprathreshold stochastic resonance (SSR), where the SR effect still works above threshold. I was totally gripped throughout Nigel's talk and it was the first stochastic resonance presentation that made real sense to me. From that moment onward I was an SR convert. This led to a major study now published by Cambridge University Press [20]. A rather exciting result out of this work is the bifurcation diagram in Fig. 32.1.

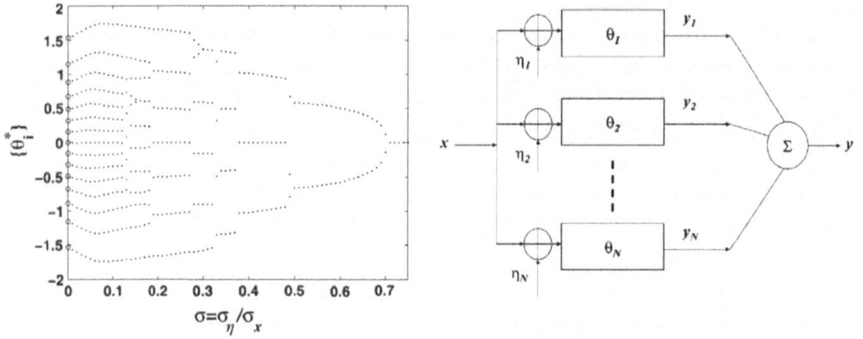

Fig. 32.1. What we see here (right) is a signal x essentially being estimated at the output y. The boxes represent different thresholds. So each box is essentially a 'voting element' asking the question if the signal is above or below its threshold. The interesting result (left) shows how to optimally distribute these thresholds (y-axis) for given amounts of noise (x-axis), in order to get the best estimate y. We see that for low noise, as expected, we must distribute the thresholds rather evenly across the signal space. However, as noise increases we need fewer thresholds that eventually collapse to one for very high noise. After [20]

This is an interesting setting that can be used to optimize thresholds in situations from sensor networks through to nanoelectronics. Also there are obvious connections to the Mulloch-Pitts neural model. However, consider this bifurcating threshold diagram as applying to human behavior. Imagine each box is a person voting on whether an ambiguous color is red or orange (for example). If the population was homogeneous it would be like having one threshold. If we had two populations, one with a predisposition toward red and one with leaning toward orange it would be like a two-threshold system. We can simulate noise in the system by dimming the lights. We may then find that for a certain light level a diverse population gives a more reliable estimate than a homogenous population.

Do we gain anything if political voters all had the same bias? The above example suggests political voting systems are meaningful in an information-theoretic sense, because we have left and right biases in the population. Homogeneity would give a monotonous result. *Vive la différence.* Going back to the voters guessing a color, we could have a homogenous population but give half the population rose colored spectacles—this introduces bias in a controlled way in order to carry out the experiment. The take home message is that bias can be a good thing.

In conclusion, SR shows us noise can be good, Parrondo's games show us that asymmetry can be good, and finally SSR shows us that bias can be good. Shles, I wish you a noisy, asymmetric, and biased 70th birthday!

32.5 The Paradox of Unanimity

We've discussed the Parrondian paradigm where two wrongs can make a right, but is an anti-Parrondo effect possible, i.e. can too much good be bad? Absolutely. A surprising case of this is described in the Talmud where if you are sentenced to death unanimously by all 23 Sanhedrin judges, you are acquitted! How do we make sense of such a counterintuitive ruling? In any decision where there are a complex set of circumstances there is incomplete information and a reasonable amount of uncertainty. So if all 23 judges are ostensibly forming independent opinions, a unanimous agreement would be rather like tossing a coin 23 times on a row and obtaining heads every time. But a run of 23 heads is so unlikely that if it happens you are probably going to suspect the coin is biassed.

Looking at it this way, we can say that in cases where there is uncertainty, we should definitely expect agreement to not be unanimous. A unanimous result is questionable in terms of bias or breakdown of independence in the system. In our Bayesian analysis of this effect we demonstrate that in a large ensemble of (say) 1000 trails, if only one of is them corrupted, then our then confidence in a unanimous decision dramatically drops—unanimity any greater than 15 in agreement rapidly drops below a 95% confidence interval [18].

32.6 Conclusion

This Chapter has been a personal journey looking the interplay between various stochastic phenomena in nonlinear systems, namely, Parrondo's paradox, stochastic resonance, the Allison mixture, and the paradox of unanimity together with the influence Michael F. Shlesinger has played. To further honor Mike's 70th birthday, a brief biography is contained in the following appendix.

Appendix

Michael F. Shlesinger (born August 8, 1948, Brooklyn, New York) is a physicist notable for his work in the area of nonlinear dynamics. He is the founder of the journal *Fractals*. His pioneering work in statistical predictions and descriptions of random and deterministic processes has influenced the physics of amorphous solids and glasses, classical mechanics, and biophysics. He is known as a proponent of fractal time and is also known for his work on fractal stochastic processes related to areas such as disordered materials and turbulence.

In 1970, he obtained his BS degree in physics and mathematics from State University of New York at Stony Brook, and then obtained his MA in 1972 from the University of Rochester. In 1975, he obtained his PhD from the University of Rochester under Elliott Waters Montroll for a thesis entitled *A Stochastic Theory of Anomalous Transient Photocurrents in Certain Xerographic Films and of the 1/f Noise in Neural Membrane.*

Initially he worked at the University of Maryland, College Park, then in 1983 he joined Office of Naval Research (ONR) and started their nonlinear dynamics program in 1984. He subsequently went on to head their physics division, before being named ONR's chief scientist for nonlinear science. His contributions to nonlinear dynamics and statistical physics include the publication of over 200 papers, editorship of over 20 books, and the organization of over 30 conferences. In 2008, he took up the Kinnear Chair in Physics at the US Naval Academy, Annapolis, United States.

He was elected to Fellow of the American Physical Society in 1993. In 2004, he received the Presidential Rank Award. In 2006, he received ONR's Saalfeld Award for outstanding lifetime achievement. In 2008, the conference *Nonlinear Dynamics at ONR* was held on Amelia Island, Florida, July 20–22, 2008, in honour of his 60th birthday.

While residing in Rockville, Maryland, he once saved the life of a woman who was being mauled by a pack of rottweilers. For putting his life at risk, the city of Rockville presented him with a medal for heroism.

His middle name is the single initial 'F'—he does not have a full middle name; another famous example being the middle initial of Harry S. Truman.

References

1. R. Clayton, J. Algar, *The GEC Research Laboratories 1919–1984* (Peter Peregrinus Ltd., London, 1989)
2. K. Von Klitzing, G. Dorda, M. Pepper, New method for high-accuracy determination of the fine-structure constant based on quantized Hall resistance. Phys. Rev. Lett. **45**(6), 494–497 (1980)
3. V. Anishchenko, A. Neiman (eds.), *International Conference on Nonlinear Dynamics and Chaos in Physics Medicine and Biology (ICND-96), 8–14 July 1996, Saratov, Russia*: Int. J. Bifurcat. Chaos **8**(4) (1998)
4. C.R. Doering, L.B. Kiss, M.F. Schlesinger (eds.), *Unsolved Problems of Noise* (Szeged, Hungary, 1996; World Scientific, Singapore, 1997)
5. G.P. Harmer, D. Abbott, Losing strategies can win by Parrondo's paradox. Nature (London) **402**(6764), 864 (1999)
6. D. Abbott, B.R. Davis, J.M.R. Parrondo, The problem of detailed balance for the Feynman-Smoluchowski engine (FSE) and the multiple pawl paradox, in *Proceedings of the Second International Conference Unsolved Problems of Noise and Fluctuations, (UPoN '99)*, vol. 511, ed. by D. Abbott, L.B. Kish (Adelaide, Australia, 11–15 July 1999), pp. 213–218
7. G.P. Harmer, D. Abbott, P.G. Taylor, J.M.R. Parrondo, Parrondo's paradoxical games and the discrete Brownian ratchet, in *Proceedings of the Second International Conference Unsolved Problems of Noise and Fluctuations, (UPoN '99)*, vol.

511, ed. by D. Abbott, L.B. Kish, (Adelaide, Australia, 11–15 July 1999), pp. 89–200
8. J. Almeida, D. Peralta-Salas, M. Romera, Can two choatic systems give rise to order? Phys. D **200**, 124–132 (2005)
9. F.A. Reed, Two-locus epistasis with sexually antagonistic selection: a genetic Parrondo's paradox. Genetics **176**, 1923–1929 (2007)
10. A.P. Flitney, D. Abbott, An introduction to quantum game theory. Fluct. Noise Lett. **2**(4), R175–R188 (2002)
11. A.P. Flitney, J. Ng, D. Abbott, Quantum Parrondo's games. Phys. A **314**, 35–42 (2002)
12. J. Rajendran, C. Benjamin, Playing a true Parrondo's game with a three-state coin on a quantum walk. Europhys. Lett. **122**(4), 40004 (2018)
13. G. Smith, J. Yard, Quantum communication with zero-capacity channels. Science **321**(5897), 1812–1815 (2008)
14. M.F. Barnsley, *Fractals Everywhere* (Springer, New York, 1988)
15. D. Abbott, Asymmetry and disorder: a decade of Parrondo's paradox. Fluct. Noise Lett **9**(1), 129–156 (2010)
16. D.G. Luenberger, *Investment Science* (Oxford University Press, Oxford, 1997)
17. A. Allison, C.E.M. Pearce, D. Abbott, Finding keywords amongst noise: automatic text classification without parsing. in *Proceedings of the SPIE Noise and Stochastics in Complex Systems and Finance, Florence*, vol. 6601 (Italy, 2007), pp. 660113
18. L.J. Gunn, F. Chapeau-Blondeau, M.D. McDonnell, B.R. Davis, A. Allison, D. Abbott, Too good to be true: when overwhelming evidence fails to convince. Proc. R. Soc. Lond. A **472**, 20150748 (2016)
19. A. Allison, D. Abbott, Stochastic resonance in a Brownian ratchet. Fluct. Noise Lett. **1**(4), L239–L244 (2001)
20. M.D. McDonnell, N.G. Stocks, C.E.M. Pearce, D. Abbott, *Stochastic Resonance* (Cambridge University Press, Cambridge, 2008)
21. N.G. Stocks, Suprathreshold stochastic resonance. Stochaos **1999**(502), 415–421 (2000) (Ambleside)
22. L.J. Gunn, F. Chapeau-Blondeau, A. Allison, D. Abbott, Towards an information-theoretic model of the Allison mixture stochastic process. J. Stat. Mech. **5**, 054041 (2016)
23. M.D. McDonnell, N.G. Stocks, C.E.M. Pearce, D. Abbott, Optimal information transmission in nonlinear arrays through suprathreshold stochastic resonance. Phys. Lett. A **352**(3), 183–189 (2006)

Chapter 33 Classification and Analysis of Chimera States

Neelima Gupte(✉) and Joydeep Singha

Indian Institute of Technology Madras, Chennai 600036, India
gupte@iitm.ac.in, joydeep@physics.iitm.ac.in

Abstract. We study the existence of different types of chimera states in a globally coupled sine circle map lattice with different strengths of intergroup and intragroup coupling. Some of the typical chimera phase configurations that can be observed in this system are aperiodic chimera states, splay chimera states and chimera states with spatiotemporally intermittent behaviour in the desynchronised group. These states are seen in different regions of the parameter space for three distinct kinds of initial conditions. We obtain the phase diagram containing the third type of chimera state, viz. the one with spatiotemporally intermittent regions, using complex order parameters. We construct an equivalent cellular automaton (CA) and reproduce the phase diagram in the region of interest by solving the mean field equation obtained for the CA.

33.1 Introduction

The chimera phase pattern in spatially extended systems is a remarkable spatiotemporal phenomenon, and has been extensively discussed for systems of coupled phase oscillators. In this context, the chimera state of a group of oscillators is defined to be a state where a synchronous subgroup of oscillators coexists with a desynchronised subgroup of oscillators. This spatiotemporal behaviour was first discovered in non-locally coupled complex Stuart–Landau oscillators [1] and has been further analysed for diverse systems such as a ring of phase oscillators [2–5], Stuart–Landau oscillators [6], networks of Kuramoto oscillators [7], coupled chemical oscillators [8–10] and mechanical oscillator networks [11]. Here, we study a system of coupled maps which is a discrete analog of systems of coupled phase oscillators, and is hence easily amenable to theoretical and numerical analysis. The specific CML used here, is of the form used in Refs. [12,13] and consists of two populations of globally coupled identical sine circle maps where the strength of the coupling within each population and that between maps belonging to distinct populations take different values.

Here, we show that the CML under consideration can support various types of chimera states with distinct spatial and temporal behaviours which can be

V. In et al. (Eds.): *Proceedings of the 5th International Conference on Applications in Nonlinear Dynamics*, Understanding Complex Systems, https://doi.org/10.1007/978-3-030-10892-2_33

obtained using distinct initial conditions, and parameter regimes. Aperiodic and stable chimera states with a synchronized subgroup and a subgroup with random phases were seen for this system [12] for a certain class of initial condition in a region of the parameter space. Oscillator systems also support a splay state, i.e. a state where the phase difference between consecutive oscillators is a constant. Our system can exhibit splay chimera states wherein a phase synchronised group coexists with a group that consists of splay phase configurations and also supports a phase kink, for a special initial condition with a system wide splay phase configuration. The switching of synchrony and de-synchrony between the two groups is also observed for this case with a variation of parameters. Again, for the same system, a general initial condition with random phases evolves to chimera states where the space time variation of the phase desynchronised group shows spatiotemporally intermittent behaviour. We analyze this case in detail. Using the global coupling topology of the CML, we define appropriate conditional probabilities, which identify the transition between the laminar and burst sites as the system evolves in time, and calculate these probabilities numerically from the space time variation of the phases of the maps of the CML.

33.2 The Model

We study a globally coupled lattice consisting of identical sine circle maps which is divided into two groups with different strengths of intergroup and intragroup coupling. A simple schematic of the coupling topology of the system is shown in Fig. 33.1.

The evolution equation for a single sine circle map is given by the equation

$$\theta_{n+1} = \theta_n + \Omega - \frac{K}{2\pi}\sin(2\pi\theta_n) \mod 1 \tag{33.1}$$

where θ_n is the phase of the map at the nth time step and Ω and K are respectively the frequency ratio and nonlinearity parameters. The two parameter space of the single sine circle map shows mode locking structures or Arnold tongues, that are organised by the Farey sequence, interspersed with regions of quasiperiodic behavior. This map exhibits both the quasi-periodic and period doubling routes to chaos [14, 15]. The evolution equation for the phase, $\theta_n^\sigma(i)$ of the ith coupled sine circle map in group σ at time step n of our CML has the form,

$$\begin{aligned}\theta_{n+1}^\sigma(i) = \theta_n^\sigma(i) + \Omega - \frac{K}{2\pi}\sin(2\pi\theta_n^\sigma(i)) + \sum_{\sigma'=1}^{2}\frac{\varepsilon_{\sigma\sigma'}}{N_{\sigma'}} \\ \times\left[\sum_{j=1}^{N_{\sigma'}}(\theta_n^{\sigma'}(j) + \Omega - \frac{K}{2\pi}\sin(2\pi\theta_n^{\sigma'}(j)))\right] \mod 1\end{aligned} \tag{33.2}$$

Here, the symbols, σ and σ' indicate the groups and each take values 1, 2, and n is the discrete time label, as before. The total number of maps in a given group σ is denoted by N_σ. For our study we use $N_\sigma = N_{\sigma'} = N$. The coupling

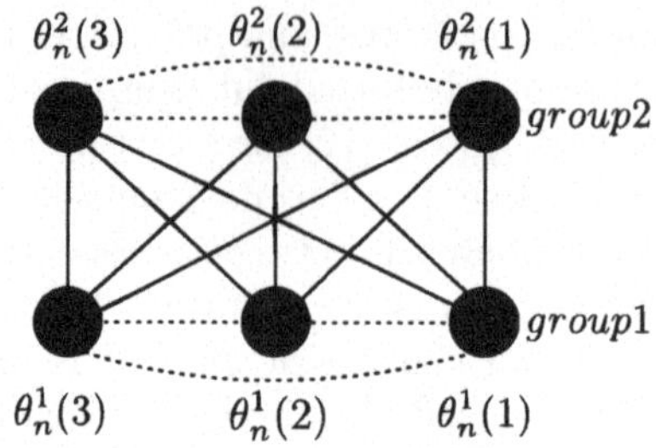

Fig. 33.1. The diagram illustrates the connection scheme of the globally connected network. The circles represent the maps in each group with 3 maps in each group. Each map in the system is coupled to all the other maps in its own group via a coupling constant ε_1 (represented by dotted edges) and to the maps in the other group via a coupling constant ε_2 (represented by solid edges)

strength parameters are, $\varepsilon_{11} = \varepsilon_{22} = \varepsilon_1$ and $\varepsilon_{12} = \varepsilon_{21} = \varepsilon_2$ and they are constrained by, $\varepsilon_1 + \varepsilon_2 = 1$. Thus our CML in Eq. (33.2) is controlled by three independent parameters, K, Ω, ε_1. We restrict these parameters to lie within the interval $[0 : 1]$ for our analysis. We show that this simple model exhibits diverse dynamical behaviours depending on the parameters, and different classes of initial conditions.

Earlier studies of this system [12] showed the existence of chimera states after the evolution of an initial condition where all of the maps in one of the groups were assigned identical phases, and random values between zero and one were assigned to the maps of the other group. Such initial conditions evolve into chimera states with phase synchronised and a phase desynchronised groups of maps for some parameter regions Fig. 33.2a. Clustered chimera states where phase synchronised clusters coexist with the phase desynchronised maps within same group along with the purely synchronised group (see Fig. 33.2b), can also be found in this system using the same initial condition. In addition to this, multiclustered phase states and globally phase synchronised state also exist at other parameter values (see Ref. [12]).

The temporal behaviour of the chimera state can be understood via the complex order parameters, $R_n^\sigma = \frac{1}{N}|\sum_{j=1}^{N} \exp i2\pi\theta_n^\sigma(j)|$ which are defined for each group, i.e. for $\sigma = 1, 2$ and the global order parameter $R_n = \frac{1}{2N}|\sum_{\sigma=1}^{2}\sum_{j=1}^{N} \exp(i2\pi\theta_n^\sigma(j))|$. Clearly R_n^σ will be one if all the phases are identical in group σ and zero if the phases are uniformly distributed between zero. Hence, for a chimera phase configuration at time step n, $R_n^\sigma \approx 1$ while $R_n^{\sigma'} \approx 0$ $(\sigma \neq \sigma')$. Using these order parameters, we find that in the temporal variation of the chimera states in Fig. 33.2a, b, the average phase of the desynchronised group in both cases evolves aperiodically (see Fig. 33.2c, d).

Chimera states of the type seen in Fig. 33.2a appear at all values of Ω and ε_1 at $K = 0$. Additionally, the phase diagram in the $K - \Omega$ space, obtained at

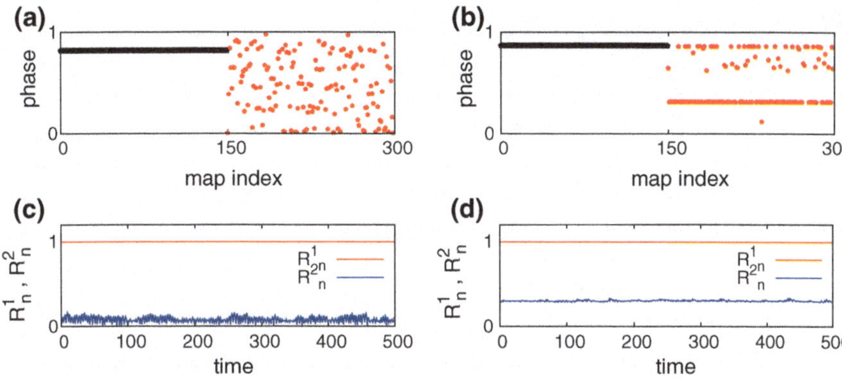

Fig. 33.2. Snapshots of the chimera state seen at the parameter values $K = 0.2$, $\Omega = 0.15, \varepsilon_1 = 0.9, N = 150$, and the clustered chimera state seen at the parameter values $K = 0.07, \Omega = 0.01, \varepsilon_1 = 0.9, N = 150$. The order parameters R_n^1 and R_n^2 is shown for **c** the chimera phase state and **d** the clustered chimera state

$\varepsilon_1 = 0.9$, shows multiple phase clustered states, two phase clustered states and fully phase synchronised states along with both types of chimera states shown in Fig. 33.2a, b. However these chimera states disappear in the $\varepsilon_1 - \Omega$ parameter space for $K = 1$ (see Ref. [12]). We note that this system can support some additional classes of chimera states as identified by their temporal and spatial properties, on evolution from other types of initial conditions.

33.3 Splay-Chimera State

To see the splay chimera states, we use an initial phase configuration where the entire system of $2N$ lattice sites is oriented in a single spatial splay state where the phase difference between any two consecutive maps is given by $\frac{1}{2N}$ (see Ref. [13]). We note here, that the phases are placed on a $1 - d$ lattice where the site labels run consecutively, so two consecutive maps means maps at adjacent sites as indicated by site labels, e.g sites i and $i + 1$. Using the splay phase initial condition if the system is evolved via Eq. (33.2) then pure splay phase states, splay chimera states and globally synchronised states are obtained as K is increased from zero to one, with the remaining parameters fixed at $\varepsilon_1 = 0.01, \Omega = 2/7$.

(i) The two copy splay states appear in the range $0 < K < 10^{-7}$ (Fig. 33.3a). (ii) On further increase of K to 10^{-4} we see a special chimera phase structure where all the maps in group two are spatially synchronised whereas in group 1, the phases of some maps are part of a splay-like state (roughly between sites 1–100 and 130–150) whereas the remaining maps show a jump (roughly between sites 100 and 130) in their phases (see Fig. 33.3b). (iii) If we increase K to 10^{-2} we observe a flip in the phase configuration of the chimera state seen in figure. In this splay chimera structure, all the maps in group one are spatially synchronised

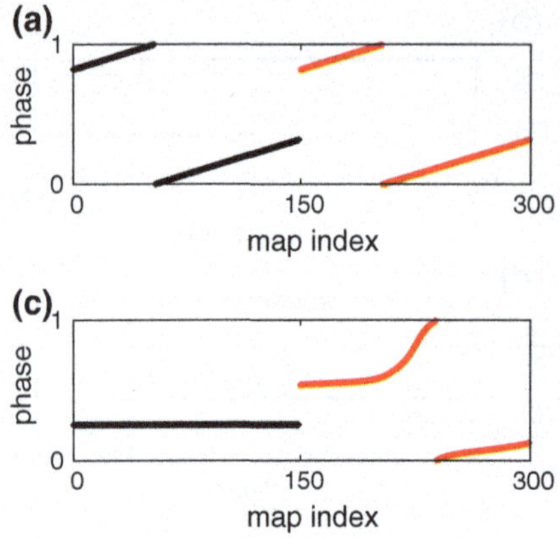

Fig. 33.3. **a** Two copy splay states were obtained at parameter values $K = 10^{-10}$. **b** If we increase K to 10^{-4} then the maps in group 2 synchronise completely while splay like structures along with a phase kink are seen. **c** A structure similar to that seen in **b** is observed for $K = 10^{-2}$. However the behaviour between group 2 maps and group 1 maps is interchanged. **d** The snapshot of the globally synchronised state at $K = 0.01$. The rest of parameters of the system are fixed at $\Omega = 2/7, \varepsilon_1 = 0.01, N = 150$

and maps at sites 150–200 and 250–300 show splay-like diagonal structure and the sites 200–250 show a phase jump (see Fig. 33.3c). (iv) As we increase K to even higher values we observe that the system settles to global synchronisation (Fig. 33.3d).

Figure 33.4a shows that R_n^1 and R_n^2 remain constant with time for the splay chimera state in Fig. 33.3b, implying that the temporal variation of the phases of maps in synchronised group and desynchronised group remains structurally stable with time while the variation of R_n^2 in Fig. 33.4b which is the order parameter for the desynchronised group in the splay chimera state shown in Fig. 33.3c indicates its aperiodic nature. The largest Lyapunov exponents calculated for both types of splay chimera states in Fig. 33.3b, c are 0.693 which indicates that they are temporally chaotic. A detailed stability analysis of splay phase configurations and their bifurcation to splay chimera states can be found in Ref. [13].

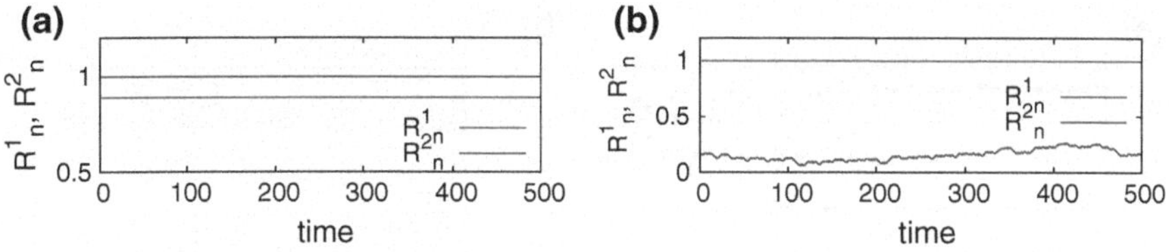

Fig. 33.4. The complex order parameters R_n^1, R_n^2 are plotted for the chimera states in **a** ($K = 10^{-4}$) and **b** ($K = 10^2$). The CML is iterated via Eq. (33.2) for 3×10^6 steps and then the order parameters are calculated for the next 300 time steps which are shown here. The parameters $\Omega = 2/7, \varepsilon_1 = 0.01, N = 150$ are the same for both the figures

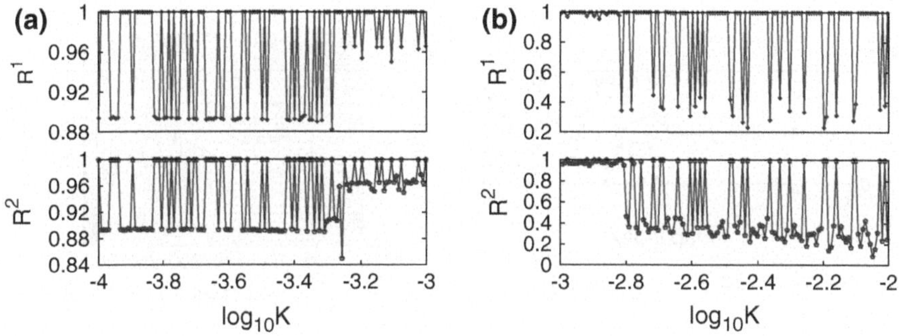

Fig. 33.5. R_n^1 and R_n^2 are plotted between the range **a** 10^{-4} and 10^{-3} and **b** 10^{-3} and 10^{-2}. The parameters $\Omega = 2/7, \varepsilon_1 = 0.01, N = 150$ are kept fixed during the variation of the K. At all the values of K we use the systemwide splay phase initial condition

We have seen that switching occurs between the two types of chimera states in Fig. 33.3b, c where the synchronisation and desynchronisation is interchanged between the groups one and two. Using the order parameters we show that this interchange of the phase synchronisation and de-synchronisation of the groups occurs in an intermittent fashion with the variation of K in the range 10^{-4} and 10^{-2}. The order parameters R_n^1 and R_n^2 are plotted in Fig. 33.5 for K values that vary between 10^{-4} and 10^{-2}.

We note that splay states are observed in a variety of experimental systems, such as crystal oscillators, and hence splay chimera states can occur in such systems as well. It would be interesting to see if such states are seen in these systems and to explore their consequences for quantities like output power.

33.4 Chimera States with Spatiotemporally Intermittent Behaviour

Now we explore a third type of chimera state (see Figs. 33.6 and 33.7) where the space time variation of the phases of the maps in the desynchronised group show

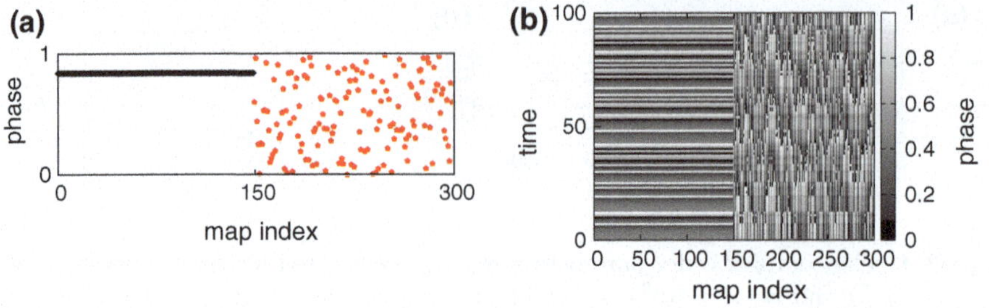

Fig. 33.6. **a** The snapshot and **b** the space time plot of the chimera state where group 1 is completely synchronized. The parameters are $K = 10^{-5}, \Omega = 0.27, \varepsilon_1 = 0.82, N = 150$

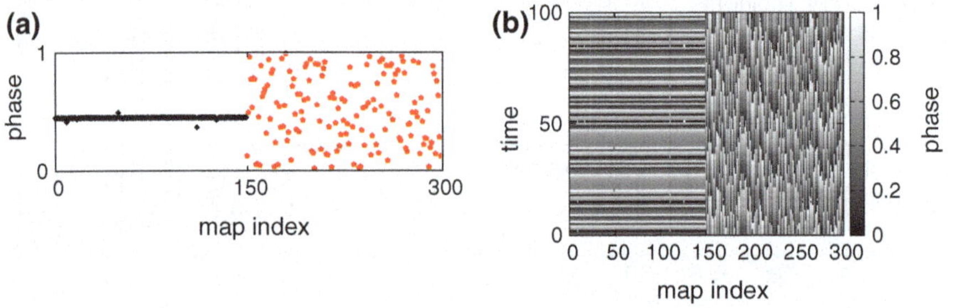

Fig. 33.7. **a** The snapshot and **b** the space time plot of the chimera state with partial synchronisation in group one. The parameters are $K = 10^{-5}, \Omega = 0.27, \varepsilon_1 = 0.93, N = 150$

spatiotemporal intermittent behaviour, as synchronised islands in the shape of cones can be observed within the desynchronised phases (see Figs. 33.6 and 33.7). This type of chimera state, which also evolves from a completely random initial condition, can have a purely phase synchronised group as one subgroup of the chimera (Fig. 33.6) (case 1), or this subgroup can be partially phase synchronised, where some defects can be seen in the phase synchronized part (Fig. 33.7) (case 2) depending on the parameters of the system.

Using the order parameters, R^1, R^2 as defined previously, we obtain a phase diagram (see Fig. 33.8) within the region $10^{-8} < K < 10^{-2}$ and $0.65 < \varepsilon_1 < 1$ where the chimera states of the kind shown in Figs. 33.6, 33.7 and fully phase desynchronised states are seen. These show that the fully desynchronised state as identified by the order parameter values ($R^1 \approx 0, R^2 \approx 0$) between $10^{-5} < K < 10^{-4}$ and $0.65 < \varepsilon_1 < 0.8$ transforms to a chimera state signalled by the values $R^1 \approx 1, R^2 \approx 0$ at $\varepsilon_1 = 0.8$. The fully phase desynchronised states (($R^1 \approx 0, R^2 \approx 0$)) which appear between $10^{-8} < K < 10^{-5.5}$ and $0.8 < \varepsilon_1 < 1$ transform to chimera states as K increases beyond 10^{-5}.

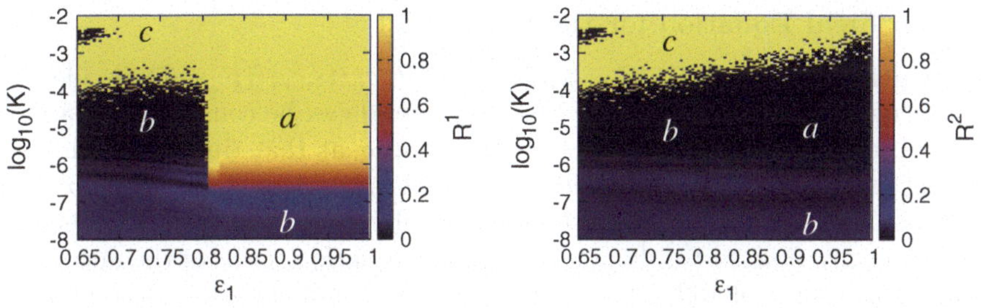

Fig. 33.8. (Left) The complex order parameters R_n^1 and (Right) R_n^2 are plotted between the region $10^{-8} < K < 10^{-2}$ and $0.65 < \varepsilon_1 < 1$. The states that can be found in the above plots are **a** chimera states ($R^1 \approx 1, R^2 \approx 0$), **b** fully synchronised states ($R^1 \approx 0, R^2 \approx 0$), **c** globally phase synchronised states and two phase clustered states ($R^1 \approx 1, R^2 \approx 1$). We have fixed the parameter values $\Omega = 0.27$ and $N = 150$ to obtain these plots

33.4.1 Construction of an Equivalent Cellular Automata

The existence of spatiotemporally intermittent behaviour in the desynchronised group of the chimera states can be analyzed by the construction of an equivalent cellular automaton. To achieve this, we identify laminar and burst stages of a lattice site during its space time evolution when the system settles in any of the chimera states shown in Figs. 33.6 and 33.7. In order to do this, we consider the phases of the maps at any two consecutive time steps n and $n+1$ and calculate the quantity $\Delta_{ij} = \left|\frac{1}{2}\left|\exp(2\pi i\theta_t^\sigma(i)) + \exp(2\pi i\theta_{t'}^{\sigma'}(j))\right| - 1\right|$ for all combinations of $i, j = 1, 2, \ldots, N$, for every $\sigma, \sigma' = 1, 2$ with $t, t' = n, n+1 (i \neq j$ if $\sigma = \sigma'$ and $t = n)$. The lattice sites considered are labelled laminar if $\Delta_{ij} < \delta$ where δ is a preassigned value. A similar check is carried out for $t = t'$. Once all the laminar sites are identified in a given space time plot of the chimera states, the rest of the sites are labelled as burst sites.

Next we assign a state variable $s_n^\sigma(i)$ which takes the value 1 if the map at the ith site at time step n is laminar and it is assigned 0 for the burst state. The global coupling topology of the system implies that the dynamics of $s_n^\sigma(i)$ or the transition probabilities to construct the CA depends on the total number of laminar sites in groups one and two at the time step n. We calculate $P(x_1, x_2)$ which is the probability of occurrence of x_1 and x_2 laminar sites in groups one and two respectively. Based on this, the transition probability is defined as $P^{x_1,x_2}(s_{n+1}^\sigma(i)|(s_n^\sigma(i))$ which is the transition probability that a lattice site i chosen at random in group σ at time step n having value $s_n^\sigma(i)$ transforms to $s_{n+1}^\sigma(i)$ at time step $n+1$, given that there are x_1 and x_2 laminar sites in groups one and two respectively. Hence there are four possibilities for each combination of x_1 and x_2, which are $P^{x_1,x_2}(0|0)$, $P^{x_1,x_2}(1|0)$, $P^{x_1,x_2}(0|1)$, $P^{x_1,x_2}(1|1)$. Using these probabilities, we obtain a mean field equation for the CA model.

A Mean Field Equation for the CA Model

By our definition, the transition probability $P^{x_1,x_2}(s_{n+1}(i)^{\sigma}|s_n^{\sigma}(i))$ is identical irrespective of the choice of i at a time step n and can be considered as a mean field which is the same at all sites in the group σ at that time step. This also implies that we have two mean fields for the CA for each of the values of σ'. Now let us assume that $m_\sigma(t)$ be an arbitrary initial value of the fraction of laminar sites for the given attractor dynamics. A linear equation terms of the mean fields or the transition probabilities and the fraction of laminar sites, $m_\sigma(t)$, can be written following the prescription by Mikkelsen et al. [16] as,

$$\begin{aligned} m_\sigma(t+1) = \sum_{x'_\sigma=0}^{N} \Bigg[& P(0,x_{\sigma'})P^{0,x_{\sigma'}}(1|0) + P(N,x_{\sigma'})P^{N,x_{\sigma'}}(1|1) \\ & + \sum_{x_\sigma=1}^{N-1} \Big(P(x_\sigma,x_{\sigma'})P^{x_\sigma,x_{\sigma'}}(1|0)(1-m_\sigma(t)) \\ & + P(x_\sigma,x_{\sigma'})P^{x_\sigma,x_{\sigma'}}(1|1)m_\sigma(t) \Big) \Bigg] \end{aligned} \tag{33.3}$$

This is a linear equation of the form $m_\sigma(t+1) = f(m_\sigma(t)) = a_\sigma m_\sigma(t) + b_\sigma$ where, a_σ, b_σ are given by,

$$\begin{aligned} a_\sigma &= \sum_{x_{\sigma'}=0}^{N} \sum_{x_\sigma=1}^{N-1} \Big(P(x_\sigma,x_{\sigma'})P^{x_\sigma,x_{\sigma'}}(1|1) - P(x_\sigma,x_{\sigma'})P^{x_\sigma,x_{\sigma'}}(1|0) \Big) \\ b_\sigma &= \sum_{x_{\sigma'}=0}^{N} \Bigg[P(0,x_{\sigma'})P^{0,x_{\sigma'}}(1|0) + P(N,x_{\sigma'})P^{N,x_{\sigma'}}(1|1) + \sum_{x_\sigma=1}^{N-1} P(x_\sigma,x_{\sigma'})P^{x_\sigma,x_{\sigma'}}(1|0) \Bigg] \end{aligned} \tag{33.4}$$

The fixed points of Eq. (33.3) in the m_1, m_2 space are given by,

$$\begin{aligned} \widetilde{m}_1 &= \frac{b_1}{1-a_1} \\ \widetilde{m}_2 &= \frac{b_2}{1-a_2} \end{aligned} \tag{33.5}$$

We find that a_σ and b_σ must satisfy the conditions, $a_1 + b_1 \le 1$, $a_2 + b_2 \le 1, a_1, a_2 \ne 1$ and $b_1, b_2 \ge 0$ since $\widetilde{m}_1, \widetilde{m}_2 \in [0:1]$. The Jacobian for the set of equations given by Eq. (33.3) is written as,

$$J = \begin{bmatrix} a_1 & 0 \\ 0 & a_2 \end{bmatrix}_{\widetilde{m}_1,\widetilde{m}_2} \tag{33.6}$$

Hence the fixed points, $\widetilde{m}_1$ and $\widetilde{m}_2$ with eigenvalues $\lambda_1 = a_1$ and $\lambda_2 = a_2$ are stable, if both $|a_1|, |a_2| < 1$. In that case any arbitrary initial value of $m_\sigma(t)$ must converge to the values of average fraction of laminar sites in the two groups of the CML. We verify this for the chimera states and list the values of $\widetilde{m}_\sigma$. in Table 33.1.

Table 33.1. The table lists the values of $\widetilde{m}_1$ and $\widetilde{m}_2$ for the parameter values $K = 10^{-5}$, $\Omega = 0.27, N = 150$. We use $\varepsilon_1 = 0.82$ for the chimera state of case 1 and $\varepsilon_1 = 0.93$ for case 2

Chimera states	a_1	b_1	a_2	b_2	$\widetilde{m}_1$	$\widetilde{m}_2$	$\widetilde{m}_1$(numerical)	$\widetilde{m}_2$(numerical)
Case 1	0.0	1.0	0.277	0.247	1.0	0.342	1.0	0.337
Case 2	0.637	0.357	0.268	0.257	0.983	0.35	0.983	0.347

We calculate $\widetilde{m}_1$ and $\widetilde{m}_2$ for group one and two, for the range of parameters between $10^{-8} < K < 10^{-2}$ and $0.65 < \varepsilon_1 < 1$ for $\Omega = 0.27$ in Fig. 33.9a, b. A comparison between Figs. 33.8 and 33.9 show that our mean field analysis reproduces accurately the phase diagram of the CML in the region of interest. Moreover, Fig. 33.9 shows that chimera states with defects in the synchronised group appear for increasing values of ε_1 within the region $0.8 < \varepsilon_1 < 1$ and $10^{-5.5} < K < 10^{-4}$.

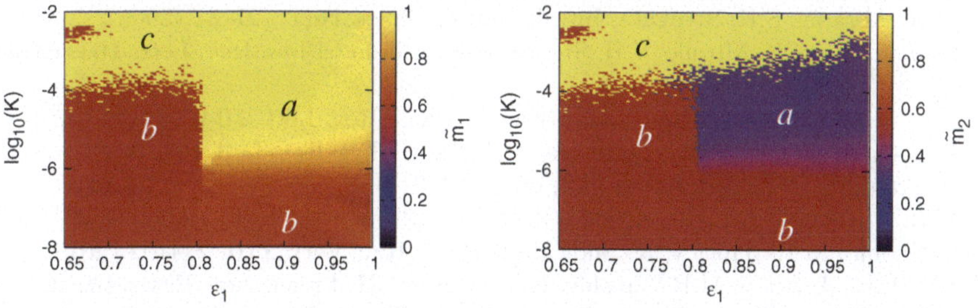

Fig. 33.9. The fractions (left) $\widetilde{m}_1$ and (right) $\widetilde{m}_2$ calculated using the values of a_1, b_1, a_2, b_2. The transition probabilities required are extracted from the behavior of the CML using the parameters $10^{-8} < K < 10^{-2}, 0.65 < \varepsilon_1 < 1, \Omega = 0.27, N = 150$. The states that can be found in this regime are **a** chimera states ($\widetilde{m}_1 \approx 1, \widetilde{m}_2 \approx 0.35$), **b** fully synchronised states ($\widetilde{m}_1, \widetilde{m}_2 \approx 0.55$), **c** globally phase synchronised states and two phase clustered states ($\widetilde{m}_1 \approx 1, \widetilde{m}_2 \approx 1$)

33.5 Conclusion

To summarise, we have shown here that a coupled map lattice having two groups of sine circle maps, connected via different intergroup and intragroup coupling strength, shows a variety of spatiotemporal behaviours depending on the regions of parameter space, and also on different initial conditions. Different classes of chimera states can be seen here. Aperiodic and stable chimera states appear in this system for an initial condition having identical phases in group one and random phases between zero and one in group two. Splay chimera states appear for a system wide splay phase as an initial condition and a switching between synchrony and de-synchrony between groups one and two for these states can be seen with the variation of the K. Chimera phase states which consist of a synchronized group and a phase desynchronised group which shows spatiotemporally intermittent behaviour are seen in a certain region of the parameter space using random initial conditions. For this case, we construct an equivalent cellular automaton to reproduce the space time evolution of these laminar and burst sites. A mean field equation is set up whose solutions give the values of the fraction of laminar/burst sites that match with the numerical calculation of these quantities. We reproduce the phase diagram of system in the parameter region of interest using the solutions of the mean field equation of the CA. We hope our techniques will find wider applications in situations where chimera states are found.

References

1. Y. Kuramoto, D. Battogtokh, Nonlinear Phenom. Complex Syst. **5**, 380 (2002)
2. D.M. Abrams, S.H. Strogatz, Phys. Rev. Lett. **93**, 174102 (2004)
3. D.M. Abrams, S.H. Strogatz, Int. J. Bifurc. Chaos **16**(1), 21–37 (2006)
4. D.M. Abrams, R. Mirollo, S.H. Strogatz, D.A. Wiley, Phys. Rev. Lett. **101**, 084103 (2008)
5. E.A. Martens, C.R. Laing, S.H. Strogatz, Phys. Rev. Lett. **104**, 044101 (2010)
6. O.E. Omel'chenko, Y.L. Maistrenko, P.A. Tass, Phys. Rev. Lett **100**, 044105 (2008)
7. H. Wang, X. Li, Phys. Rev. E **83**, 066214 (2011)
8. M.R. Tinsley, S. Nkomo, K. Showalter, Nat. Phys. **8**, 662–665 (2012)
9. S. Nkomo, M.R. Tinsley, K. Showalter, Phys. Rev. Lett. **110**, 244102 (2013)
10. J.F. Totz, J. Rode, M.R. Tinsley, K. Showalter, H. Engel, Nat. Phys. **14**(3), 282–285 (2018)
11. E.A. Martens, S. Thutupalli, A. Fourriére, O. Hallatscheck, PNAS **110**(26), 10563–10567 (2013)
12. C.R. Nayak, N. Gupte, AIP Conf. Proc. **1339**, 172 (2011)
13. J. Singha, N. Gupte, Phys. Rev. E **94**, 052204 (2016)
14. M.H. Jensen, P. Bak, T. Bohr, Phys. Rev. Lett. **50**, 1637 (1983)
15. E. Ott, *Chaos in Dynamical Systems* (Cambridge University Press, Cambridge, 1993)
16. R. Mikkelsen, M. van Hecke, T. Bohr, Phys. Rev. E **67**, 046207 (2003)

Author Index

V. In et al. (Eds.): *Proceedings of the 5th International Conference on Applications in Nonlinear Dynamics*, Understanding Complex Systems, https://doi.org/10.1007/978-3-030-10892-2

The manufacturer's authorised representative in the EU is Springer Nature Customer Service Centre GmbH, Europaplatz 3, 69115 Heidelberg, Germany. If you have any concerns regarding our products, please contact ProductSafety@springernature.com

Printed and bound by CPI Group (UK) Ltd, Croydon, CR0 4YY
05/10/2026
02228130-0010